New Perspectives on Construction in Developing Countries

Developing countries face the challenge of maintaining economic growth and socio-economic development, at the core of which sits the construction industry. Much research on construction in developing countries took place in the 1970s and 1980s, but little since; a gap which this book fills.

Including contributions from prominent academics and practitioners in Singapore, South Africa, Portugal, the Netherlands, Switzerland and the UK, this is a truly international analysis of a subject of global interest. The most insightful and relevant recent research, on topics such as the Millennium Development Goals, the informal construction sector, human resource development, technology and social change, is addressed in the context of the construction industry in the developing world.

While the challenge has grown and the needs have become even more pressing, the research to date has rarely presented effective solutions. Focusing on those aspects of the construction industry most crucial to development, this is a much needed up-to-date study that sheds new light on a variety of concepts and issues.

This is essential reading for researchers, professionals and students interested in the construction industry in developing countries. Readers of this book will be interested in its companion volume: *Contemporary Issues of Construction in Developing Countries*.

George Ofori is Professor at the National University of Singapore. He is a Fellow of the Chartered Institute of Building, Royal Institution of Chartered Surveyors and Society of Project Managers. His research is on construction industry development, international project management and leadership. He has been a consultant and adviser to several international agencies and governments.

About CIB and about the series

CIB, the International Council for Research and Innovation in Building and Construction, was established in 1953 to stimulate and facilitate international cooperation and information exchange between governmental research institutes in the building and construction sector, with an emphasis on those institutes engaged in technical fields of research.

CIB has since developed into a world-wide network of over 5000 experts from about 500 member organisations active in the research community, in industry or in education, who cooperate and exchange information in over 50 CIB Commissions and Task Groups covering all fields in building and construction related research and innovation.

http://www.cibworld.nl/

This series consists of a careful selection of state-of-the-art reports and conference proceedings from CIB activities.

Open & Industrialized Building A. *Sarja*
ISBN: 0419238409. Published: 1998

Building Education and Research J. *Yang* et al.
proceedings
ISBN: 041923800X. Published: 1998

Dispute Resolution and Conflict Management P. *Fenn* et al.
ISBN: 0419237003. Published: 1998

Profitable Partnering in Construction S. *Ogunlana*
ISBN: 0419247602. Published: 1999

Case Studies in Post-Construction Liability A. *Lavers*
ISBN: 0419245707. Published: 1999

Cost Modelling M. *Skitmore* et al.
(allied series: Foundation of the Built Environment)
ISBN: 0419192301. Published: 1999

Procurement Systems S. *Rowlinson* et al.
ISBN: 0419241000. Published: 1999

Residential Open Building S. *Kendall* et al.
ISBN: 0419238301. Published: 1999

Innovation in Construction A. *Manseau* et al.
ISBN: 0415254787. Published: 2001

Construction Safety Management Systems S. *Rowlinson*
ISBN: 0415300630. Published: 2004

Response Control and Seismic Isolation of Buildings M. *Higashino* et al.
ISBN: 0415366232. Published: 2006

Mediation in the Construction Industry P. *Brooker* et al.
ISBN: 0415471753. Published: 2010

Green Buildings and the Law J. *Adshead*
ISBN: 90415559263. Published: 2011

New Perspectives on Construction in Developing Countries G. *Ofori*
ISBN: 0415585724. Published: 2012

Contemporary Issues in Construction in Developing Countries G. *Ofori*
ISBN: 0415585716. Published: 2012

Culture in International Construction W. *Tijhuis* et al.
ISBN: 041547275X. Published: 2012

New Perspectives on Construction in Developing Countries

Edited by George Ofori

Routledge
Taylor & Francis Group

LONDON AND NEW YORK

First published 2012 by Spon Press

2 Park Square, Milton Park, Abingdon, Oxfordshire OX14 4RN
52 Vanderbilt AVenue, New York, NY 10017

Routledge is an imprint of the Taylor & Francis Group, an informa business

First issued in paperback 2019

British Library Cataloguing in Publication Data
A catalogue record for this book is available from the British Library

Library of Congress Cataloging-in-Publication Data
New perspectives on construction in developing countries / edited by George Ofori.
p. cm.
Includes bibliographical references.
1. Construction industry—Developing countries. 2. Building—Economic aspects—Developing countries. I. Ofori, George.
HD9715.D44N49 2011
338.4'7624091724–dc22
2011000189

ISBN13: 978-0-415-58572-9 (hbk)
ISBN13: 978-0-367-35586-9 (pbk)

Typeset in Sabon by Prepress Projects Ltd, Perth, UK

Contents

Figures

Tables

Boxes

Contributors

Akintola Akintoye, PhD, is Chair Professor and Head of School of Built and Natural Environment at University of Central Lancashire. He is a Chartered Surveyor and a Chartered Builder. He has gained international recognition for his work on risk management, procurement, estimating and modelling, and construction inventory management. His current research is on Public and Private Partnership and Private Finance Initiative.

Yaw A. Debrah, PhD, is Professor of Human Resource and International Management at University of Wales, Swansea. He has held academic positions at Brunel University, Cardiff University and Nanyang Technological University, Singapore. He has also worked in Africa, Canada and the USA. He has published numerous articles on Human Resource Management (HRM) in Asia and HRM and International Business/Management in Africa. He is on the editorial boards of several journals.

Emilia van Egmond, PhD, is a Senior Lecturer and Researcher on Innovation and International Technology and Knowledge Transfer for Sustainable Construction at Eindhoven University of Technology. She has professional and academic experience in countries in Africa, Asia, Europe and Latin America and is an advisor to the Dutch government and several international organisations. She was director of a Dutch government project on Capability Building in Ghanaian Construction.

Jorge Lopes, PhD, is Professor and Head of the Department of Construction and Planning at the Polytechnic Institute of Bragança, Portugal. He has written several publications in construction economics, construction management and valuation, and is a member of the editorial board of the *International Journal of Strategic Property Management*. He is a consultant of the Road Institute of Cape Verde.

Rodney Milford, PhD, is Programme Manager: Construction Industry Performance at the Construction Industry Development Board (cidb) in South Africa, and previously Director of CSIR Building and Construction Technology. He is a past-President of the International Council for Research and Innovation in Building and Construction (CIB), a past-President of the South African Institution of Civil Engineering and a previous member of the Board of the cidb.

Patricia Lim Mui Mui holds a BSc (Building) from the National University of Singapore. She is a Contracts Manager with the National Art Gallery, Singapore, overseeing cost and contract matters for the gallery redevelopment project. She is a member of Singapore Institute of Surveyors and Valuers. She has worked as a quantity surveyor on both private- and public-sector projects.

Richard B. Nyuur, MSc, is currently studying for a doctoral degree in Business Management at the School of Business and Economics, Swansea University. His research focuses on foreign direct investment in developing African countries, corporate social responsibility and human resource issues in the construction industry in Ghana.

George Ofori, PhD, DSc, is Professor at the National University of Singapore. He is a Fellow of the Chartered Institute of Building, Royal Institution of Chartered Surveyors and Society of Project Managers. His research is on construction industry development, international project management and leadership. He has been a consultant and adviser to several international agencies and governments.

Ellis L. C. Osabutey, PhD, is a lecturer in International Business and Knowledge Management at Middlesex University Business School. His research interests include strategy and international business, innovation and technology transfer management, international human resource management and cross-cultural management.

Suresh Renukappa, PhD, is a Research Associate in sustainable procurement and economics at the University of Central Lancashire. After his doctoral research on managing change and knowledge associated with sustainability initiatives for improved competitiveness, he has worked in research and consultancy in the area of sustainability in India, the UK and the USA. He has also worked as a management consultant in the software industry. He has authored several papers which have been published in journals, book chapters, and reports on economic, social and environmental sustainability issues.

Les Ruddock, PhD, is Professor of Construction and Property Economics in the School of the Built Environment at the University of Salford. He is Coordinator of the CIB Working Commission (W55) on Construction Industry Economics and has written extensively on the economics of the construction and built environment sectors.

Steven Ruddock, PhD, is a Research Fellow at Salford Centre for Research and Innovation (SCRI), where his research interest focuses on economic appraisal. He is also a futurist, working with Built Environment Shaping Tomorrow (BEST) and specialising in the relationship between the construction sector and the macro-economy.

Winston M. Shakantu, PhD, is a Professor of Construction Management at the Nelson Mandela Metropolitan University in Port Elizabeth. He is a member of the Chartered Institute of Building. He has held academic positions at institutions including the University of Cape Town, South Africa, and has published widely.

Ron Watermeyer, DEng, has been at the forefront of many development initiatives in South Africa since the early 1990s. In 2010 he was awarded the ICE International Medal for contributions over time to the procurement and delivery of infrastructure. He is a Director at Soderlund and Schutte (Pty) Ltd, Consulting Engineers, Johannesburg.

Jill Wells, PhD, has more than 35 years' experience in development work, much of it focusing on construction. Her expertise lies in economic, social and labour issues related to the development of the construction sector. She has worked as 'construction specialist' at the International Labour Office in Geneva, spent considerable periods in academia and now works for Engineers Against Poverty.

Abbreviations

ACES	Association of Consulting Engineers Singapore
AES	architectural and engineering services sector
AIC	advanced industrial country
B-BBEE	broad-based black economic empowerment
BCA	Building and Construction Authority
BRIC	Brazil, Russia, India and China
C21	Construction 21
CCA	Chief Construction Adviser (UK)
CEO	chief executive officer
CETA	Construction Education and Training Authority
CIB	International Council for Research and Innovation in Building and Construction
CIC	Construction Industry Council
CIDA	Construction Industry Development Agency (Australia)
cidb	Construction Industry Development Board
CIJC	Construction Industry Joint Committee
CIRC	Construction Industry Review Committee (Hong Kong)
CONQUAS	Construction Quality Assessment System
CORENET	Construction and Real Estate Network
CoreTrade	Construction Registration of Tradesmen
CPSC	Construction Policy Steering Committee (Australia)
CRB	Contractors Registration Board (of Tanzania)
CRS	Contractors Registration System
CVA	construction value added
D&B	design and build
DAC	Development Assistance Committee
DPW	Department of Public Works
e-BDAS	E-Buildable Design Appraisal System
EAP	Engineers Against Poverty
ECDP	Emerging Contractor Development Programme
EDB	Economic Development Board
EPEC	Electronic Procurement Environment for the Construction Industry

EPWP	Extended Public Works Programme
EU	European Union
FDI	foreign direct investment
FIDIC	International Federation of Consulting Engineers
GDP	gross domestic product
GFCF	Gross Fixed Capital Formation in Construction
GLC	government-linked company
GMR	Global Monitoring Report
GNI	gross national income
HDB	Housing and Development Board
HDE	historically disadvantaged enterprise
HDI	Human Development Index
HIE	high-income economy
HR	human resource
HRD	human resource development
HRM	human resource management
IAS	Investment Allowance Scheme
ICE	Institution of Civil Engineers
ICT	information and communications technology
IEG	Independent Evaluation Group
IES	Institute of Engineers Singapore
IFC	International Finance Corporation
IGS	Infrastructure Gateway System
ILO	International Labour Office
IMF	International Monetary Fund
ISO	International Organisation for Standardisation
KPI	key performance indicators
LDC	less developed country
LEAD	Local Enterprise and Association Development Scheme
LEED	Leadership in Environmental Engineering and Design
LIC	low-income country
LIE	low-income economy
LMIC	lower-middle-income country
MBA/NSW	Master Builders Association in New South Wales
MBSA	Master Builders South Africa
MDG	Millennium Development Goal
MIC	middle-income country
MIE	middle-income economy
MPI	Multidimensional Poverty Index
NABCAT	National Black Contractors and Allied Trades Forum
NAMC	National Association of Minority Contractors
NatBACC	National Building and Construction Committee
NCDP	National Contractor Development Programme
NEC	New Engineering Contract
NGO	non-governmental organisation

NIC	newly industrialised country
NSF	National Skills Fund
NTT	National Trade Test
NVQ	National Vocational Qualifications
ODA	official development assistance
OECD	Organisation for Economic Co-operation and Development
OGC	Office of Government Commerce
PDP	previously disadvantaged person
PPI	Private Participation in Infrastructure
PPP	public–private partnership
PSIB	Process and System Innovation in Building and Construction
R&D	research and development
RDP	Reconstruction and Development Programme
REDAS	Real Estate Developers Association
SAICE	South African Institution of Civil Engineering
SCAL	Singapore Contractors Association Limited
SCOT	Social Construction of Technology
SDA	Skills Development Act
SIA	Singapore Institute of Architects
SIBL	Singapore Institute of Building Limited
SIP	Singapore Institute of Planners
SISV	Singapore Institute of Surveyors and Valuers
SME	small and medium-sized enterprise
SMMEs	small, medium and micro-enterprises
SNA	Systems of National Accounts
SOE	state-owned enterprise
SPM	Society of Project Managers
SRI	socially responsible investment
TD&I	technology development and innovations
UMIC	upper-middle-income country
UN	United Nations
UNCHS	United Nations Centre for Human Settlements
UNDP	United Nations Development Programme
UNEP	United Nations Environment Programme
VAT	value-added tax
WBCSD	World Business Council for Sustainable Development

1 The construction industries in developing countries

Strategic review of the book

George Ofori

Introduction: a paradox

The construction industry plays an important role in economic growth and in long-term national socio-economic development. This is only one of the reasons why it is necessary to study, and gain a good understanding of, the construction industries of developing countries. Two other closely inter-related reasons may be outlined. First, governments tend to use their investments in construction to introduce changes in the national economies (Hillebrandt, 2000). This was evident in many of the 'stimulus packages' which were launched in a number of industrialised countries to address the global economic and financial crisis in 2008–2009. Thus, it is important for these policy decisions to be informed by sound knowledge of (i) the nature of the industry, and the circumstances in which the proposed solutions would work; (ii) the type of projects where the desired change is most likely to be achieved; and (iii) the action that needs to be taken by the construction industry itself to enable it to play the role assigned to it. The last point is in stark contrast to the reality, where there is a lack of knowledge on the construction industry owing to lack of interest of governments and other stakeholders. For example, in most countries, the industry has a poor image and low priority is given to issues concerning it.

The second reason why studies on the construction industries in developing countries must be undertaken is that it is vital for construction to perform well the tasks that are assigned to it. This is because constructed items are important inputs into other productive activities, and hence time is of the essence with regard to the execution of projects. From the cost perspective, effective and efficient construction will save money and provide value for the clients and for society. With respect to quality and durability, constructed items form a significant part of any nation's savings. From the social angle, constructed items improve the quality of life of the people in any society. In particular, in the developing countries, they provide the basic needs of housing, water, health and education. It is appropriate here to relate construction activity to the attainment of the Millennium Development Goals (MDGs).

Perhaps without directly knowing it, the world's leaders expect construction to help improve the standard and quality of life of their people.

There was much research on the construction industries in developing countries in the late 1970s and throughout the 1980s. These works resulted in a large volume of literature on key aspects of the industries. The aspects covered included (i) the role of the construction industry in economic development; (ii) features of the construction industries including their needs and problems; (iii) the factors which influence their performance; and (iv) particular parts of the industries such as contractor development, materials development, technology development and institution building. The authors of the works offered many recommendations for developing the construction industries and enhancing their performance and prospects. There is much evidence of the researchers at that time being aware of the earlier work, and building on their findings and conclusions.

If the research work outlined in the previous paragraph had continued, there would have been much greater understanding of the nature of the construction industry and its needs, problems, challenges and prospects in the context of socio-economic development. What is the position now? This issue will be taken up again at the end of this chapter.

Brief historical review of research on construction industry development

Duccio Turin and his colleagues in the University College Environmental Research Group (UCERG) and Built Environment Research Unit (BERU) at University College London undertook studies on the relationship between construction activity and economic growth and development. Turin's (1973) was arguably the most significant of these works. This was followed by works on particular countries by the team, including Andrews and colleagues (1972). Several of the studies were undertaken upon commissioning by clients such as the World Bank. An international team of experts undertook a comprehensive study of the construction industry in Tanzania and produced *The Local Construction Industry Report* (Ministry of Works Tanzania, 1977), one of the finest works on the industry in a developing country for several decades. It is pertinent to note that many of the points made in that report are still relevant today. For example, it considered 'productivity development' to be at the heart of the industry's performance, including cost levels. It concluded that poor productivity in Tanzania's construction industry was due to lack of materials, equipment and spare parts; lack of skilled manpower, especially supervisors; lack of working capital; lack of incentives for the workers; and little genuine interest among managers and professionals in improving project planning, site management and cost control (p. 13). The authors of the report recommended that a separate agency be established to improve co-ordination and communication in the construction sector. This was a new thing. The National Construction Council of Tanzania was formed soon after this, and was the first of the construction industry development agencies to be established.

Under the Construction Management Programme of the International Labour Office (ILO), which covered some 20 countries in Africa, a number of country reports, and those on particular aspects of construction such as project management and procurement, were produced. From these reports, some useful books were published. First, *Foundations for Change* (Edmonds and Miles, 1984) was a compilation of chapters on various relevant subjects, including country studies on Ghana and Sri Lanka. Second, again, arguably the most comprehensive work on construction contractor development was published by the ILO (Relf, 1986). This book was informed by field work by consultants commissioned by the ILO in a number of developing countries. It was followed by the 'Improve Your Business' programme for developing small and medium-sized enterprises, which the ILO implemented in many countries until well into the 1990s (see, for example, Miles and Ward, 1991). Third, Miles and Neale (1991) considered case studies of four construction industry development institutions in the developing countries, and proposed a '12-point action plan' for such agencies: (i) general industry promotion; (ii) influencing policy and investment decisions; (iii) undertaking research; (iv) development of local construction capability; (v) improving the regulatory framework; (vi) training and management development; (vii) promoting construction management as a discipline; (viii) promoting computer based management information systems; (ix) consultancy and advisory services; (x) internal training and staff development; (xi) creating natural linkages; and (xii) creating international linkages.

The World Bank (1984) produced another seminal work. Its title was simple: *The Construction Industry: Issues and Strategies in Developing Countries*. One of the main features of this slim but useful volume is an outline, in the appendix, of possible terms of reference for a study on a construction industry in a developing country, entitled 'Framework Terms of Reference for Studies of the Construction Industry in Developing Countries'. This is still relevant today, and Ofori (2007) laments that it is not being used by researchers and consultants on the construction industry. Another publication of the World Bank which merits greater recognition and attention is that by Kirmani and Baum (1992) on the consulting professions in the developing countries.

The then United Nations Centre for Human Settlements (UNCHS) sponsored several studies on the construction industries in developing countries. These include (i) works on the role of the industry in human settlements in general (UNCHS, 1981a,b); (ii) a study of technology development (UNCHS, 1991) which included one of the pioneering studies on the impact of the construction industry on the physical environment; and (iii) a study of contractor development (UNCHS, 1996). In May 1993, the UNCHS and the United Nations Industrial Development Organization (UNIDO) held their first consultation on the industry in development in Tunisia.

The issues addressed included the following 'constraints' (UNIDO, 1993): (i) big volume of need together with poorly organised construction industry unable to meet the increasing demand; (ii) fluctuations in demand and levels of activity, which have adverse impact on the building up of capacity and experience in the industry, with as a result a high level of risk exacerbated by lack of data and information for proper analyses; (iii) limited technology development and innovation, and lack of utilisation of computer technology; (iv) skills development, which also hampers the local construction firms in competition with foreign ones; (v) reliance on foreign firms and imported plant and equipment; (vi) lack of access to finance; and (vii) poor performance on projects characterised by high costs, delays and poor productivity. These are issues that are still of concern to the developing countries today.

Books on the construction industries in the developing countries which have been published include those of Wells (1986) and Ofori (1993a). Other books which have considered particular aspects of construction include Abbott (1985) on technology transfer and Werna (1996) on small contractor development. The subject of construction in developing countries has been covered in sections, or in passing, in some of the good books on construction economics such as that by Hillebrandt (2000).

In the early 1990s, Ofori (1993b, 1994) presented a review of research on construction industry development as well as a review of practice in construction industry development. He noted that there had been a decline in the volume and pace of work on the construction industries in developing countries. In another paper, Ofori (2007) observed that work on the area had ground to a virtual halt. He gave the following reasons for this phenomenon: (i) funding from donor agencies no longer flowed to such studies; (ii) many of the good publications (such as Turin, 1973) were rather old, and no longer available or accessible; and (iii) as a result of point (ii), the field of construction in developing countries has lacked a suitable foundation on which to build and many of the works do not have much to say that is new or profound.

Recent developments

There has not been much work on the construction industries in developing countries in the past one and a half decades. At the same time, the problems confronting the industries have grown, and the need for action to address them has become even more pressing. In the absence of recent authoritative literature, the work which is currently undertaken (for example, in graduate research studies, or consultancy assignments for governments and international organisations) is often at a basic level, repeating what was done some decades ago, and reaching conclusions which have already been established. Moreover, these works tend to offer recommendations which are not new, and some of which had failed to work when they had been tried earlier.

There is much evidence of the researchers of the 1970s and 1980s being aware of the earlier work, and using their findings and conclusions as points of departure.

Not only are some of the earlier findings being ignored, but some of the significant ones are also being misassigned. This includes Turin's (1973) establishment of the relationship between construction and an economy as it grows. However, there are some bright spots on the landscape, and some encouraging signs are emerging. Two of these positive developments are now discussed.

First, some of the construction industry development agencies, such as the one in South Africa, have undertaken, or sponsored, some significant studies which have yielded a great volume of materials which help to engender understanding of the contemporary times. The work done by the Construction Industry Development Board (cidb) in South Africa on procurement and project delivery (including a toolkit), registration of contractors, contractor development, tracking of industry performance (including status reports and monitoring reports, which include periodic surveys of the state of the industry) and best practices schemes is a useful addition to the body of knowledge. (cidb publications are available at: http://www.cidb.org.za/.)

Second, several countries have realised that it is important to have policy documents on their construction industries. Again, South Africa offers lessons. The government produced a 'green paper' which set out proposals and a programme for radically restructuring and developing the industry. This was followed by a statute which provided for particular actions in the development of the construction industry, and also for the establishment of a championing agency, the cidb, in that country. Tanzania's national construction industry policy, adopted in 2003, has the following goal:

> To develop an internationally competitive industry that will be able to undertake most of the construction projects in Tanzania and export its services and products and ensure value for money to industry clients as well as environmental responsibility in the implementation of construction projects.
>
> (Ministry of Works, 2005, p. 1)

In other words, Tanzania wants to have one of the best construction industries in the world. The objectives of the national construction industry policy of Rwanda are to (i) harmonise roles of the public and private sectors; (ii) develop and strengthen local capacity for effective participation; (iii) strengthen and support the professional bodies; (iv) promote use of appropriate technology; (v) promote the participation of women in the national construction industry; and (vi) ensure that the national construction industry supports sustainable economic and social development (Ministry of Infrastructure, 2009).

This book

This book seeks to fill the gap between the early works on the construction industries in the developing countries and their improvement. It is part of a two-book set; the other book is entitled *Contemporary Issues in Construction in Developing Countries*. It also comprises chapters by prominent researchers on the various topics. It covers the implications for developing countries of issues which are often discussed without much reference to these countries.

Each of the chapters in this book covers an important aspect of the construction industries in the developing countries in the present era. After beginning with a consideration of macro-level issues of the industries, the rest of the book is devoted to a discussion of topics which are the key components of construction industry development as is widely agreed upon by authors. The broad subjects include (i) the role of the construction industry in economic development; (ii) contractor development; (iii) issues relating to finance; (iv) institution building; (v) human resource development; and (v) technology development.

The authors agreed to produce their chapters in accordance with a set of guidelines. These included the provision that each chapter would contain (i) a review of the literature; (ii) a discussion of current issues with relevant examples from several countries and inter-country comparisons; (iii) proposals for developmental action; and (iv) future directions for research and practice. The authors were urged to bear in mind that the developing countries are not a homogeneous group. They were also requested to include, in their conclusions and recommendations, policy implications of their findings and conclusions for the governments of the developing countries in relation to developing the construction industries, and enhancing their performance and prospects.

Strategic overview of the book

An attempt is now made to introduce the contents of the book by not only providing a summary of each chapter, but also relating the chapters to the key developments on the particular subject.

Construction and economic development

The first part of the book comprises three chapters on the broad subject of construction and economic development.

Chapter 2

There have been many significant changes in the global economy and society since the early works on the construction industries in the developing countries (see, for example, Turin, 1973) were undertaken at the beginning of the 1970s. Apart from the global trends such as globalisation, there have been significant changes in policy directions and inclinations in the developing countries themselves, generally towards liberalisation. Some of the manifestations of new developments include the emergence of Brazil, China and India as major economic powers. Owing to the intimate relationship between the construction industry and other sectors of the economy, these developments should have had an impact on the construction industry. It is necessary to consider these changes and their implications for the industry, in terms of both the challenges they pose and the opportunities they provide.

In the second chapter of the book, Les Ruddock and Steven Ruddock seek to answer the following key question: 'What are the meta-issues that will shape development over the coming decades and what challenges will be presented?' He considers major trends in thinking on development economics. The issues he discusses include attainment of the Millennium Development Goals (MDGs), sustainable development, rural development, climate change and technology development. The discussion in the chapter is undertaken in the context of the micro-economic challenges facing the developing countries. Ruddock makes policy proposals, with suggestions of courses of action which will prepare the construction industry to play the role expected of it, as the countries strive to develop.

Chapter 3

As noted above, Turin (1973) and his colleagues at University College London pioneered the study of the role of the construction industry in long-term national socio-economic development, and the changes that occur in a construction industry as the country develops. These studies were replicated by various other authors, who found similar relationships. However, the analyses of Turin (1973) and those researchers who repeated his studies are based on many assumptions, some of which are not easy to ignore. These include the use of data over the same period in a number of countries (cross-sectional data), instead of time-series data on the same country.

In Chapter 3 of this book, Jorge Lopes reviews these early studies on the contribution which construction makes to the economy and how its role, contributions and linkages change as nations develop. He discusses key works on economic growth and economic development as well as studies on the role of the construction industry in the process of national economic growth and development. Several aspects of this particular chapter set it apart from previous works on this subject. First, a comparative analysis of the relationship between the measures of construction output and those of the national aggregate is undertaken for two groups of countries in sub-Saharan Africa. This provides further evidence of the need for country specificity when making proposals on any aspect of the construction industries in developing countries. Second, the link between investment in construction, especially infrastructure, and the attainment of the economic and social targets under the MDGs is explored. Finally, some proposals for public policies on economic development and the role of the construction industry in it are made.

Chapter 4

The MDGs were set in 2000, and are to be attained by 2015. They have guided development policies and programmes in many countries. Governments, some businesses, international organisations and non-governmental institutions have committed themselves to the attainment of these goals. However, reports show that, whereas there have been some significant achievements, progress towards reaching many of the goals has been slow, and many of them will not be attained by the due date. It is pertinent to note that the MDGs have not been universally welcomed, and there have been some strong criticisms on many of their aspects from both the intellectual and practical points of view. The very legitimacy, relevance and accuracy of the goals are questioned. On the other hand, the proponents of the MDGs claim that they are not only necessary and achievable, but, indeed, quite modest. For example, it is often noted that, even if all the MDGs are attained by the set date, it will still leave a large number of people in the developing countries in deprived and precarious situations.

As construction plays a crucial role in the development process, it is important to understand how it contributes to the attainment of the MDGs. In Chapter 4, George Ofori explains the MDGs and the targets set to guide policy formulation and activities towards their attainment. He places the MDGs in their context within the broader Millennium Declaration in order to highlight their relevance. He reviews the literature on MDGs, and discusses the controversy on them. He considers progress towards the attainment of the MDG targets in various regions of the world. Ofori then analyses the role of infrastructure in the programmes towards attaining the MDGs. He notes that if construction plays such a key role, then superior performance on each construction project is critical. He then offers a framework of performance criteria for assessing projects in developing countries.

Construction industry development: macro-level issues

The second part of the book is made up of three chapters on the macro-level issues of construction industry development.

Chapter 5

The establishment and strengthening of the various institutions in the construction industry, both public and private, as well as the linkages and co-ordination among them, are key aspects of the development of a construction industry. In Chapter 5, Rodney Milford considers the role of these institutions. He starts from the organisations which were established in the UK and other countries such as Australia and South Africa over 100 years ago, and draws lessons from the factors which led to their promotion, as well as their achievements and continued relevance. These trade associations were established to safeguard the interests of their members and to promote the particular professions and trades. They have realised key successes in many countries such as in health and safety, and training.

Milford also notes that, since the late 1990s, many countries have set up new public institutions to spearhead the development of their construction industries, in most cases to implement national reform programmes for developing their industries. These organisations have had various forms. He gives the examples of statutory bodies in Malaysia, Singapore and South Africa, and private forums in the UK and the Netherlands. Milford discusses the objectives, roles and achievements of these institutions. He also draws out the key lessons learned from the consideration of the experiences of various countries. He uses a case study of South Africa, focusing on the activities, challenges and achievements of the cidb, to illustrate the points made in the chapter. Finally, he makes proposals for policy on construction industry development which can apply in these countries.

Chapter 6

The strategies and programmes for improving the capacity, capability and performance of the construction industries in developing countries must be effectively implemented if the objectives of industry development are to be realised. Given the complex nature of the construction industry itself, and its relationships with other sectors of the economy, various parties in government and industry will have to play particular roles in the overall effort, and the co-ordination of the actions to implement the initiatives is critical. For this reason, many of the recent review programmes have highlighted the need for championing.

The study presented by George Ofori and Patricia Lim Mui Mui in Chapter 6 aims to find out ways to implement effectively the industry development initiatives which have been formulated in many countries. The chapter starts with a review of the literature on championing. The authors next consider some of the recent industry reform initiatives in many countries, both developed and developing. They focus on Singapore, presenting a field study which was based on interviews with senior practitioners and officers who took part in the radical review of the construction industry in 1999. Ofori and Lim note that, whereas the report on that reform exercise had highlighted 'collective championing of the industry', the findings revealed that the collaboration between the public agency and the umbrella industry association which had been identified in the report to champion the development of the industry had not worked well. They reiterate that long-term partnership between the government and the industry is essential to achieve the desired radical change in construction. They suggest that the umbrella industry grouping should endeavour to realise its potential to be a think tank on the industry and the government's partner in development.

Chapter 7

The potential of the informal sector is seldom considered in the policies, strategies and programmes for developing the construction industries. This is because there is inadequate information on, and understanding of, the sector, especially in the developing countries. In Chapter 7 of this book, Jill Wells points out that, over nearly two decades, studies have described a growth of informal construction activity in the developing countries. In fact, a similar phenomenon has also been found in developed countries. Wells notes that the concept of what is informal and what is formal in the construction industry is generally not clearly defined. She suggests that it is now widely recognised that the key essence of informality is the absence of regulation.

Wells outlines four spheres of regulation which relate to (i) the terms and conditions of employment for construction workers; (ii) enterprises operating in the sector; (iii) the process of construction – the delivery system; and (iv) the product. In each of these areas, she provides evidence of the expansion of informal activity by giving a number of examples from many countries. She notes that, as the level of informality grows, the influence of the state is being reduced. She goes on to assess the factors contributing to the processes of informalisation and the linkages within and between the four different spheres which she had delineated. She ends the chapter by considering policy options. She suggests that policy makers should note that, whereas some individuals or firms may choose to work informally, other have no choice. In some cases, it might be sensible to 'formalise the informal'. In others, it might be worthwhile to support the informalisation of the 'formal'.

Construction industry development: key areas

The third part of the book is on the key areas of construction industry development.

Chapter 8

Technology development will enable construction industries to undertake the construction of complex and technologically sophisticated buildings and items of infrastructure. All things being equal, it will enable the industries to deliver higher-quality buildings and items of infrastructure at lower cost, providing value for money.

In Chapter 8 of this book, Emilia van Egmond addresses the question: 'How is technology development and innovation in the construction industries in the developing countries to be managed?' She notes that new technologies, knowledge and innovations in construction have been recognised as important tools for attaining the MDGs. She highlights two major challenges. The first is that posed to the construction industry by human needs for a sustainable built environment. The second one is that the developing countries face high levels of unemployment while they face scarcity of resources, weak infrastructure and capabilities, and high levels of housing shortages. She also points out that in the developing countries, as in other nations, the construction process has an adverse impact on the environment. These observations form the context of the discussion in the chapter.

Van Egmond suggests that technology development and innovation offer opportunities to the construction industries in the developing countries to leapfrog in development, and to provide effective solutions to the challenges she outlines. However, the development and diffusion of new technologies are complex and difficult processes with no guarantees, varying degrees of success and, indeed, a high probability of failure. Van Egmond provides a review of the literature on technology development, knowledge and innovation as well as the international flows of technology and knowledge. She provides clear definitions of the related terms to facilitate the understanding of the discussion. She considers the modes of technology and knowledge flows, and discusses the drivers and barriers. She makes recommendations on the management of innovation, technology and knowledge development for the benefit of the construction industries in the developing countries.

Chapter 9

The employment generation characteristic of the construction industry is often highlighted. To attain this potential, each country should have the required numbers and quality of a wide range of professional, technical and trade skills. Thus, human resource development is of critical importance. However, the industries in the developing countries have faced severe shortages of qualified and experienced personnel for several decades.

In Chapter 9 of this book, Ellis Osabutey, Richard Nyuur and Yaw Debrah note that the construction industry can generate employment and incomes in the location of its projects. Thus, construction projects can alleviate poverty in the developing countries as they build the large volume of buildings and infrastructure which they need to support their socio-economic development. However, owing to globalisation, many of the positions on major construction projects in these countries are taken by foreign personnel because most of these countries lack the professional and technical expertise and the skilled site workers required to construct complex buildings and infrastructure. This absence of the necessary human resources in quality, quantity and variety adversely affects the planning and implementation of construction projects in developing countries. Training facilities and skill development programmes are inadequate.

Osabutey *et al.* seek to improve the understanding of governments, practitioners and researchers, and others, of human resource development trends in developing countries. They review the theory and practice and also examine what has been done on the subject in research. They note that human resource management and human resource development can significantly improve the performance of construction firms, but that they are ignored by most firms. In particular, those in the developing countries are not in a position to apply them owing to their sizes and the scale of their activities. Thus, governments have a role to play. Osabutey and colleagues discuss what needs to be done, and suggest an agenda for future research.

Chapter 10

The development of construction firms has been recognised as an important component of programmes for improving the performance of the construction industries in the developing countries for several decades. As noted above, the ILO undertook several studies on it, and had programmes in the field on it as well. The World Bank has also implemented programmes on it, and continues to support it in its procurement policies and procedures, for example, by granting bidding preferences to local firms. Authors note that, whereas all construction industries have a pyramid structure, with a few large companies and a large number of small ones, the latter are fragile and face an unfriendly operating environment. Thus, many countries have launched programmes to develop their construction firms.

In Chapter 10, Winston Shakantu suggests that contractor development, in some regards, is viewed as a sound rationale for promoting enterprises associated with a high labour absorption capacity. That is why governments, especially in developing countries, are actively promoting the development of an efficient and effective construction industry that uses resources better, reduces waste and transforms the working environment of the people it engages. Shakantu notes that, whereas a large volume of literature has been developed on the role of the government in the construction industry, not much work has been done on the issue of contractor development. He explains contractor development, especially in developing countries. After a review of works on the subject, he uses cases from South Africa to illustrate both the national programme and how some particular contractors have benefited from it. He offers suggestions on how developing countries can promote contractor development, including how the needs of the firms can be identified and addressed with appropriate interventions at both the local and national levels.

Chapter 11

On a construction project, the consultants undertake the planning, prepare budgets and feasibility studies, and undertake design. Thus, they set the tone for the development, and determine all the key aspects of it, including the materials and technologies to be used. In turn, these have an influence on cost, time, quality and safety on the project, depending on the fit between what is specified and the context in which the work will be undertaken. Moreover, the consultants supervise activity on the site, and can also play a role in the operation of the built item.

In Chapter 11, Ron Watermeyer considers the consulting sector in the construction industries of developing countries. He notes that consulting firms can contribute to both the 'green' (earth agenda) and 'brown' (people agenda) aspects in developing countries throughout the life cycle of a project, from planning to deconstruction. He suggests that consulting firms can play a key role in the major challenge of the day, which he considers to be acceleration of infrastructure delivery and the improvement of the quality of life. He explains the nature of a consulting firm, and considers the fundamentals of construction works, explaining the workflow on them. He considers the nature of a profession and the types of professions involved in the various aspects of the development of construction works. He also considers the manner in which a profession is regulated.

A useful aspect of the chapter by Watermeyer is the detailed consideration of the various options for procuring construction works and professional services and their implications for the nature of the services of the professional. He notes that the issues relating to consulting firms in developing countries can then be considered against this background, namely (i) the location, distribution and usage of professional resources within the construction industry; (ii) emerging and existing opportunities for consulting firms; and (iii) approaches to overcoming capability and capacity constraints facing consulting firms in developing countries within a procurement context. He makes many proposals for developing the consultancy sectors in these countries.

Chapter 12

In the construction industry, funds are required for many purposes. Considering the high expense of constructed items, the sums involved are huge. In particular, the client, contractor and end purchaser require substantial sums of money to meet their obligations on each project. Thus, access to finance is a critical issue for a construction firm. Again, owing to the large sums involved, the cost of this finance is also important. Several factors have an influence on the financial situation of a construction firm. Most of these stem from the nature of construction activity, the terms on which the contractor is paid, the nature of the financial sector in the country, the availability of credit and the cost of borrowing, many of which are also related to the track record of the construction firm.

In Chapter 12, Akintola Akintoye and Suresh Renukappa discuss finance for construction in developing countries. They consider the sources of finance available to organisations at the levels of the industry, the company and the project, and the challenges which the construction industries in developing countries face in gaining access to finance. They discuss the different meanings of 'construction project financing'. This is followed by consideration of infrastructure investment in developing countries. They discuss various funding strategies and financial packages and instruments and the characteristics, merits and demerits of each form. They note that the well-being of a client as well as a construction company depends on its ability to raise and manage funds. However, in the developing countries, the need for finance is acute as the construction firms are under-resourced, and the financial institutions are not willing to provide them with assistance. They end the chapter by outlining the key points to be considered in raising finance for construction projects in developing countries.

Conclusion

To end the strategic review of this book, it is appropriate to quote from the construction policy of Rwanda. One of the objectives of the policy is:

> Develop & Strengthen Local Capacity for Effective Participation: Government shall develop and strengthen the capacity of the local contractors, consultants, suppliers and manufacturers for effective participation in the construction industry. The support shall include creating a Rwanda National Association of Building and Civil Engineering Contractors and a Rwanda Association of Consulting Engineers to continue the promotion of contractors and consulting engineers' interests respectively, nationally and internationally. The Government shall ensure participation of local consultants in the procurement process and in the implementation of all consultancy service contracts in the public sector by requiring that they increasingly take on lead roles. Where international competitive bidding is a condition for the funding of a project, local consultants would be encouraged to undertake partnerships or joint ventures with foreign consultants in order to help develop local experience and capacity.
>
> (Ministry of Infrastructure, 2009, pp. v–vi)

It is important that researchers provide countries such as this, which seek to improve the performance of their construction industries, with the knowledge which will provide the basis of sound policies, strategies, programmes and initiatives.

References

Abbott, P. G. (1985) *Technology Transfer in the Construction Industry*. The Economist Publications: London.

Andrews, J., Hatchett, M., Hillebrandt, P., Jaafar, A.J., Kaplinsky, S. and Oorthuys, H. (1972) *Construction in Overseas Development: A Framework for Research and Action*. University College London Environmental Research Group: London.

Edmonds, G. A. and Miles, D. W. J. (1984) *Foundations for Change: Aspects of the Construction Industry in Developing Countries*. International Technology Publications: London.

Hillebrandt, P. (2000) *Economic Theory and the Construction Industry*, 3rd edition. Macmillan: Basingstoke.

Kirmani, S. S. and Baum, W. C. (1992) *The Consulting Professions in Developing Countries: A Strategy for Development*. Discussion Paper No. 149. World Bank: Washington, DC.

Miles, D. and Neale, R. (1991) *Building for Tomorrow: International Experience in Construction Industry Development*. International Labour Office: Geneva.

Miles, D. and Ward, J. (1991) *Small-Scale Construction Enterprises in Ghana: Practices, Problems and Needs*. Construction Management Programme, International Labour Office: Geneva.

Ministry of Infrastructure (2009) *Rwanda National Construction Industry Policy*. http://mininfra.gov.rw/index2.php?option=com_docman&task=doc_view&gid=113&Itemid=263, accessed on 17 December 2010.

Ministry of Works Tanzania (1977) *The Local Construction Industry Report: General Report*. Ministry of Works: Dar es Salaam.

Ministry of Works (2005) *Implementation Action Programme for Construction Industry Policy, Draft*. Ministry of Works: Dar es Salaam.

Ofori, G. (1993a) *Managing Construction Industry Development: Lessons from Singapore's Experience*. Singapore University Press: Singapore.

Ofori, G. (1993b) Research on the construction industry at the crossroads. *Construction Management and Economics*, 11, 175–185.

Ofori, G. (1994) Practice of construction industry development at the crossroads. *Habitat International*, 18 (2), 41–56.

Ofori, G. (2007) Construction in developing countries. *Construction Management and Economics*, 25, 1–6.

Relf, C. (1986) *Guidelines for the Development of Small-scale Construction Enterprises*. International Labour Office: Geneva.

Turin, D. A. (1973) *The Construction Industry: Its Economic Significance and Its Role in Development*, 2nd edition. University College Environmental Research Group, University College London: London.

UNCHS (1981a) *Development of an Indigenous Construction Sector*, Report of the Ad Hoc Expert Group Meeting on the Development of the Indigenous Construction Sector. UNCHS: Nairobi.

UNCHS (1981b) *The Construction Industry in Human Settlements Programmes*. UNCHS: Nairobi.

UNCHS (1991) *Technology in Human Settlements: Role of Construction*, HS/262/91E. UNCHS: Nairobi.

UNCHS (1996) *Policies and Measures for Small Contractor Development in the Construction Industry, HS/375/95E*. UNCHS: Nairobi.

UNIDO (1993) *Prospects for Development of the Construction Industry in the Developing Countries*. Issue Paper I, First Consultation on the Construction Industry, UNIDO and UNCHS, 3–7 May, Tunis.

Wells, J. (1986) *The Construction Industry in Developing Countries: Alternative Strategies for Development*. Croom Helm: London.

Werna, E. (1996) *Business as Usual*. Avebury: Aldershot.

World Bank (1984) *The Construction Industry: Issues and Strategies in Developing Countries*. World Bank: Washington, DC.

Part I

Construction and economic development

2 Changes in societies and economies
New imperatives

Les Ruddock and Steven Ruddock

Introduction

The world is at a turning point in development thinking and practice and this chapter focuses on the consequent challenges facing developing countries and the consequences for the construction industry. Charles Gore of the United Nations Conference on Trade and Development (Gore, 2009) has propounded the notion of 30-year cycles for the developing world. The period of 1950–1980 was an era of 'national developmentism' and liberation of peoples; 1980–2010 represented global integration and liberation of economies. What now, and what are the new paradigms for the next decades? Gore's (2009) view is that the next global development cycle will be based on the principle: *from global integration to global sustainable development*. If such a change is to manifest itself, the construction industry must play a role in the pursuit of the goal of the provision of a sustainable environment. This chapter seeks to answer the following key questions: if developing countries are at that watershed, what are the meta-issues that will shape development over the coming decades and what challenges will be presented?

It is estimated that in 2010 the working population of the developing world exceeded 3 billion for the first time and the total will continue to grow. It is expected to reach about 4 billion within 30 years or so (United Nations, 2009). Therefore, the developing countries are now at the beginning of a demographic window, when the working-age population will be proportionately highest and the potential for economic growth will be at its peak. A number of inter-related factors – such as government policy and global economic conditions – will determine whether or not this change will necessarily result in economic growth. The achievement of economic growth will require investment in infrastructure, education, training and research, and technology development, and the construction industry will need to play its part in the skilful management of environmental factors associated with that growth and development, and those arising from them.

The Millennium Development Goals (MDGs) represent a framework for measuring progress in development, particularly in the context of environmental sustainability (World Bank, 2010a). The MDGs were endorsed by world leaders in the Millennium Declaration, which was translated into a roadmap setting out eight time-bound and measurable goals to be reached by 2015:

Goal 1   Eradicate extreme poverty and hunger.
Goal 2   Achieve universal primary education.
Goal 3   Promote gender equality and empower women.
Goal 4   Reduce child mortality.
Goal 5   Improve maternal health.
Goal 6   Combat HIV and AIDS, malaria and other diseases.
Goal 7   Ensure environmental sustainability.
Goal 8   Develop a global partnership for development.

The construction industry can play a significant role in the efforts to attain several of the MDGs, through the building of schools, hospitals and other items of infrastructure, but, arguably, it is mainly in the pursuit of Goal 7, 'Ensure environmental sustainability', that the industry has the most important part to play. This becomes clear when one considers the specific MDG7 targets of improving the lives of at least 100 million slum dwellers by 2020 and halving the proportion of people without access to safe drinking water and basic sanitation, as well as reversing the loss of environmental resources. The *Global Monitoring Report* 2010 (World Bank, 2010a) noted that significant progress has been made on some MDGs in many countries. There have been significant reductions in poverty globally; there has also been progress in reducing child and maternal mortality, and towards ensuring environmental sustainability.

A key challenge in future will be the achievement of sustainable urban development. Rapid urban growth throughout the developing world is outstripping the capacity of most cities to provide adequate services for their residents. Over the next 30 years, virtually all of the world's population growth is expected to be concentrated in the urban areas in the developing countries (UNFPA, 2007).

One of the most critical challenges is to find ways of mitigating and adapting to climate change, whilst at the same time realising the development aspirations and unrealised potential of millions of people in the developing countries. Coupled with the imperative of climate change mitigation, the recent global financial crisis has been a significant event for developing countries; and it has coincided with the end of a 60-year global development era (Gore, 2009). In the wake of the global financial crisis, the developing world will be considered as potentially constituting an engine of recovery. In any case, whether or not the developing countries become the dominant pole of economic activity, for global recovery to endure, multi-polar growth will be essential, and growth will increasingly depend upon the developing countries. From an emphasis on globalisation to reduce poverty, derivation of the benefits from globalising in the development of productive capacities will be the main challenge in future. A high level of business activities will require several enabling systems. Among these is an adequate financial and capital system including the necessary property infrastructure. In the next stage of the development cycle (Gore, 2009), there needs to be recognition of the strong relationship between business development and the lifespan of fixed capital investment. The replacement of, and the increase in, the fund of basic capital goods (infrastructure) will require significant investment in this next stage of the cycle.

In the context of new technological paradigms, there is a mismatch between the socio-institutional matrix and the emerging technological paradigm. Silberglitt and colleagues (2006) identify the technologies that could have significant global impacts by 2020 and that countries with low scientific capability could acquire. The role of the construction industry in the development and provision of cheap autonomous housing is one example of such technologies.

The 1990s and the pre-global financial crisis era was an 'age of abundance' (Lindsey, 2007). In most countries, incomes were rising, capital markets were processing very large flows of money, and investment and technological gains meant that ever more information was available ever more cheaply. The 2010s will be an age of relative scarcity of resources as the new industrial giants such as China and India compete for access to oil, minerals and farmland.

This introduction has identified the main issues facing developing countries in the context of the current challenges that will be faced by their economies and societies. The rest of the chapter develops these issues and considers the contribution that the construction industry will be asked to make in the pursuit of progress in national development over the next decades.

Development cycles

The global financial crisis marks the end of a 60-year global development era. During this period, there were two periods of development thinking and practice, which were each about 30 years long. What began in 1950 and continued until 1980 was a period of 'national developmentalism', which focused on the liberation of peoples and economic growth. In Africa, this period marked the early years of nationhood after independence. Economic development during this era was understood as an essential aspect of the building of new nations, and this process was supported through foreign aid. The period from 1980 to 2010 was an era in which the focus of policy was global integration. This was prompted by the economic crisis of the early 1980s and the ideological revolution driven forward in the USA and UK, in particular. To manage the global fiscal and debt crisis, developing countries adopted structural adjustment programmes of liberalisation, privatisation and stabilisation. In each of these periods, there was an inflection in the later stages of the 30-year period, with a shift towards redistribution with growth in the 1970s, and a focus on poverty alleviation and the MDGs after 2000. However, each time, the fundamental orientation – towards national developmentalism in the first period and global integration in the second period – remained at the heart of development thinking and practice.

The global financial crisis, which is considered to have started in spring 2007 as international securitisation markets started to close down, has been a significant event for developing countries. According to estimates by the World Bank (2010a), 55 million more people were pushed into poverty as a result of the economic crash and the decline in growth rates in developing countries (excluding China and India). The rate of economic growth in developing countries fell to 1.6 per cent in 2009. This came on top of the existing chronic unemployment and underemployment in these countries. Other key issues were rising numbers of hungry people and the imperative of climate change, which requires the deployment of sustainable energy technologies globally (World Bank, 2010a).

If the financial crisis is understood as marking the end of a long cycle of development, a sustained and inclusive improvement in the outlook will depend on the widespread deployment of new transport and communications infrastructure and new energy infrastructure, which will act as carriers of a new wave of production and process innovations, and of new patterns of economic development. Energy innovations are particularly critical if, as is often predicted, the point of 'peak oil' (the moment when the global oil supply starts to decline) will be reached during the next decade (Newman, 2008).

Macro-economic challenges for developing economies

No single definition of the term 'developing country' is accepted internationally; the level of development varies widely among the nations which are commonly referred to as developing countries. However, it is pertinent to distinguish the emerging economies, notably Brazil, Russia, India and China (the BRIC economies), whose economies are more advanced and developed than those of the other developing countries, but which do not yet have the full features of a developed country. Their emergence as major players in the world economy has been recognised for several years. For example, *The Economist* (2006) noted that:

> Emerging economies are driving global growth and having a big impact on developed countries . . . As these newcomers become more integrated into the global economy and their incomes catch up with the rich countries, they will provide the biggest boost to the world economy since the industrial revolution.

There is concern about the extent to which the global economic and financial crisis of 2007–2008 has knocked many countries off their tracks of solid growth. Financial markets for developing countries are improving and capital flows are returning but the volume of foreign direct investment (FDI) may remain low as long as economic activity in the developed world continues to be at low levels, particularly as many developed countries will struggle to pursue policies to deal with sovereign debt problems for several years. Table 2.1 shows that, whereas in the advanced countries deficits are likely to continue rising, in the developing world, although fiscal deficits increased in 2009 and 2010, they are projected to fall over the period 2010–2014 (IMF, 2010).

The current economic crisis has led to the deepest recession globally since the Second World War and there may be long-lasting damage to global growth prospects but the short- and medium-term growth prospects for emerging and developing countries are still positive, as can be seen from the IMF (2010) projections presented in Table 2.2, which indicate a variety of recovery rates.

The crisis has greatly affected disposable incomes in many countries, where a contraction in real activity was reinforced by a deterioration in the terms of trade (World Bank, 2010b). In 2008–2010, about one-third of emerging and developing countries were experiencing declines in disposable incomes. This has potentially serious adverse effects on levels of poverty. Macro-economic policy responses have differed but most developing countries first focused on containing the impact of the financial market crisis. Exports by emerging economies to other emerging economies have soared – for example, half of China's exports now go to other emerging countries. In many emerging markets, domestic consumption and investment have increased; over half of China's investment is in infrastructure and property. This pattern is mirrored in other countries. In the Middle East, iconic buildings, urban infrastructure,

Table 2.1 Government debt as a percentage of GDP (2010 and 2014 forecast)

Region	2006	2007	2008	2009	2010	2014
Advanced countries	78	78	83	98	106	114
Developing countries	38	38	36	39	40	35

Source: IMF (2010).

Table 2.2 Global output: percentage change (2010–2013 forecast)

Region	2007	2008	2009	2010	2011–2013
World	5.2	3.0	–0.6	4.2	4.4
Advanced economies	2.8	0.5	–3.2	2.3	2.4
Emerging and developing economies	8.3	6.1	2.4	6.3	6.6

Source: IMF (2010).

new airports and innovative developments such as eco-cities are among development projects which are being financed by petrodollars, and Mexico, Brazil and Russia have launched infrastructure projects that will take decades to complete. Other resource-rich countries also have significantly large development programmes. For example, in 2009 and 2010, Libya spent the largest per capita figures in the world on infrastructure.

Importance of developing economies in the promotion of global growth

Akın and Kose (2007) examined the changing nature of growth spillovers between developed economies and developing countries, driven by the process of globalisation. They note that profound changes have been taking place in the global economy over the past two decades as financial linkages between developed countries and their developing counterparts have become much stronger. Moreover, the emerging countries have differentiated themselves from the others by growing at an extraordinary pace while rapidly integrating themselves into the global economy. Moreover, these emerging economies have become increasingly important players in the global economy as they have begun to account for a substantial share of the world output. These changes have recently been at the centre of an intensive debate about whether emerging market economies can decouple themselves from the slowing economies in the United States and other western countries (see Helbling *et al.*, 2007; Dées and Vansteenkiste, 2007).

The last three decades represent the globalisation era, when there has been a substantial increase in the volume of trade and financial flows. By opening their trade and capital accounts during the globalisation period, emerging market economies have differentiated themselves from other developing countries. Since the mid-1980s, there has also been a concurrent shift in the comparative advantage of the emerging economies from primary commodities to a diversified range of manufactured products as the emerging economies have pursued aggressive industrial policies based on export-driven growth strategies (Weiss, 2005). In particular, both groups have mainly been exporters of primary commodities while manufactured products have constituted the bulk of their imports. However, in the emerging economies, the share of commodities has declined to below 20 per cent while the share of manufactured exports has rapidly increased to three-quarters of the total volume of exports during the globalisation period.

Decoupling

In an era of globalisation, the notion that a recession in the major economies of the developed world must inevitably be followed by similar economic downturns in the developing world seems an obvious one. Developing countries have become much more integrated into the world economy, with the levels of exports of their output increasing from 25 per cent in 1990 to

50 per cent in 2010. The perception of the developing world as an engine for global recovery is one which has been the subject of much debate since the onset of the global financial and economic crisis. A traditional view has been that the economies of developing countries are dependent on the vicissitudes in the economies of their developed counterparts. The issue concerns the possibility of the existence of decoupling from this traditional pattern. Particularly in an era of globalisation, as economies have become more intertwined through trade and finance, business cycles might be expected to become more synchronised, yet there is considerable evidence that decoupling will be the pattern of the future.

It is clear that the situation in the emerging economies is now much different from what it has been in recent decades after economic crises. Since the Second World War, there had been a pattern in which a slowdown in the developed world led to a worsening of the economic conditions in the developing world but, by 2007–2008, decoupling was evident as the US and western European economies slowed but the emerging economies (particularly the BRIC countries) continued to grow. Table 2.3 shows that the estimated economic growth rates of the emerging and developing countries outstrip the world average and the figures for the advanced economies. The rates for China and India are even higher.

This distinction between the growth rates of the developed and the developing world is a continuation of the pattern, prevalent in the years immediately prior to the recession, when the growth rates in the emerging economies were between 4 and 8 per cent higher than those in the developed economies, followed by a recessionary period which was deeper and more prolonged in the developed economies (see Figure 2.1).

Decoupling has advantages for both the emerging economies and the developed ones in a period of economic downturn as the recession-hit countries require an outlet for their exports. However, on the negative side, emerging market-centred growth could make for higher interest rates in

Table 2.3 Estimated growth rates in GDP for 2010 and 2011

Region	Year on year percentage growth in output	
	2010	*2011*
World	4.2	4.3
Advanced economies	2.3	2.4
Emerging and developing economies	6.3	6.5
China	10.0	9.9
India	8.8	8.4

Source: IMF (2010).

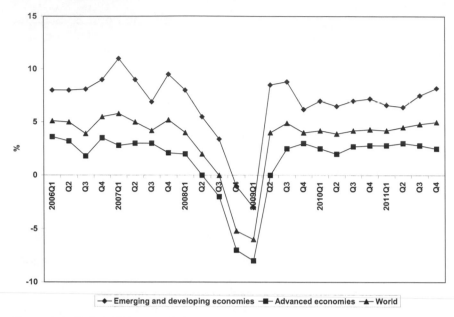

Figure 2.1 Global GDP growth (per cent, quarter to quarter, annualised). Source: IMF staff estimates.

advanced countries, and push up the prices of oil and other commodities when the developed world can least afford it.

Since 2008, government regulators in western countries have been forced to help to save collapsing banks in order to avert a meltdown of the financial system, and sovereign wealth funds from the developing world have been taking large stakes in venerable western banks such as Citibank and UBS in order to keep them liquid – thus tying up valuable funds. The price of food has also been spiralling; the prices of rice and other grains doubled between 2007 and 2008 (Von Braun, 2008), leading to food riots in some developing countries. The prices of most other agricultural products have also been increasing in recent years. In fact, the prices of almost all commodities, including those used for energy and construction, as well as those meant for consumption, have been rising rapidly. The price rises have been fuelled by demand from the emerging countries, particularly the BRIC nations, which, together, have a population of nearly 3 billion people. In order to maintain their high rates of growth and help lift more of their populace out of poverty, they require increasingly greater volumes of commodities. However, a bigger worry is whether the natural resources exist to meet these burgeoning demands. A similar crisis was faced in the 1970s. After a period of strong global economic growth, when the world economy was recording an average of 5 per cent a year of gross domestic product (GDP) increases, the world hit supply constraints in oil and food. For the next 15 years, global GDP growth slowed to an average of 3.2 per cent per year (IMF, 2010). This became known as the 'stagflation' era. Growth opportunities were limited, but prices continued to rise as supply continued to be short.

Economic growth and construction activity

The relationship between the construction industry and national income in the context of economic development has been the subject of many studies in recent decades (see, for instance, Ofori, 1990; Lopes *et al.*, 2002). In these studies, the classical approach in economic growth theory, in which capital formation (particularly physical infrastructure) is the main engine of economic growth and development, has been considerably validated. Construction is an important sector in a developing economy, in terms of its contribution to total output, and the number of workers and proportion of the workforce it employs (Ruddock and Lopes, 2006), as well as the high level of backward linkages through the supply chain.

The developing economies now account for approximately half of all global construction spending and, although the USA remains the largest national construction market, it will soon be overtaken by China, which, with India, continues to be a high-growth economy and leads the way in increases in volumes of construction activity, as illustrated in Figures 2.2 and 2.3. Figure 2.3 shows the rise of China and the decline of the two industrial giants, Japan and the USA, in terms of global construction spending over the

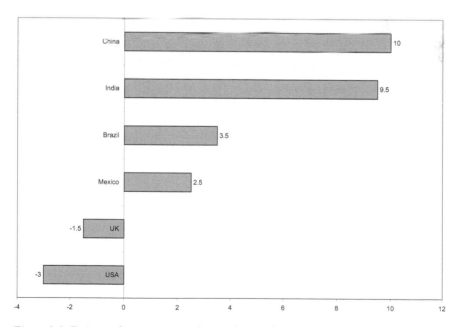

Figure 2.2 Estimated average annual growth (%) of construction spending in selected countries (2008–2011). Source: adapted from Davis Langdon and Seah International (2009).

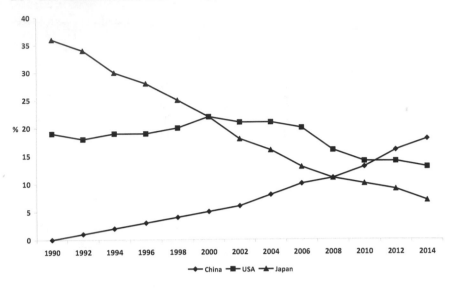

Figure 2.3 Shares of global construction spending. Source: IHS Global Insight (2010).

past two decades. It also presents a projection of spending in these countries for the period up to 2014.

Finance

At a micro level, the World Bank (2000) has proposed a set of strategies for attacking poverty: promoting opportunity, facilitating empowerment and enhancing security. Microfinance is regarded as an essential tool for poverty reduction as it offers poor people access to basic financial services such as loans, savings and money transfer services. The potential for consumption and investment in developing countries will be unprecedented, if the millions of the 'unbanked' are able to join the formal banking system. However, the removal of restrictive regulations, such as the requirement for banking licences, must be expedient if microsavings programmes are to reach the poor.

With global demographic changes and developments in communication technologies, the consequences for financial markets will be far-reaching. Currently, London and New York are the world's only two genuinely global financial centres but other cities will join those two as global centres within a generation. The scale of some of the demographic trends can be difficult to absorb. For example, China's middle class is now bigger than the entire population of the United States and a well-functioning financial sector will be a *sine qua non* for any emerging economy. As a result, the centre of gravity of the world's financial map is beginning to move east and south.

Financing infrastructure

Investment in infrastructure is a fundamental building block of a prosperous economy but, since the onset of the global economic and financial crisis, risk aversion among private equity and infrastructure funds and the ambivalent state of public capital markets have led to a paucity of fresh capital. For instance, the World Bank (2010b) estimates that sub-Saharan Africa is facing a total funding gap of US$31 billion over the next 10 years as traditional finance models for infrastructure have faltered in the wake of the global economic crisis. The market for infrastructure finance has changed and will continue to evolve. The emerging economies have a strong demand for capital. Therefore, a change in capital and investment inflows as profound as the one which occurred after the Asian economic crisis of the late 1990s, when there was a big accumulation of foreign exchange by China and other countries to protect themselves against future calamity, will be called for. The credit crunch has convinced Chinese policy makers that they cannot rely much longer on exports to the west to support economic growth (IMF, 2010). Instead, domestic demand must take up the baton and capital flows will, as a consequence, go into reverse; the money will flow back into emerging domestic demand instead. Having absorbed the west's technology and know-how, the developing world will need its capital too. The growth of international financial flows has overshadowed that of trade flows in the era of globalisation. This unprecedented change has been mainly associated with the rapid liberalisation of capital account regimes since the mid-1980s (IMF, 2010). As a consequence, the composition of capital flows, in particular to the emerging economies, has changed rapidly, and portfolio-equity and FDI inflows have become more prominent.

Major changes in development economics thinking are needed in the context of infrastructure funding. Rogers (2010) suggests that there is a need for private-sector-driven growth, but the state will have to play a facilitating role to promote engagement with the global economy in ways that advance development. In many developing economies, private investment in physical infrastructure would benefit from the fast-tracking of reforms of the regulatory framework.

Sustainable urban development

The building booms of the nineteenth century helped to create the distinctive fabric of European cities, and the twentieth-century building cycle shaped North American cities. The iconic urban forms of the twenty-first century have been, and will continue to be, under construction in Asia. The property cycle has a vital function as a driver of innovation and growth but Barras (2009) warns of the dangers of the inherent instability of the property market and its multiplicity of inter-connections with the wider economy: 'Within the crucible of the worst economic crisis since the Second World

War, the instruments of the next recovery are already being forged. If the price of growth is instability, then out of instability there comes growth.' (Barras, 2009, p. 344).

According to United Nations projections (United Nations, 2009), nearly two-thirds of the people in the developing world will live in cities by 2025. This high rate of urbanisation is occurring primarily in countries which lack the economic, social and physical infrastructure to foster supportive sustainable environments. Slum dwellers, who currently make up one-third of the world's urban population, are at least as poor as rural people, who are the traditional focus of the poverty alleviation programmes implemented in the developing world. The shift in population from the rural to the urban areas has continued at a rapidly increasing rate in the last decade. The rate of population growth for the developing world as a whole is 1.9 per cent but the urban growth rate is much greater, at 3.5 per cent (United Nations, 2009). The cities in the developing world are growing rapidly, placing great strain on infrastructure, such as utilities and roads. Many authors note that there is a disparity in incomes and access to basic facilities in the developing countries (United Nations, 2009). Nowhere is the divide between rich and poor more evident than in the widespread presence of large areas of slum dwellings in the large cities of developing countries. The most rapid economic growth and urban development, as well as the most massive and volatile flows of investment in buildings, are now to be found in Asia. According to the Asian Development Bank (2010), 44 million people are added to the populations of Asian cities each year and every day sees the construction of 20,000 new dwellings and 200 kilometres of new roads.

Managing urban transition

Many developing countries in Africa and Asia are in the early stages of urban transition. According to recent projections, Africa's urban population is expected to grow by over 900 million people in the first half of this century, whilst those of urban areas in Asia will grow by over 2 billion (United Nations, 2010). Figure 2.4 compares levels of urbanisation for the regions of Latin America, Africa, Asia and Europe.

The level of urbanisation in Latin America overtook that of Europe in the 1990s and the region has completed a rapid urban transition over the last 20 years. The trajectory of urbanisation in Latin American countries such as Brazil (with levels of urbanisation higher than those of many European countries) may provide lessons for other countries currently undergoing such transition, according to a study undertaken for the International Institute for Environment and Development (Martine and McGranahan, 2010). The authors stress that, at the beginning of the planning and management of the urbanisation process, the authorities must accept that the poor have the right to be in cities, and then the authorities need to prepare ahead of time for their land and housing needs within a constantly updated

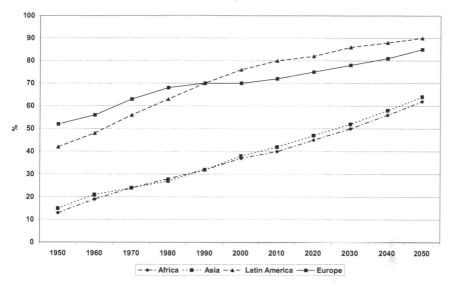

Figure 2.4 Percentage of the population in urban areas (1950–2050). Source: United Nations (2010).

vision of sustainable land-use in the ongoing management of the process. The most effective way of achieving this is to plan ahead and provide land and services, rather than taking remedial actions in future, which are likely to be more costly to the poor city dwellers and the city itself (Angel, 2008; UNFPA, 2007). A critical issue is matching the provision of infrastructure to future sites of urban growth with the rate at which the growth occurs, and recognising that infrastructure has an influence on where that growth takes place.

The prevalence of the informal settlements in which the poor people live makes it difficult for them to gain access to infrastructure and services. A study by Feler and Henderson (2008) found evidence of systematic practices to prevent or discourage poor people from joining the urban communities – a policy of keeping the poor in bad living conditions appears to be considered as an effective way of resisting in-migration and urban growth. As Feler and Henderson observe:

> The development of unserviced informal housing sectors has immediate effects beyond restraining migration: inequality in living conditions and development of unhealthy neighbourhoods with high negative externalities . . . These policies have implications for the future as countries develop and undertake investments to make cities more liveable, which will involve catch-up investments. Building water and sewer infrastructure long after the development of dense neighbourhoods can be very costly, requiring extensive spatial reconstruction and reconstruction of

neighbourhoods. We find evidence consistent with strategic behaviour and that such policies affected population growth rates of localities.

(pp. 1–2)

As a country experiencing a high population drift from the countryside to its rapidly growing cities, China is struggling to develop the social infrastructure necessary to support its large, growing and increasingly urban population. The country also faces the demographic phenomenon of an ageing population – by 2050, China will have more than 300 million people aged 65 or over. (On the other hand, several of the poorest developing countries in Africa will have the world's largest youth populations over the next two decades, and with fertility rates remaining high, pressure will be put on housing, and on education, health care and sanitation infrastructure.)

In the context of sustainability, the most important factor is the construction and use of new buildings as they account for a large part of developing Asia's carbon emissions. This figure may be 30 per cent in the case of China, where nearly half of the world's new floor space is being built each year. Also, the buildings do not age well, as many which were built in the 1990s are already being pulled down and replaced (*The Economist*, 2010). In China, the challenge is being met by green codes which require energy-saving standards to be met in heating, cooling and lighting new buildings. However, many buildings are designed first and greened later; this is a cheaper but less effective approach. Developers are ill-versed in thinking about energy, water and sewage as a seamless whole. State utilities put down power, water and sewerage mains and developers bid for the rights to build blocks with specified numbers of housing units, schools, offices, shops, green space and so on, and the developer builds the block and plugs it into the centralised utilities grid and other elements of the physical infrastructure. The consequential environmental impact of 'superblock' development can be a negative one. It requires a proactive attitude from enlightened developers to put up greener buildings without waiting for governments to issue the regulations and set the tone.

Rural development

Complementary to the need for sustainable urban development is a demand for new or improved infrastructure in rural areas. The food price crisis of 2008 accelerated the acquisition of farmland in developing regions by other countries (Hallam, 2009). Countries which are rich with capital but face land constraints (such as the Gulf states) and those with large populations or food security concerns (such as China, India and South Korea) are seeking to produce food overseas. This is not a new phenomenon; China started leasing land for food in Cuba and Mexico in the 1990s (IFAD, 2009). The form of contracts differ considerably – from land leases and concessions to outright purchases abroad – but the need to secure food supplies through contract farming and investment in rural infrastructure requires civil engineering and construction activity, including the putting in place of agricultural infrastructure and roads. The benefits from such investments include the creation of a significant number of jobs not only in the agricultural sector but also indirectly in the construction sector through the development of rural infrastructure and poverty-reducing improvements such as the construction of new housing, schools and medical facilities. In these regards, foreign investment can provide key resources for rural development including the development of necessary infrastructure and social facilities for the community but it is in the long-run interests of investors, host governments and local people to ensure that practices are sustainable and benefits are shared.

Climate change

The majority of developing countries are in tropical or sub-tropical regions – the regions predicted to be most adversely affected by the impacts of climate change – and the environmental and socio-economic impacts of climate change pose a serious threat to economic development and poverty reduction in these countries. Tackling the causes of climate change and minimising the consequences are inherently linked processes, and both are essential. Adaptation to the impacts of climate change to reduce vulnerability and to moderate the risk of climatic impacts on lives can be achieved through sustainable development, and under the Gleneagles Plan of Action (2005), the G8 group of industrialised nations agreed to assist developing countries to adapt to climate change.

Whereas developing countries are the greatest victims of climate change, the scale of their contribution to the problem, through economic development, is also considerable. According to the *World Development Report 2010* (World Bank, 2010c), poor and middle-income countries account for just over half of total carbon emissions. If developing countries follow the model adopted by developed countries, their energy use and associated emission of greenhouse gases will continue, spurred by population and economic growth. Stern (2006) estimated that a 2°C rise in global temperature will cost about 1 per cent of world GDP but the relative cost for developing countries will be much higher: about 4 per cent of GDP for Africa and 5 per cent for India. Although the carbon footprints of the citizens of developing countries are smaller per capita, they bear a greater burden of the environmental costs.

Natural disasters

Developing countries are more vulnerable to natural disasters than the richer nations for various reasons, one of which is the poor quality of the existing housing and physical infrastructure in disaster-prone regions in the former countries. This is exemplified by the 2004 tsunami, when countries such as Indonesia and Sri Lanka faced one of the worst natural disasters in their history; a large proportion of the loss of property was in housing and infra-structure. After successful emergency relief operations, Sri Lanka initiated post-tsunami reconstruction of housing and infrastructure, which resulted in a massive boost in construction activities in the country. This construction spending caused ripple effects to be transmitted throughout the economy because of the high multiplier effects resulting from the growth in construc-tion activity through the linkages between the industry and other sectors of the economy (Ruddock *et al.*, 2010).

Rising sea levels, due to global warming, present a problem, as the devel-oping countries are particularly prone to flooding, and 10 of the world's 15 largest cities, including Shanghai, Mumbai and Cairo, are in low-lying areas which are vulnerable to rising water levels. The developing countries need to improve the defences against rising sea levels but, to put the costs involved and the funding required into context, whereas the Netherlands spends $100 per person a year on flood defences, in Bangladesh, that figure is a quarter of the average person's annual income.

The biggest vulnerability to climate change is that the weather gravely affects the main economic activities of developing countries, particularly farming. Since two-thirds of the land area of Africa is desert or arid, the continent is particularly exposed as the Sahara encroaches southwards upon previously fertile land.

Water security

The scarcity of water already affects large parts of the world. Consequently, there is an issue of the availability of fresh water, and there is a focus on water security, which refers to people's access to enough safe and affordable water to satisfy their needs for household use, food production and liveli-hoods (Molden, 2007).

It is predicted that temperatures will almost certainly rise – model predictions for a range of emission scenarios suggest an increase in global temperatures of between 1.1°C and 6.4°C by the end of the twenty-first century – and there is much evidence to suggest that fresh water resources could be strongly affected by climate change (Parry *et al.*, 2007).

New technological paradigms

A study undertaken by the RAND Corporation (Silberglitt *et al.*, 2006), to inform the US National Intelligence Council's '2020 Project' and help pro-vide US policy makers with a view of how world developments could evolve, identified technology applications and performed a rough net assessment of these applications according to the technical feasibility, implementation feasibility and whether the application has the potential for global diffusion or whether diffusion would be moderated by the economic and social status of particular countries.

The ability of developing economies to benefit from new technological developments varies considerably and, basically, developing countries fall into three levels of technology capacity (Silberglitt *et al.*, 2006): 'Scientifically proficient' countries (such as China and India) will benefit from many, but not all, of the most sophisticated technology advances; 'Scientifically developing' countries (such as Brazil, Chile, Colombia, Indonesia, Mexico, South Africa and Turkey) will be poised to take advantage of modestly sophisticated technology advances; but 'Scientifically lagging' countries (most developing countries fall into this category) will need to make concerted efforts to eliminate barriers to, and support efforts towards the acquisition of, simple technology advances. The principal drivers of the implementation of advanced technology in developing countries will be human development needs, effective management of resources and prevention of pollution and environmental damage. The most challenging barriers will be effective governance, political stability and matching of the available technology applications to the social and cultural conditions and local needs such as remote power and communications, clean water, and food and shelter. On the basis of the future technologies that could most feasibly be implemented in developing countries, the most significant technology applications that are of direct interest to the construction industry are:

- cheap solar energy: solar energy systems inexpensive enough to be widely available to developing countries as well as to economically disadvantaged populations that are not on existing power grids;
- cheap autonomous housing: self-sufficient and affordable housing that provides shelter adaptable to local conditions as well as energy for heating and cooling.

The emerging economies, if they can address the barriers to implementation, will be able to use technology applications to support continued economic growth and human development of their populations. As emerging technological powers, China and India will have the best opportunity to approach the ability of the scientifically advanced countries to use applications to achieve their national goals. The scientifically lagging countries will face the most severe problems: disease, lack of clean water and sanitation, and environmental degradation. They will also probably lack the resources to address these problems. Consequently, they stand to gain the most from implementing the 2020 technology applications. However, to do so, these nations will need to make substantial progress in building institutional, physical and human capacity. The sponsorship of international aid agencies and developed countries can provide assistance in these efforts, but the developing countries will have to improve governance and achieve greater stability before they will be able to benefit from the application of the available innovations.

Conclusion

In recent years, important human development indicators have been improving. For instance, in the developing world, life expectancy has increased and many of those countries are starting to close the gap between them and the developed world. Child mortality has also decreased in every developing region of the world. Much has been achieved but many countries still remain off-track in the efforts to attain the MDGs. Climate change is hampering efforts to deliver the MDG promise, and will undermine international efforts to combat poverty. Looking to the future, the danger is that the MDG initiative will stall and then reverse progress built up over generations, not just in cutting extreme poverty but also in health, nutrition, education and other areas (UNDP, 2008).

A particularly critical challenge for global and national economic governance in developing countries is to find ways of pursuing sustainability (and mitigating and adapting to climate change) whilst at the same time reducing global inequalities and realising the development aspirations and unrealised human potential of millions of people in those countries. With a focus on sustainable development, an emphasis on the development of productive capacities and a central role for science, technology and innovation, the construction industry will be called upon to play an essential role in achieving those objectives. In the past two decades, there has been widespread enthusiasm for private sector participation in the provision of infrastructure, particularly through various forms of public–private partnership arrangements. There is now a more tempered acceptance that, in the foreseeable economic climate, most of the infrastructure will have to be provided directly from public funds, with recognition of the importance of understanding how to improve the efficiency and quality of service delivery in housing, health facilities, education and infrastructure in other areas. The recent financial and economic crises have shown that both markets and governments can fail spectacularly. Although markets must be the drivers of growth, good government is necessary to provide the conditions for markets to work well and, in the area of public spending, there will need to be an emphasis on efficiency in spending on infrastructure in the social sector.

The form of leadership can be a vital factor in determining the implementation of appropriate strategies, and the existence of a strong 'hard state' government, having the will and authority to create and maintain policies that lead to long-term development that will help all the citizens, has been construed to be a prerequisite to successful development. It is noted that in 1957 South Korea had a lower per capita GDP than Ghana but by 2008 Korea's indicator was 17 times as high as that of Ghana (IMF, 2008). This is an often quoted example of the divergent development paths associated with different styles of government. There is evidence to show that acceleration of progress is possible when strong governmental leadership, effective policies and institutional capacity for scaling up public investments exist and are complemented by adequate financial and technical support from the international community. Some examples can be used to illustrate the success of such supportive measures: Afghanistan's 'Basic Package of Health Services' focused on the construction of health centres and hospitals, training of health workers and large-scale vaccinations. Although this programme was carried out in the midst of the conflicts, the mortality rate for children under five years declined significantly between 2002 and 2004. Similarly, Ethiopia promoted small and medium-sized enterprises and community-based urban works programmes and constructed over 80,000 public housing units to address the 60 per cent rise in the population of slum dwellers between 1990 and 2008 (IMF, 2008).

However, this is not to underestimate the potential role of the private sector in assisting the development process. In particular, multi-national corporations should be regulated so that they follow reasonable standards for wages and labour conditions, pay reasonable taxes and reinvest some of their profits in the country. Whether investment comes from private, public or agency funding, good infrastructure will continue to be a prerequisite for economic and social development in developing countries.

References

Akın, C. and Ayhan Kose, M. (2007) *Changing Nature of North–South Linkages: Stylized Facts and Explanations.* IMF Working Paper 07/280, Washington, DC.

Angel, S. (2008) Preparing for urban expansion: a proposed strategy for intermediate cities in Ecuador. In G. Martine, G. McGranahan, M. Montgomery and R. Fernandez-Castilla (eds) *The New Global Frontier: Cities, Poverty and Environment in the 21st Century.* IIED/UNFPA: London.

Asian Development Bank (2010) *Sustainable Transport Initiative Operational Plan: Policies and Strategies.* Asian Development Bank: Manila.

Barras, R. (2009) *Building Cycles: Growth and Instability.* Wiley-Blackwell: Chichester.

Davis Langdon and Seah International (2009) *World Construction 2009.* David Langdon: London.

Dées, S. and Vansteenkiste, I. (2007) *The Transmission of U.S. Cyclical Developments to the Rest of the World.* ECB Working Paper 798, European Central Bank: Frankfurt.

The Economist (2006) Leader. 16 September.

The Economist (2010) Asia's alarming cities. 1 July.

Feler, L. and Henderson, J. V. (2008) *Exclusionary Policies in Urban Development: How Under-Serving of Migrant Households Affects the Growth and Composition of Brazilian Cities.* Working Paper 14136, National Bureau of Economic Research, Cambridge.

Gleneagles Plan of Action (2005) *Climate Change, Clean Energy and Sustainable Development.* Prime Minister's Office: London.

Gore, C. (2009) *Development: What's Next and What Are We Going to Do about It?* Presentation to UKCDS Board, 21 October 2009, Wellcome Trust, London.

Hallam, D. (2009) *Issues, Policy Implications and International Response*. Global Forum on International Investment, OECD: Paris.

Helbling, T., Berezin, P., Ayhan Kose, M., Kumhof, M., Laxton, D. and Spatafora, N. (2007) Decoupling the train? Spillovers and cycles in the global economy. *World Economic Outlook*, April, 121–160. International Monetary Fund: Washington, DC.

IFAD (2009) *The Growing Demand for Land*. Discussion paper, International Fund for Agricultural Development, United Nations, Washington, DC.

IHS Global Insight (2010) http://wwwIHSGlobalInsight.com/Construction

IMF (2008) *World Economic Outlook*. International Monetary Fund: Washington, DC.

IMF (2010) *World Economic Outlook*. International Monetary Fund: Washington, DC.

Lindsey, B. (2007) *The Age of Abundance: How Prosperity Transformed America's Politics and Culture*. HarperCollins: New York.

Lopes, J., Ruddock, L. and Ribeiro, F.L. (2002) Investment in construction and economic growth in developing countries. *Building Research and Information*, 30, 152–159.

Martine, G. and McGranahan, G. (2010) Brazil's early urban transition: what can it teach urbanizing countries? Special issue of *Urbanization and Emerging Population Issues*, vol. 4.

Molden, D. (2007) *Water for Food, Water for Life: A Comprehensive Assessment of Water Management in Agriculture*. Earthscan: London.

Newman, S. (2008) *The Final Energy Crisis*, 2nd edition. Pluto Press: London.

Ofori, G. (1990) *The Construction Industry: Aspects of Its Economy and Management*. Singapore University Press: Singapore.

Parry, M., Canziani, O., Palutikof, J., van der Linden, P. and Hanson, C. (2007) *Climate Change 2007 Impacts, Adaptation and Vulnerability*. Cambridge University Press: Cambridge.

Rogers, F. H. (2010) *The Global Financial Crisis and Development Thinking*. Policy Research Working Paper 5353, Development Research Group, World Bank: Washington, DC.

Ruddock, L. and Lopes, J. (2006) The construction sector and economic development: the 'Bon' curve. *Construction Management and Economics*, 24, 717–724.

Ruddock, L., Amaratunga, D., Wanigaratne, N. and Palliyaguru, R. (2010) Post-tsunami reconstruction in Sri Lanka: assessing the economic impact. *International Journal of Strategic Property Management*, 14, 219–232.

Silberglitt, R., Anton, P. S., Howell, D. R., Wong, A., Gassman, N., Jackson, B. A., Landree, E., Pfleeger, S. L., Newton, E. M. and Wu, F. (2006) *The Global Technology Revolution 2020, In-Depth Analyses: Bio/Nano/Materials/Information Trends, Drivers, Barriers and Social Implications*. TR-303-NIC, RAND Corporation: Santa Monica, CA.

Stern, N. (2006) *The Economics of Climate Change*. HM Treasury: London.

UNDP (2008) *Human Development Report 2007/2008*. United Nations: New York.

United Nations (2009) *World Population Prospects*. United Nations Publications: New York.

United Nations (2010) *World Urbanization Prospects, 2009*. Population Division: New York.

UNFPA (2007) *State of the World Population 2007: Unleashing the Potential of Urban Growth*. United Nations Population Fund: Washington, DC.

Von Braun, J. (2008) *Assessing the Challenges of a Changing World Food Situation: Preventing Crisis and Leveraging Opportunity*. US Agency for International Development Conference: Washington.

Weiss, J. (2005) *Export Growth and Industrial Policy: Lessons from the East Asian Miracle Experience*. Working Paper, Asian Development Bank Institute.

World Bank (2000) *World Development Report 2000/2001: Attacking Poverty*. Oxford University Press: New York.

World Bank (2010a) *The MDGs after the Crisis: Global Monitoring Report*. World Bank: Washington, DC.

World Bank (2010b) *Global Economic Prospects 2010*. World Bank: Washington, DC.

World Bank (2010c) *World Development Report 2010: Development and Climate Change*. World Bank: Washington, DC.

3 Construction in the economy and its role in socio-economic development

Jorge Lopes

Introduction

The contribution of the construction industry to the national economy and its role in socio-economic development has been addressed by various authors and international organisations, and many of them have focused on developing countries. A particular feature is that construction is the only sector of the economy that appears twice in the national accounts of any country: as a major component of fixed capital formation and as a sector that contributes to the gross domestic product (GDP) (Hillebrandt, 2000). Historical reviews point to the importance of construction in the process of industrialisation and urbanisation that followed the advent of the industrial revolution in western and northern Europe and other parts of the globe. As regards the association between construction and economic growth, several writers (Strassman, 1970; Turin, 1973; World Bank, 1984; Wells, 1986; Bon, 1990) have analysed the changing role of the construction industry at various stages of economic development and presented a development pattern for the industry based on the stage of development of a country's economy. One of the main features drawn from these works is the common assumption that directly relates the measures of construction output to a country's level of economic development. Furthermore, in this view, the construction sector, as a major component of a country's physical capital, plays a determinant role in the development process.

The positive association between construction (indeed physical infrastructure) and economic growth has been the subject of debate on the part played by the proponents of endogenous growth theory and international organisations such as the World Bank in the structural adjustment programme for Africa. Indeed, in the aftermath of the 1979–1980 oil-shock and the international financial crisis that followed in 1981, most of the sub-Saharan African countries experienced decreasing growth in per capita income until the late 1990s (see Table 3.1) despite heavy investment in construction and other physical capital over the period 1970–1980. The World Bank (1994) posited that, rather than the quantity of infrastructure, the main concern in developing countries should be the improvement of the quality of infrastructure. According to this reasoning, it could be argued that this would be achieved through adequate maintenance and upgrading of existing infrastructure stocks and by prioritising investments that modernise production and enhance international competitiveness.

The argument above brings about the other side of the construction industry and ancillary industries. Construction projects, particularly public infrastructure projects, require large amounts of national resources. In less developed countries, quite often, the 'real costs' of a major construction project are understated if one considers the figures presented in the national accounts tables. Technical assistance (usually paid for in foreign currency) and some other unexpected costs can significantly inflate the costs of a construction project. Thus, it could be argued that in the resource-constrained developing countries, part of the scarce resources which are devoted to investment in construction projects could alternatively be used in other important sectors of the economy (such as health, education and agriculture).Therefore, African governments and their development partners shifted economic policies away from infrastructure investment to macro-economic stabilisation accompanied by social intervention.

In the early 2000s, international organisations and development agencies started to become aware of the important role infrastructure would play in the efforts to attain the Millennium Development Goals (MDGs) in sub-Saharan Africa (Organization of African Unity, 2001; Commission for Africa, 2005, cited in World Bank, 2009a). An important question which should be the concern of the construction economics research community and national and international development agencies is how a well-functioning construction industry could contribute to sustainable economic growth and development.

The structure of this chapter is as follows: the next section discusses the concepts of economic growth and economic development, and explores relevant issues in the area of development economics. The main approaches in growth theory are discussed, and a review of the literature on the role of the construction industry in the process of economic growth and development is undertaken. The third section presents quantitative analyses of the relationship between the measures of construction output and those of the national aggregate in two groups of countries in sub-Saharan Africa according to their stages of economic development: low-income countries (LICs) and middle-income countries (MICs). The statistical sources and data are presented and commented upon, and the analysis and discussions of the results are elaborated upon. The fourth section explores the link between construction investment and economic and social targets related to the MDGs. A concluding section finalises the analysis presented in this chapter.

Economic growth and development: an overview

The study of modern economic growth can be traced in the literature to Adam Smith's *Wealth of Nations*, published in 1776. From then on, economists have tried to understand what makes national economies progress. For about as long, they have analysed how income distribution and growth are connected (Todaro, 1992; Thirlwall, 1994). This section discusses the main features of the determinants, measurement and comparability of economic growth and development. Then it presents a review of the literature on growth theory and the role of capital formation, and a discussion of the role of construction in economic development.

Determinants, measurement and comparability of economic growth and development

The major determinants in the process of economic growth and development are increase in population; increase in per capita and total national output; and external relations among countries. These have determined the common historical trends in the process of development of nations, particularly of the advanced industrial countries (Kuznets, 1968). However, the role of capital formation is central in any discussion of economic growth and development. Indeed, as Maddison (1987, p. 656) pointed out in his study of six advanced industrial countries, the close correlation between capital and output movements over the long run is the reason why simple regressions find capital such a powerful explanation of growth. It is worth pointing out that the concept of capital has been evolving. The World Bank (1998) identified four types of capital: (i) human capital (the stocks of knowledge and skills in the population); (ii) physical capital (the plant, machinery, equipment and economic infrastructure); (iii) natural capital (natural resources and the environment); and (iv) social capital (the shared values and institutions which give cohesion to society). More recently, Hess (2010), in a study of the determinants of growth in developing economies, used the concept of 'adjusted net saving rate', which incorporates not just physical capital depreciation, but natural capital depletion and environmental damage.

'Economic development' and 'economic growth' are terms often used interchangeably in much of the economics literature (Low, 1994). However, economic growth can be defined as the steady process by which the productive capacity of the economy is increased over time to bring about high levels of national income (Todaro, 1992). In his study, *Economic Growth in the Third World*, Reynolds (1985, pp. 7–8) suggests that the process of economic growth generally comprises two phases, both of which deserve special study: (i) a period of *extensive growth*, a situation in which population and output are growing at roughly the same rate, with no secular rise of per capita output; (ii) a period of *intensive growth* in which, after a certain point in time, a sustained rise in a country's per capita income occurs. The beginning of this sustained rise in per capita income is called the *turning point*. It should be noted that the reaching of the turning point is not a guarantee for growth in the long-term future, and there are cases where growth has changed to stagnation or decline (p. 8). This concept of economic growth appears to be appealing, and it also seems to be shared by Kuznets (1968), so that it has the merit of placing population analysis in a prominent place in the development process. With respect to the turning point, the development pattern of most sub-Saharan African countries since the 1960s appears to have added value to Reynolds's (1985) development paradigm.

Development, and to a certain extent economic development, is a more ambiguous concept among policy makers and social scientists. The World Bank (1989) refers to 'sustainable development', that is, a strategy of development in which growth in real output must be accompanied by measures that address poverty alleviation, and does not compromise the welfare of the next generations. Goulet (1971, cited in Thirlwall, 1994, p. 8) distinguishes three components in his wider view of development, which he calls life-sustenance, self-esteem and freedom. It follows that, in this view, development is a process that pursues continuous improvement in these issues. In a similar vein, Todaro (1992, p. 487) defines development as the process of improving the quality of human lives through three equally important aspects: (i) raising people's incomes and consumption levels of food, medical services, education and so on; (ii) creating conditions conducive to the growth of people's self-esteem through the establishment of social, political and economic systems and institutions that promote human dignity; and (iii) increasing people's freedom to choose by enlarging the range of their possible choices in consumer goods and services. Mabogunje (1989), adopting a multi-dimensional perspective with regard to the concept of development, points out that any theory of development will ultimately reflect the social, historical and national background of its author.

The issue of economic development not only connotes different implications for different people, but also defies exact measurement (Low, 1994). However, there is an increasing consensus among authors concerned with the issue of development that any notion of strictly economic progress must, at a minimum, look beyond growth in per capita income to the reduction of poverty and greater equity, to progress in education, health and nutrition, and to the protection of the environment (World Bank, 1991). From this definition, it is clear that the issues of poverty, growth and the protection of the environment underlie the development agenda as reflected in the UN's Millennium Declaration (UN, 2000).

The main indicators of the national output of a nation are gross national income (GNI) and gross domestic product (GDP). Basically, the latter indicator measures what is produced within a country's borders, and is a better measure of growth in productive capacity. GNI measures how much of what is produced belongs to residents of the country, and is more closely related to changes in welfare. In the UN Systems of National Accounts (SNA), GNI is derived from the GDP adjusted to the net factor incomes (labour and capital) with the rest of the world. Alternatively, one can consider GNI per capita and GDP per capita when the variable, population, is introduced. In modern history, since the Second World War, growth in output per capita, generally, runs 0.5 to 1.5 per cent below growth in output in developed countries, and 2.0 to 2.5 per cent below in developing countries, so that the population growth rates are typically in these ranges for developed and developing countries, respectively (see UN *Demographic Yearbook*, various years).

GNI per capita has been utilised by institutions such as the World Bank and Organisation for Economic Co-operation and Development (OECD) to compare living standards among their member countries. In order to facilitate international comparisons, a common currency is used, usually the US dollar. The World Bank publishes annually the *World Development Indicators* in which member countries are ranked and classified according to GNI per capita measured in US dollars. The income groups are as follows: low-income economies (LIE); middle-income economies (MIE); and high-income economies (HIE). The middle-income group is further divided into two sub-groups: lower-middle-income and upper-middle-income. The value limits for each group tend to vary annually upwards, because the overall size of the world's economy tends to generally grow.

From the discussion above, it can be suggested that the development process is not merely an increase in the national output per capita. Furthermore, the weighted average of the growth of income of different groups of people pays no regard to the distribution of income. One of the measures used to express income inequality in any economy is the *Gini coefficient of distribution* (for the method for calculating this index, see, for example, Thirlwall, 1994), which varies from 0 (complete equality) to 1 (complete inequality). According to Kuznets (1965, cited in Thirlwall, 1994, pp. 13–14), the gap between rich and poor would initially increase in less developed countries, and only in later stages of industrialisation will the degree of inequality tend to decrease. Deininger and Squire (1996) examined the relationship between growth and inequality in all regions in the world. Their findings seem to contradict those of Kuznets. They calculate the average Gini coefficients for the world's regions for the period 1960 to 1990, and their main results show that, although differences in inequality among regions are very large (from around 0.5 in Latin America and slightly less than 0.5 in sub-Saharan Africa to a value of about 0.35 in high-income countries), they have changed little over time. Since, in that period, different regions of the world have experienced varied growth rates (an upswing in East Asia and stagnation in sub-Saharan Africa), the stability of the Gini coefficient would suggest that economic growth does not necessarily increase inequality. However, Barro (2000), using the framework for the determinants of growth that was developed in his earlier studies, offers some support to Kuznets's findings. The *Kuznets curve* – whereby inequality first increases and later decreases during the process of economic development – emerges as a clear empirical regularity. However, this relationship does not explain the bulk of variations in equality across countries over time (Barro, 2000, p. 29).

Other measures of the welfare of a nation are health-related and education indicators. In the former, the indicators which are most often used to find out the stage of development of a nation are the number of people per doctor, life expectancy at birth, infant mortality rate and the percentage of people with access to safe water and sewerage. With regard to education, the most common measures used in the statistics are the percentage of the age groups enrolled in primary, secondary and tertiary educational institutions. Since 1990, the United Nations Development Programme (UNDP) publishes annually the *Human Development Report* (UNDP, 1990–), which gives alternative measures of the welfare of nations that do not necessarily accord with the standard measure of the level of income per capita. Thus, the UNDP constructs a Human Development Index (HDI) which takes into account the measures of income per head (adjusted for international comparisons) combined with measures of life expectancy and literacy. Countries are then ranked by the index, from lowest (0) to highest (1), and compared with the ranking of per capita income. In the low- to middle-income range, some countries rank low by per capita income and high (relatively) by the HDI, and vice versa. In the high-income economies, although there may be relative variations in a country's ranking using the two indices, a high income per capita tends to generally correspond to a high HDI.

Another attempt to improve comparison of living standards of different economies is through the *purchasing power parity* among countries, especially between nations at different stages of development. It is well known that the exchange rates are, in the main, determined by the demand and supply of goods and services which are traded on the international market. The rationale for the determination of the purchasing power parity is that the national output of a country comprises not only traded goods but also non traded goods (and also government services) in which prices are determined by unit labour costs, which tend to be lower, the poorer the country. There are several methods for determining purchasing power parity ratios and prices (see UN, 1985; Summers and Heston, 1991) to make international comparisons in which the currency of a country (usually international US dollars) or a group of countries acts as the unit of account without altering the ratios of living standards between countries. This methodology has been utilised by the OECD in the compilation of the national accounts statistics of its members.

Apart from the World Bank classification of the economies of the world (HIEs, MIEs and LIEs), other country group classifications have been used in much of the social science literature. The UN and its agencies, and also the World Bank, use the terminology of 'developing countries' to represent a wide group of countries at varied stages of development, excluding the former Union of Soviet Socialist Republics (USSR) and OECD countries. Another criterion for classifying economies is according to the degree of industrialisation. In this view, economies are classified as advanced industrial countries (AICs), newly industrialised countries (NICs) or less developed countries (LDCs). Basically, the HIEs correspond to the AICs, the upper-middle-income economies correspond to the NICs, and lower-income economies added to the lower-middle-income economies represent the LDCs (Bon, 1992; Ruddock and Lopes, 2006).

Growth theory: a brief review

Growth theories have been evolving from a classical approach, focused primarily on physical capital accumulation as the main engine of growth, to one where technology and knowledge play an increasing role in the development process. The recent literature on economic growth has emphasised the importance of both human capital and the quality of institutions for economic development. A debate has emerged about the relative importance of these two factors, and the extent to which institutions cause human capital accumulation and growth, or vice versa (Eicher *et al.*, 2009).

The classical approach to growth theory envisaged that output per capita would be stationary as the rate of output declined, with diminishing improvements in productivity. Rosenstein-Rodan (1943) postulated 'the big push' by which output would grow in proportion with capital and then an economy would propel itself into self-sustaining industrialisation and growth. Along the same lines, Lewis (1954, p. 150) stated:

> the central problem in the theory of economic development is to understand the process by which a community which was previously saving 4 or 5 per cent of its income converts itself into an economy where voluntary saving is about 12 to 15 per cent of national income. This is the central problem because the central fact of economic development is rapid capital accumulation including knowledge and skills with capital.

Rostow (1960, 1963) proposed the theory of stages by which it is possible to identify stages of development and to classify societies according to those stages. He distinguished five stages: traditional, transitional, take-off, maturity and high mass consumption. Central in Rostow's thesis is the transition from take-off to maturity, that is the beginning of industrialisation. According to Rostow (1960), the take-off is a short stage of development during which economic growth becomes self-sustaining. Thus, investment must rise to a level of over 10 per cent of national output in order for per capita income to rise sufficiently to ensure adequate levels of investment. Another condition is that one or more 'leading growth sectors' must emerge. Investment in the transport infrastructure was singled out as an important component for facilitating this successful transition. Rostow's propositions were the subject of some debate. Kuznets (1963) argued that Rostow's findings lacked statistical evidence, and the analysis of take-off and preconditions of the stages neglected the effect of historical heritage and the time of entry into the process of modern economic growth (Kuznets, 1963, p. 40). It is also worth noting that De Long and Summers (1991) distinguish two types of capital (machinery and equipment, and construction investment), which have different influences in the growth process. They analysed data on 61 countries of all continents over the period 1960–1985, and found no significant association between construction investment and economic growth. In contrast, they reported that economic growth was strongly associated with investment in machinery and equipment. Despite these criticisms, Rostow's preconditions for take-off have highlighted the need for investment in construction projects and its role in the development process (Low, 1994).

Based on the growth experience of the United States, Solow (1956) developed the neo-classical model of economic growth. In this approach, the permanent rate of growth of output per unit of labour input is independent of the investment rate and depends entirely on technological progress. Some figures from the World Bank (1991) illustrate Solow's (1956) model. The total output of the United States in the first part of the twentieth century grew at about 3 per cent per year. Its capital stock also grew at about 3 per cent per year, whereas the labour input grew at only about 1 per cent per year. In the capital–labour mix, capital accounts generally for about one-third, and labour two-thirds. Thus, inputs were rising at 1.7 per cent per year. The residual (i.e. increase in total factor productivity), which Solow called 'technological progress in the broadest sense', accounted for the major part of the growth in output. However, most of technological progress finds its way into actual production only with the use of new and different capital equipment (Solow, 1988, 1994). That is, countries would grow only through exogenous technological changes embodied in machinery and equipment. As an implication of diminishing returns, this model also postulates that growth rates would be expected to converge across countries.

However, the growth rates of developing countries have diverged particularly since the first oil-shock in 1973–1974. This seemed to contradict the expectation of convergence, and has been a major concern among development economists and development agencies, but in practice the path of technological change has not been similar. Neither has it been transmitted in most developing countries. It has been argued (see, for example, Agarwala, 1983) that excessive industrial protection, tariffs and other import restrictions, and market distortions have encouraged an inefficient pattern of production and growth. Thus, development economists have felt the need to go beyond the neo-classical theory to understand the experience of developing countries (Romer, 1989, p. 203).

The endogenous economic growth approach (Lucas, 1988; Romer, 1989, 1990) envisages that investment in human capital produces increasing externalities in the development process. Central in Romer's thesis is the role played by research and innovation in the process of development. Romer's (1990) model uses a simplified version of the production function which combines labour input and capital input with a variable that represents technology. However, technology corresponds to the number of intermediate capital inputs for which satisfactory designs have been developed. The model postulates that growth in the variable technology can sustain long-term growth of per capita income and explains why growth rates have been diverging, particularly since the post-Second World War period. Thus, technology and knowledge are taken as factors of production and are not left aside in the peripheral concept of the 'residual' in the neo-classical approach. The role of the government, then, is to ensure a macro-economic environment that favours these dynamics of permanent innovation (Romer, 1994). Barro (1991), using a cross-sectional study of data on almost 100 countries for the period 1960–1985, found that the growth rate of GDP per capita was positively related to the initial human capital and the average rate of physical investment, and negatively related to the initial level of GDP per capita in the beginning of that period. Barro (1997) extended the sample period to 1995 and reported similar results. O'Neil (1995) seems to confirm these findings and suggests that the divergence in growth rates is a consequence of different levels of human capital across countries. The spectacular growth process of some Eastern Asian economies in the period after the Second World War offers some support to the endogenous economic growth approach.

Since the founding article of the endogenous growth theory (Romer, 1986), various versions of the endogenous growth model have been proposed (see Madsen, 2008, for an extensive review of the literature). Some of the studies can be seen as a revival of the neo-classical approach (see, for example, Mankiw *et al.*, 1992). Jones (1995), using a modified version of the endogenous growth model, concludes that, although economic growth is generated endogenously through research and development, the long-run growth rate depends only on parameters that are exogenous, including the rate of growth of population.

Another strand of the literature on economic growth has been revolving around the notion that heterogeneity of a country's growth process is of fundamental importance in the study of economic growth (Owen *et al.*, 2009). A number of studies which apply cross-country growth regression models (see, for example, Barro, 1991) treat all countries of a continent as belonging to a homogeneous group, after conditioning on other relevant variables (such as investment ratios, initial GDP levels and school enrolment ratios). These models typically use a parameter of heterogeneity that depends upon the geographical grouping of countries, and includes interactions of terms of continent dummies with explanatory variables. Regarding sub-Saharan Africa, explanations for the continent's long-term stagnation that have been put forward in the literature are diverse. They include geographical location; ethnic diversity; choice of political and economic institutions; insufficient infrastructure; limited openness to international trade; and the lack of social capital (Paap *et al.*, 2005). Some of these works point to the existence of a 'poverty trap' (i.e. zero growth in a long-term perspective) in developing countries and could be considered a revival of the Big Push theory (Bloom *et al.*, 2003). According to Easterly (2006), this renewed interest in the classical approach of growth theory as a framework for shaping development policy making is partly motivated by the international effort to meet the MDGs in developing countries by the year 2015. It is worth nothing that Easterly (2006) found that there was no clear evidence of a poverty trap and that the evidence of divergence between rich and poor nations in 1960–2002 was associated more with the quality of institutions than with initial per capita income. However, a report of the UN Millennium Project (UN, 2005, cited in Easterly, 2006, p. 290) argued that poor countries were in a poverty trap. Escaping the trap requires:

> A big push of basic investments between now and 2015 in public administration, human capital (nutrition, health, education), and key infrastructure (roads, electricity, ports, water, and sanitation), accessible land for affordable housing, environmental management.

Other writers (see, for example, Paap *et al.*, 2005; Owen *et al.*, 2009) have contended that group membership is not determined by geography and that the quality of institutions is an important factor that sorts countries into different growth regimes. Paap and colleagues (2005) analysed a data set on growth rates of real GDP per capita in 69 countries in sub-Saharan Africa (34), Latin and Middle America (14), Asia (13) and North Africa and the Middle East (8) over the 40-year period from 1961 to 2000. They found that there were three clusters of growth rates in that sample (low, medium and high growth rates) and that one out of four countries in sub-Saharan Africa presented a development pattern similar to those of many Asian and Latin American countries. These results are in line with the findings of a recent study (Radelet, 2010, cited in Center for Global Development, 2010) that the steady economic growth achieved by 17 emerging countries in sub-Saharan Africa since the mid-1990s has been driven by five fundamental changes: (i) more democratic and accountable governments; (ii) more sensible economic policies; (iii) the end of the debt crisis and changing relationships with donors; (iv) the spread of new technologies; and (v) the emergence of a new generation of policy makers, activists and business leaders.

Role of construction in economic development

Historically, the construction industry has been linked with the process of economic growth and development. It has been noted above that the economic growth of a nation can be defined as a sustained increase in its population and product per capita. With growth in population, there is a need for more products 'that permit us to feed, clothe, and shelter ourselves – the structures in which our goods are produced and stored, over which goods are shipped to market and in which goods are consumed' (Lange and Mills, 1979, p. 1). The increase in per capita and global product is related to the construction industry in the sense that various activities of the industry provide the facilities indispensable for undertaking activities in, and thus for developing, other sectors of the economy, and the construction industry has direct links with the manufacturing industry – the main partner of construction in the process of economic growth and development.

Although the direct contribution of the construction industry to economic growth is significant, it also contributes to the basic objectives of development including employment creation and income generation and redistribution. The extent to which growth and employment creation should be balanced depends largely upon technical, economic and social conditions. Construction could play an important role in resolving this conflict because it is technologically flexible, implying that many of its operations can be more or less labour-intensive depending upon technical conditions and available resources in the country at the time (Moavenzadeh and Koch Rossow, 1976).

Duccio Turin (1966) was one of the first authors to analyse the relationship between the construction industry and the macro-economy in economic development. Turin (1973, p. 1) highlighted the purpose of a later study as:

> to provide guidance to the policy-making bodies responsible for the development of the construction industry by drawing their attention to the nature of the construction process . . . to the steps that could be taken to remove some of the existing and future constraints in the vital areas of materials, manpower, financial resources, organisation and management, institutional set-up and statutory requirements.

Paul Strassman was also one of the pioneers in the study of the macro-economics of the construction industry and its role in socio-economic development. Strassman (1970, cited in Han and Ofori, 2001, p. 190) argued that construction was a major force replacing the manufacturing industry to drive economic growth after the initial stage of development, and he postulated the 'middle-income country bulge' concept.

Turning back to Turin's (1973) purpose, it covered all levels of economic activity and concerned all vital players in the construction industry. The sample analysed by Turin (1973) consisted of 85 countries in all continents representing all stages of economic development, in the period 1955–1965. Countries were ranked in five groups in decreasing order of average per capita product, from a maximum (at the time) of US$3,130 (North America) to a minimum of US$130 for some countries of Asia and Africa. The observations made by Turin (1973) can be summarised as follows:

1   Construction value added (CVA) was 3 to 5 per cent of GDP in developing countries and 5 to 8 per cent in industrialised countries.
2   Capital formation in construction was 6 to 9 per cent of GDP in developing countries and 10 to 15 per cent in industrialised countries. In all countries, construction gross output accounted for 45 to 60 per cent of gross capital formation.

3 Using cross-country comparisons, there exists a direct relationship between the level of GDP per capita and the level of the construction industry activity (measured by the share of CVA in GDP).

4 The construction industry bought between 50 and 60 per cent of its non-primary inputs from other sectors of the economy. The building materials sub-sector accounts for most of these inputs.

5 Developing countries directed 30 to 50 per cent of construction investment to civil engineering whereas the developed nations devoted 25 to 30 per cent.

The World Bank (1984) corroborated the findings of Turin's (1973) study by considering data for the period 1970–1980, and ranked the construction industry in developing countries as fourth out of 20 sectors of the economy in terms of inter-sectoral linkages. The importance of the construction industry in economic growth in the developing world has long been recognised by the World Bank and its affiliates. This is not surprising given the nature of the World Bank since its conception at the Bretton Woods Conference just before the end of the Second World War. Although it has the name of a bank, its original aim was the promotion of economic and financial co-operation among member states. Later, with the formation of other international and regional groupings concerning the richer countries of the world (such as the then European Economic Community and the OECD), the World Bank diverted its attention from its original concern and started to address the financing and monitoring of development projects in less developed countries and the newly independent states of Africa and Asia. According to its own statistics, 44 per cent of the total cost of projects approved for assistance by the World Bank and its affiliates in the three-year period fiscal 1980–1982 went to construction work. Besides addressing the macro level of the construction industry, the World Bank (1984) went further. It suggested directions for future actions to promote the development of its member states and proposed a set of measures for all levels of construction industry activities (demand-side, supply-side, institutional set-up and research activities) to improve the efficiency of this important sector of the national economy. It should be noted that earlier and further works commissioned or supported by other international development agencies addressed the same issue, with a particular concern on developing economies (see, for example, Moavenzadeh and Rossow, 1976; UNCHS, 1982; UNIDO, 1985). The first two works address the contribution of construction to socio-economic growth and development; the last-mentioned deals with the role of the building materials sub-sector in a developing country's industrialisation strategy.

Wells (1986) analysed a sample of more than 100 countries in all continents, representing all stages of economic development, and followed closely the methodology adopted by Turin (1973). However, an important contribution from her work was the establishment of a mathematical model relating different measures of construction activity – construction value added, gross construction output, and employment in construction – to the level of GDP per capita. It should be noted that, compared with Turin's 1973 study, the work by Wells benefited from a longer period of consideration in the analysis (1960–1980) and the continuous improvement in statistical coverage, not to mention the number of separate countries which then existed (by 1980). Again, the countries were grouped according to the level of income: from group 1 with an average income over US$2,000 to group 4 with an average income under US$350 in 1980. The main findings of Wells (1986) can be stated as follows: (i) the construction output as a percentage of GDP is related to GDP per capita in an increasing form of income level; and (ii) if the relationship between countries at different income levels at a fixed point in time also occurs within any country over time, then construction output increases, in relative terms, with increasing per capita GDP in any country over time. The changes in this ratio would be faster for countries in the middle-income range.

Turin's (1973) analysis, and those of the researchers who adopted methods similar to his, have been challenged on a number of grounds: the reliability of the data, the limitations of the coverage, the appropriateness of the methods of analysis employed and the assumption of construction as an engine of economic growth and development. Regarding the last-mentioned point, Drewer (1997, cited in Han and Ofori, 2001, p. 191) suggested that, at best, construction could be an effective motor of economic growth in a limited context, and over a short period, whereas, at the other extreme, economic growth led by uncontrolled expansion of the construction industry might lead to disastrous economic consequences.

Bon (1990) analysed the relationship between the construction industry and the national economy, and presented an overview of several global trends concerning the changes in construction employment and sectoral share in GDP over six continents for the period 1970–1985. The analyses on the changing role of the construction industry at various stages of economic development and the development pattern of the construction industry within the historical processes of industrialisation and urbanisation were further developed in Bon (1992). The main aspect of the propositions was that, in the first stages of economic development, the share of construction in GDP increases, and this proportion decreases in the last stages of development. At some stage, construction volume will decline, not only relatively, but also absolutely. However, Ruddock and Lopes (2006) found that there was no clear evidence of an absolute decline of construction activity in advanced industrialised countries.

Other researchers on the development process might put particular emphasis on the expansion of other sectors of the economy based on their perceptions of the most appropriate strategy for development, but they generally tend to agree on the need for the construction sector to grow (Ofori, 1990, 1993). Some of the authors focus on the issue of employment creation

(Ganesan, 1979, 1994); others emphasise the multiplier effect on other sectors of the economy (Currie, 1974; Bon, 1991). Currie (1974) proposed a development strategy for a developing country based on the 'leading sectors', by which an export-led promotion strategy added to an expansion in the building industry would lead a country to sustained economic growth and development. In this view, for a sector to qualify as a 'leader', it should have two characteristics: an unexplored or latent demand that can be actualised and a sufficiently large demand to cause its satisfaction to have a significant impact on the whole economy; another qualification is that an increase in the sector's growth can be exogenous and occur independently of the current overall growth rate of the economy (Currie, 1974, p. 6). In a similar vein, Low (1994, p. 7) proposes a strategy that focuses on the roles construction and marketing play in the development process, and argues that there is a need for these two factors to be balanced and synchronised for economic development to proceed effectively.

Existing paradigms on the structural change in the construction industry, as a national economy develops over time, tend to be based on cross-sectional data across countries rather than longitudinal studies based on one country's time-series statistics. However, a number of longitudinal studies pertaining to developing countries have been developed (Lopes, 1998; Han and Ofori, 2001; Lopes *et al.*, 2002). Lopes (1998) analysed the role of construction in economic development of countries in sub-Saharan Africa. The development patterns of construction and related sectors were modelled on the basis of data for the period 1980–1993 and a sample of 15 countries exhibiting two different patterns of growth in that period. He argued that the share of construction in GDP and GDP per capita grow at the same rate only in a declining economy, and that, in a growing economy, construction volume, typically, would not grow faster than the rest of the economy. Han and Ofori (2001) studied data on the geographical distribution of construction activity among the provinces of China over the period 1990–1998. They reported that there was a positive correlation between economic growth and growth rate of construction value added, and an inverse relationship between the growth of GDP and the share of construction in GDP.

With the availability of long and more reliable time-series data and the development of econometric methodology related to the study of economic relationships between variables, a new set of studies has emerged. Some of these studies (Green, 1997; Lean, 2001; Tse and Ganesan, 1997; Yiu *et al.*, 2004; Wong *et al.*, 2008) have applied econometric analyses within Granger's (1969) framework to test the causality link between construction output and GDP. For example, Yiu and colleagues (2004) found that, for Hong Kong, the real growth of the aggregate economy leads the real growth of the construction output and not vice versa, at least in the short term. On the other hand, Wong and colleagues (2008), using more recent data covering a longer period of Hong Kong's high-income status, concluded that the direction of the causality is from the construction sector, particularly the civil engineering sub-sector, to GDP. Along the same lines, Anaman and Osei-Amponsah (2007) analysed the relationship between the construction industry and the macro-economy in Ghana, based on time-series data from 1968 to 2004, and found that the construction industry leads economic growth in that country. Chen and Zhu (2008) analysed provincial data on housing investment in three main regions of China and found that there was a bi-directional Granger causality between GDP and housing investment for the whole country, while the impact of housing investment on GDP behaves differently in the three regions.

The contrasting empirical results on the direction of causality (if any) show that the relationship between the growth of the construction industry and that of the national aggregate is not yet fully understood, even with the use of advanced econometric methods (Lopes *et al.*, 2011). Thus, the question that remains to be answered is whether or not the construction industry is an engine of economic growth and development.

The next section considers the relationship between CVA (as a measure of the construction activity) and GDP in two groups of countries in sub-Saharan Africa according to their stages of economic development.

Quantifying the relationship between construction output and GDP

Statistical sources and methodology of data collection

The main statistical sources used in this analysis are the 2010 edition of the *Yearbook of National Accounts Statistics: Main Aggregates and Detailed Tables* (UN, 2010a), *Africa Development Indicators 2008–2009* (World Bank, 2008) and *World Development Report 2010* (World Bank, 2009b). The internet site of the UN Statistical Office presents data on GDP and its components in both the expenditure and production approaches. It presents various sets of economic series detailing the evolution of GDP and its components in different statistical formats over the long period 1970–2008, in the world, the regions of the world, and individual countries: at current prices in national currencies, constant 1990 prices in national currencies, current prices in US dollars and constant 1990 prices in US dollars.

The indicators of economic activity which are analysed in this chapter are GDP and CVA. Unfortunately, data on gross fixed capital formation in construction (GFCFC) are not provided in the UN publication. Thus, it is not possible to compare GFCFC with gross national income (GNI) (which, as discussed above, are both 'total' figures). For this reason, CVA is used as a proxy for the analysis of the pattern of evolution of construction investment across the sub-Saharan African region. It is compared with GDP as both are measures of value added. In order to facilitate international comparison as well as for aggregation purposes, constant 1990 prices in US dollars are used. With respect to the investigation of the relationship between the construction industry and the economy according to the stage of economic development of a country (or group of countries), GNI per capita for the benchmark year, 2008, has been chosen. The data for this are provided by the World Bank (2009b). *The World Development Report 2010* presents the following definitions of the income groups of countries, categorising the economies according to the 2008 GNI per capita:

a   low-income countries (LICs), US$975 or less;
b   lower-middle-income countries (LMICs), US$976–$3,855;
c   upper-middle-income countries (UMICs), US$3,856–$11,905; and
d   high-income countries (HICs), US$11,906 or more.

Data on the evolution of economic indicators in the period 1980–2006 were obtained from the *Africa Development Indicators 2008–2009* (World Bank, 2008). Thus, because of data consistency, the quantification of the relationship between the construction sector and GDP is analysed only for the period 1980–2006.

Data

As referred to earlier, the indicator used as a proxy for construction investment is CVA. CVA is calculated in the same way as that for any other sector, but includes only the activities of the construction activity proper. For example, it excludes the building materials industry, which is accounted for in the manufacturing sector. The main indicator of economic activity used in this study is GNI per capita. It adjusts the growth in the economy with the growth in population.

Using data obtained from the *UN Yearbook of National Accounts Statistics* (United Nations, 2010a), data are presented for the share of construction in GDP (at constant 1990 US dollars) for the period 1980–2006. GNI per capita is presented for the year 2008. The evolution of basic indicators, of sub-Saharan Africa as a whole, as well as excluding two important economic players of that region, South Africa and Nigeria, is presented for the period 1980–2006 (Table 3.1).

Table 3.1 GDP per capita in sub-Saharan Africa (SSA) in 1980–2006

	Constant prices (2000 US$)				Average annual growth (%)		
	1980	1990	2000	2006	1980–1989	1990–1999	2000–2006
SSA	593	532	508	580	–1.0	–0.6	2.4
SSA excluding South Africa	371	339	332	368	–0.9	–0.3	2.6
SSA excluding South Africa and Nigeria	348	331	323	379	–0.3	–0.2	2.3

Cross-matching sources, data are available for 45 countries and these can be split into two groups according to the level of GNI per capita in 2008: LICs and MICs. Tables 3.2 and 3.3 and Figures 3.1 and 3.2 illustrate these two groups. Only Equatorial Guinea could, in theory, be considered a HIC, owing to its high GNI per capita.

Table 3.2 GNI per capita and share of CVA in GDP (%) for selected years (LICs)

Country	GNI *per capita,* current US$ (2008)	CVA/GDP (1980)	CVA/GDP (1990)	CVA/GDP (2000)	CVA/GDP (2006)
Benin	690	3.65	3.11	3.56	3.91
Burkina Faso	480	2.90	4.67	5.05	5.49
Burundi	140	3.29	3.35	4.37	3.11
C. African Rep.	410	1.77	2.81	2.57	2.97
Chad	530	1.02	1.69	1.32	1.31
Comoros	374	9.39	3.17	5.38	6.01
Congo, D. Rep.	150	3.65	5.00	3.27	4.24
Gambia, The	390	4.86	4.51	4.05	4.77
Ghana	670	3.30	3.30	3.48	3.56
Guinea	390	10.23	10.20	11.49	12.41
Guinea-Bissau	250	13.71	9.99	6.76	10.63
Kenya	750	3.74	2.92	2.66	2.53
Liberia	170	3.89	3.33	2.07	3.45
Madagascar	410	1.69	1.11	1.45	2.83
Malawi	290	6.81	4.96	3.84	5.10
Mali	580	2.00	2.91	5.14	4.92
Mauritania	840	3.42	4.78	6.27	8.95
Mozambique	370	9.65	5.19	8.40	7.37
Niger	330	2.63	2.45	2.30	2.55
Rwanda	410	6.82	6.78	7.97	9.18
Senegal	970	2.82	3.29	3.94	5.13
Sierra Leone	320	3.31	2.12	2.07	3.96
Tanzania	430	3.57	4.76	6.84	7.81
Togo	400	5.84	5.14	4.44	5.20
Uganda	420	3.51	4,84	6.39	7.58
Zambia	950	2.71	2.56	2.15	3.40
Zimbabwe	340	3.88	2.86	2.22	0.86

Table 3.3 GNI per capita and share of CVA in GDP (%) for selected years (MICs)

Country	GNI per capita, current US$ (2008)	CVA/GDP (1980)	CVA/GDP (1990)	CVA/GDP (2000)	CVA/GDP (2006)
Angola	3,450	4.64	2.92	2.73	3.45
Botswana	6,470	9.11	7.28	5.88	5.13
Cameroon	1,150	6.92	4.59	3.57	3.58
Cape Verde	3,130	10.39	11.92	8.47	8.76
Congo, Rep.	1,970	7.23	2.99	4.44	5.01
Côte d'Ivoire	980	3.63	1.79	4.04	3.29
Djibouti	1,130	4.19	9.62	5.89	6.84
Equatorial Guinea	14,980	7.25	4.51	3.05	3.38
Gabon	7,240	5.70	6.71	6.37	6.54
Lesotho	1,080	9.63	14.69	13.84	12.45
Mauritius	6,400	5.74	5.62	5.88	5.96
Namibia	4,200	6.51	2.30	2.33	3.20
Nigeria	1,160	3.90	1.69	2.14	2.40
Seychelles	10,290	9.26	4.79	8.51	9.14
S. Africa	5,820	3.82	2.98	2.29	2.62
Sudan	1,130	5.56	6.04	5.01	5.00
Swaziland	2,520	5.84	2.49	6.39	6.70

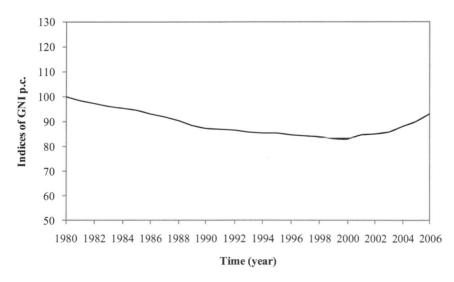

Figure 3.1 Indices of GNI per capita in LICs (mean average; 1980 = 100).

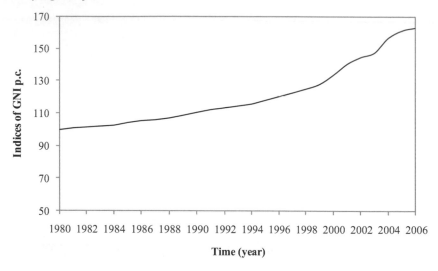

Figure 3.2 Indices of GNI per capita in MICs (mean average; 1980 = 100).

Analysis and discussion

Table 3.1 shows the evolution of GDP per capita in sub-Saharan Africa as well as that of sub-Saharan Africa excluding South Africa, and then sub-Saharan Africa excluding South Africa and Nigeria. The division shown in Table 3.1 is a reflection of the influence those two countries wield in the sub-Saharan African economy. According to the World Bank (2009b), the GDP of South Africa and Nigeria in nominal prices constituted over half (51.4 per cent) of sub-Saharan Africa's GDP. The reasons for this dominance are not the same for the two countries. Nigeria plays a big role because it is, by far, the most populous country in the region, whereas South Africa is important owing to its unmatched industrial structure and technological development that makes it the economic pole of sub-Saharan Africa.

Lopes (1998) studied data on the economy of sub-Saharan Africa for the period 1980–1992 and pointed out that 'most of the countries are not just standing still: for the last fifteen years, they have been moving dramatically backwards' (p. 637). Although some successes have been noticeable in a number of countries, this is still the development pattern of the majority of sub-Saharan African countries. It can be seen that both the region and its subdivisions, in terms of GDP per capita, experienced decreasing growth in the period 1980–2000 and a reasonable upturn in the period from 2000 onwards. This is in line with the evolution illustrated in Figure 3.1. The LICs, as a group, experienced dramatic decreasing growth in the period 1980–2000 and an average annual rate of growth of almost 2 per cent in the period 2000–2006. The striking aspect worthy of note concerning the LICs is illustrated in Figure 3.1. GNI per capita in the LCIs in 2006 (measured as an average for the group) was lower than that in 1980. On the other hand, the MICs, in terms of GDP per capita, grew slightly in the period 1980–2000, with an average annual growth rate of about 1 per cent, and notched up a spectacular rate of growth, more than 4 per cent annually on average in the period 2000–2006. As illustrated in Figure 3.2, GNI per capita for the MICs in 2006 was about 1.65 times that in 1980.

Now, considering the relationship between the construction sector and the national economy, Tables 3.2 and 3.3 and also Figures 3.1 to 3.4 show that the evolution pattern of the share of CVA in GDP in the developing countries of sub-Saharan Africa is markedly different according to the country's stage of economic development as determined by GNI per capita. The share of CVA in GDP in the LICs, despite differences across countries as well as taking into account annual fluctuations, varied, in general, from 4 per cent to 5 per cent of GDP, as is illustrated in Figure 3.3. In terms of the evolution in the period, the share of that indicator was in line with the development pattern of GNI per capita: it decreased in the period 1980–1990, remained practically stagnant in the period 1990–2000, and grew at a reasonable rate in the period 2000–2006. It is worth noting that, in the later years of the period, the share of CVA in GDP was higher than in the earlier years of the same period. That is, in the first stages of economic development, and in an increasing growth pattern, the construction industry tends to grow faster than national output. Conversely, in an economic downturn, the industry tends to decrease not only absolutely but also relatively.

Regarding the MICs, Table 3.3 and Figure 3.4 show that the share of CVA in GDP varied, in general, from 5.0 per cent to 6.5 per cent in the period

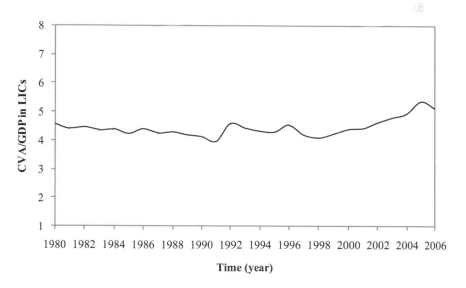

Figure 3.3 Evolution of the share of CVA in GDP (%) in LICs (mean average).

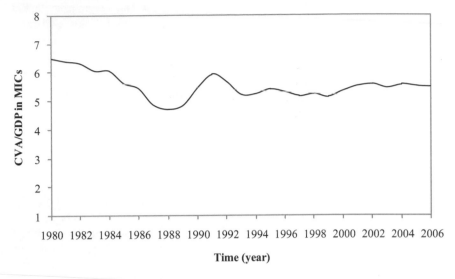

Figure 3.4 Evolution of the share of CVA in GDP (%) in MICs (mean average).

1980–2006, also disregarding differences across countries as well as annual fluctuations. Figures 3.2 and 3.4 also show that a small increase in GNI per capita corresponded to a fairly significant decrease in the share of CVA in GDP in the period from 1980 to the mid-1990s. From then onwards, the share of construction in GDP remained practically stagnant at around 5.5 per cent of GDP. The pattern experienced by the MICs is worthy of note: despite a significant increase in national income per capita, particularly in 2000–2006, the share of CVA in GDP in the later years of the period did not reach the value attained in the earlier part of the period. These results presented here seem to corroborate those of a previous work concerning the developing countries of Africa (Lopes, 1998) that found that in those countries that have middle-income status or are in a sustained process of reaching it, and achieve a certain level of construction activity (say 5 to 6 per cent of GDP, depending upon the year taken as a basis), the proportion of construction in GDP tends to remain stagnant. In other words, the rate of growth of the volume of construction follows that of the national economy.

Construction infrastructure and the challenge of the Millennium Development Goals

The preceding section has described the changes which are discernible in the evolution of construction and its sectoral share in GDP in two groups of countries in sub-Saharan Africa according to their levels of GNI per capita. This section considers the role of construction in socio-economic development, particularly by exploring the link between construction infrastructure and the economic and social targets related to the MDGs.

The results of the analyses in the previous sections suggest that the role of construction in the economy goes beyond its share in national output. As pointed out earlier, the construction industry has historically been linked with the process of industrialisation and development. Railway systems and canals played an important role in the connection of different regions of Europe, North America and some parts of Latin America. Transport infrastructure facilitated trade and co-operation among countries and also the diffusion of technical innovations from the most advanced to the less advanced areas of the globe (Rostow, 1963). The construction industry played a key role in the reconstruction of war-ravaged Europe: the heavy programme of construction improvement of housing and social infrastructure, besides its contribution to the national output, was also a reflection of a better redistributive economic policy in Europe after the Second World War. Following the UN Millennium Declaration in 2000, the heads of state and government of sub-Saharan Africa have emphasised the role transport infrastructure can play in enhancing inter-regional co-operation and fostering economic and social development (Organization of African Unity, 2001).

In the early 2000s, the physical infrastructure in sub-Saharan Africa was in a very poor state. The volume of external capital flows (particularly from donor countries which are members of the Development Assistance Committee of the OECD) for infrastructure in Africa had reached a historic low. As already mentioned, over the 1990s, African governments and their development partners sharply reduced the share of resources allocated to infrastructure, and in the aftermath of the Asian financial crisis, in the early 2000s, private capital flows declined sharply. In 2005, the Commission for Africa, in line with another policy shift on the part of the international development agencies, singled out infrastructure as one of the continent's central development challenges:

> Infrastructure is a key component of the investment climate, reducing the costs of doing business and enabling people to access markets; is crucial to advances in agriculture; . . . is critical to enabling Africa to break into world markets; and is fundamental to human development, including the delivery of health and education services to poor people. Infrastructure investments also represent an important untapped potential for the creation of productive employment.
>
> (Commission for Africa, 2005, cited in World Bank, 2009a)

The Group of Eight summit at Gleneagles in 2005 called for action by the major economies and multi-lateral donors in the financing of sub-Saharan African infrastructure. This led to the formation of the *Infrastructure Consortium for Africa*. This consortium would constitute a forum where major donors could work with continental and regional institutions to spearhead economic integration (World Bank, 2009a). One of the practical results of this political arrangement was the publication of the flagship report *Africa's Infrastructure: A Time for Transformation* in 2009. This publication diagnosed the infrastructure needs of sub-Saharan Africa, addressing the twin challenges of financing and sustainability, particularly the attainment of the MDGs.

Table 3.4 indicates that the estimate for the overall cost to build, maintain and operate Africa's infrastructure is US$93 billion annually over the period 2006–2015, approximately 15 per cent of sub-Saharan Africa's GDP in 2006. Of this total, about two-thirds is for investment and about one-third for operation and maintenance. In sectoral terms, about 40 per cent is allocated to the power sector. The second largest component is water and sanitation – a key sector for meeting the MDGs – with about 23 per cent of the total and the third largest share of the cost is associated with transport, which is approximately 20 per cent of the overall spending needs. In terms of regional groups, the burden of the price tag relative to the countries' GDP is markedly different across groups (World Bank, 2009a). For middle-income countries and resource-rich countries, the amount is in the range of 10 per cent to 13 per cent of their GDPs. For low-income countries, as much as 25 per cent of GDP would be needed, and for a particular sub-group of the latter – fragile states (war-ravaged countries) – the burden would be an astonishing 37 per cent of GDP. If one takes into account that the middle-income countries already spend a reasonable share of their wealth in investing in infrastructure and the spending needs are almost equally divided across groups, one can envisage the implausibility for the poorer countries in Africa to finance the funding gap of their estimated spending needs.

As Tables 3.5 and 3.6 show, the funding gap for infrastructure in sub-Saharan Africa is US$31 billion or about 5 per cent of GDP, taking into account efficiency improvements. About US$23 billion a year, or over 70 per cent of the funding gap, is for the power sector. The other significant component of the gap, representing a shortfall of US$11.4 billion, is associated with water supply and sanitation (WSS). The funding gap in the latter sector in the low-income countries, particularly in fragile states, looks like

Table 3.4 Overall infrastructure spending needs for Africa, 2006–2015 (US$ billion annually)

Sector	Capital expenditure	Operation and maintenance	Total needs
ICT	7.0	2.0	9.0
Irrigation	2.7	0.6	3.3
Power	26.7	14.1	40.8
Transport	8.8	9.4	18.2
WSS	14.9	7.0	21.9
Total	60.4	33.0	93.3

Source: World Bank (2009a).

Note
ICT, information and communication technology; WSS, water supply and sanitation.

Table 3.5 Funding gaps, by sector and country group (US$ billion annually)

Country type	Power	ICT	Irrigation	Transport	WSS	Potential for reallocation	Total
Middle-income	10.7	−0.9	0.1	−0.3	0.0	−4.1	5.5
Resource-rich	4.5	0.5	1.8	−1.4	3.7	−0.8	8.2
Low-income non-fragile	4.7	−0.2	0.7	−0.5	5.2	−0.4	9.5
Low-income fragile	2.7	0.7	0.0	2.0	3.9	0.0	9.4
Sub-Saharan Africa	23.2	1.3	2.4	−1.9	11.4	−3.3	30.6

Source: World Bank (2009a).

Note
Totals do not add up because efficiency gaps cannot be carried across country groups.

Table 3.6 Funding gaps, by sector and country group (percentage of GDP)

Country type	Power	ICT	Irrigation	Transport	WSS	Potential for reallocation	Total
Middle income	3.9	−0.3	0.0	−0.1	0.0	−1.5	2.0
Resource rich	2.0	0.2	0.8	−0.6	1.7	−0.4	3.7
Low-income non-fragile	4.2	−0.2	0.6	−0.4	4.7	−0.3	8.6
Low-income fragile	7.1	1.9	0.1	5.3	10.2	0.0	24.6
Sub-Saharan Africa	3.6	−0.2	0.4	−0.3	1.8	−0.5	4.8

Source: World Bank (2009a).

Note
Totals do not add up because efficiency gaps cannot be carried across country groups.

an unattainable target in the foreseeable future in the light of the present economic situation and prospects of the countries themselves, and the challenges posed to the development partners by the recent global financial crisis.

As pointed out above, international development agencies and bilateral donors have emphasised infrastructure (more precisely the services provided by the infrastructure) as the key factor for the attainment of the MDGs. The MDGs are the world's time-bound and quantified targets for addressing extreme poverty in its many dimensions – income poverty, hunger, disease, lack of adequate shelter, and exclusion – while promoting gender equality, education and environmental sustainability (UNDP, 2005). Following the adoption of the UN Millennium Declaration at the Millennium Summit in

New York in September 2000, eight MDGs, measured through 21 targets, were devised. Most of the MDG targets have a deadline of 2015, and 1990 is the baseline against which progress is quantified.

Restricting here the analysis to those infrastructure items which incorporate mainly construction investment (WSS, transport and electricity – note that a significant part of the investment in the power sector is for multi-purpose use), construction infrastructure has an important role to play in the process of attaining the MDGs in sub-Saharan Africa. MDG 7 – Ensure environmental sustainability – is particularly relevant to the construction industry. This goal is translated into the following targets: (i) integrate the principles of sustainable development in country policies and programmes and reverse the loss of environmental resources; (ii) halve, by 2015, the proportion of people without sustainable access to safe drinking water and basic sanitation; and (iii) have achieved by 2020 a significant development in the lives of 100 million slum dwellers.

Table 3.7 suggests that sub-Saharan Africa, as a whole, is unlikely to achieve, by the deadline of 2015, the targets related to access to safe drinking water and basic sanitation. However, some middle-income economies of Africa such as Cape Verde, South Africa, Botswana, Namibia, Seychelles and Mauritius are either on track or experiencing good progress towards achieving the targets, particularly in access to safe drinking water (WHO/UNICEF, 2010). Regarding the urban population living in slum areas, whereas some progress was noticed in relative terms, the absolute number of slum dwellers has actually been growing and will probably continue to increase in the near future.

With respect to the emission of CO_2, sub-Saharan Africa registered 0.9 metric tons per capita in 2007 compared with 12 metric tons per capita in the developed world, and, in deforestation, the rate of increase started to decrease in the period 1990–2010.

The WSS sector, besides its direct effect on the provision of water and sanitation services, has a pervasive impact on other social targets, namely in the prevention of disease, improvement in education and promotion of

Table 3.7 MDGs in sub-Saharan Africa: Goal 7 – Ensure environmental sustainability

Proportion of people using an improved water source (%)		Proportion of people with access to improved sanitation facilities (%)		Proportion of urban population living in slum areas (%)		Forested area as a percentage of land area (%)		Emission of CO_2 (billions of metric tons)	
1990	2008	1990	2008	1990	2010	1990	2010	1990	2007
49	60	28	31	70	62	31	28	0.5	0.7

Source: UN (2010b).

gender equality so that women save time when they begin using an improved water source. Transport fosters trade by reducing the cost of transporting goods and passengers, reduces child and maternal mortality and improves access to education services. Electricity enhances productivity, eradicates poverty by fostering economic growth and reduces child and maternal mortality.

One important area related to the construction industry which needs special attention on the part of government agencies is urban planning. The rate of urbanisation in sub-Saharan Africa is increasing sharply. This has its merits. For example, higher population densities lower the per capita costs of providing safe water, sewer systems, waste collection and most other infrastructure and public amenities. Moreover, sound urban planning restricts development in flood-prone areas and other hazardous places and provides critical access to services, and infrastructure developments can provide physical protection for the natural environment. On the other hand, most of the urban areas have been unable to cope with the increasing populations, and large numbers of their inhabitants have a poor quality of life. As pointed out by the World Bank (2009b), overcrowding, insecure tenure, illegal settlements sited in landslide- and flood-prone areas, poor sanitation, unsafe housing, inadequate nutrition and poor health exacerbate the vulnerabilities of the population in urban slums. These are the realities for millions of people in sub-Saharan African countries. An efficient construction industry can contribute to the efforts to tackle these problems. For example, it can address the vulnerabilities of slum dwellers by devising labour-intensive and cost-effective technologies, and by implementing practical sustainable measures in the framework of the *Agenda 21 for Sustainable Construction in Developing Countries* (International Council for Research and Innovation in Building and Construction [CIB] and United Nations Environment Programme [UNEP] 2002). This framework for a co-ordinated response to the challenges of limiting the impact of construction on the environment puts an emphasis on collaboration among different stakeholders of the construction industry. To this end, three types of interdependent and multi-dimensional enablers were identified: technological enablers, institutional enablers and enablers related to value systems. The development of these enablers requires an approach that operates simultaneously at different scales, as well as over different time horizons (Du Plessis, 2007, p. 71). Indeed, it is the flexibility of the construction industry in being able to adjust to different framework conditions that makes it such a great contributor to the process of economic development.

Concluding remarks and policy implications

The picture that emerges from the analysis of the evolutionary process of the construction industry and its role in national socio-economic development suggests that the share of construction in GDP tends to increase with the level of per capita income in the first stages of economic development. When countries reach a certain level of economic development, the construction output will grow more slowly than national output in the later stages of their development. That is, it decreases relatively but not absolutely. Thus, it is reasonable to assume that, when a certain level is achieved (say the share of CVA in GDP at around 5–6 per cent) and countries enter into a path of sustained economic growth and development, the construction output tends to grow, in general, at the same rate as that of the broader economy.

The analysis also suggests that the role of construction in economic development is by no means a fully understood process. Two barriers appear to be particularly relevant: (i) the relative lack of attention paid to the construction industry by international organisations which aim to promote economic co-operation and development (for example, in comparison with the manufacturing and transport sectors); and (ii) the scarcity and unreliability of the data on the construction industry and related sectors which are needed to undertake longitudinal country studies, particularly in developing countries. These studies are the sub-stratum for generating good understanding of the process of the economic development of nations. The development of Brazil, China and India might, at last, provide the time-series data which are required to build appropriate models of the role of construction in national socio-economic progress, and strengthen the argument on the importance of the industry (Ofori, 2007, p. 7).

The results of the study also underlie the need to address the twin challenges of finance and sustainability in sub-Saharan Africa in the effort towards attaining the MDGs, and the situation is particularly acute in the low-income countries in the light of the countries' own economic circumstances and prospects, and the current global financial crisis. The results of the study may have some implications for public policies. Given the experience of the growth process in sub-Saharan Africa, what should be the focus of growth-enhancing policies in the two groups of countries? How can the construction industry contribute to this end, and help a country in the LIC group to move to the MIC group? It is evident that further investment in construction infrastructure might be recommended for LICs but might not necessarily be a growth priority for MICs. Most recent data indicate that there is no significant funding gap in infrastructure investment in the MICs of sub-Saharan Africa in order to achieve the economic and social targets of the MDGs. These countries should prioritise their investment projects by balancing economic and financial factors with social targets. For the low-income countries, taking into account the financial stress facing these countries, the analyses suggest that most of the effort should be directed at investment in construction projects in order to achieve a level of construction industry activity of, say, 5–6 per cent of GDP, which some studies have shown to be required for a reasonable functioning of the economy. The priority should be given to construction projects which have high multiplier effects in the economy, particularly transport and multi-purpose (power and water) infrastructure. A concerted effort to implement integrated sub-regional infrastructure projects seems also to be the way forward.

References

Agarwala, R. (1983) *Price Distortions and Growth in Developing Countries.* World Bank Staff Working Papers, No. 575. The International Bank for Reconstruction and Development: Washington, DC.

Anaman, K. and Osei-Amponsah, C. (2007) Analysis of the causality links between the growth of the construction industry and the growth of the macro-economy in Ghana. *Construction Management and Economics*, 25 (9), 951–961.

Barro, R. J. (1991) Economic growth in a cross section of countries. *Quarterly Journal of Economics*, 106, 407–443.

Barro, R. J. (1997) *Determinants of Economic Growth: A Cross-Country Empirical Study.* MIT Press: Cambridge, MA.

Barro, R. J. (2000) Inequality and growth in a panel of countries. *Journal of Economic Growth*, 5, 5–32.

Bloom, D., Canning, D. and Sevila, J. (2003) Geography and poverty traps. *Journal of Economic Growth*, 8, 355–378.

Bon, R. (1990) The world building market 1970–85. In *Proceedings of the CIB W65 Symposium: Building Economics and Construction Management*, Sydney.

Bon, R. (1991) *What Do We Mean by Building Technology?* Inaugural lecture delivered at the University of Reading, Reading.

Bon, R. (1992) The future of international construction: secular patterns of growth and decline. *Habitat International*, 16 (3), 119–128.

Center for Global Development (2010) *Emerging Africa: How 17 Countries Are Leading the Way.* Policy Brief, September.

Chen, J. and Zhu, A (2008) *The Relationship between Housing Investment and Economic Growth in China: A Panel Analysis Using Quarterly Provincial Data.* Working Paper 2008:17, Department of Economics, Uppsala University.

Currie, L.(1974) The 'leading sector' model of growth in developing countries. *Journal of Economic Studies*, 1 (1), 1–16.

De Long, J. B. and Summers, L. H. (1991) Equipment investment and economic growth. *Quarterly Journal of Economics*, 106, 445–502.

Deininger, K. and Squire, L. (1996) Measuring income inequality: a new database. *World Bank Economic Review*, 10 (3), 565–591.

Du Plessis, C. (2007) A strategic framework for sustainable construction in developing countries. *Construction Management and Economics*, 25, 67–76.

Easterly, W. (2006), Reliving the 1950s: the big push, poverty traps, and takeoffs in economic development. *Journal of Economic Growth*, 11, 289–318.

Eicher, T., Garcia-Penalosa, C. and van Ypersele, T. (2009) Education, corruption, and the distribution of income. *Journal of Economic Growth*, 14, 205–231.

Ganesan, S. (1979) *Growth of Housing and Construction Sectors: Key to Employment Creation,* Progress in Planning, Vol. 12, Part 1, London.

Ganesan, S. (1994) Employment maximisation in construction in developing countries. *Construction Management and Economics*, 12, 323–335.

Granger, C. W. J. (1969) Investigating causal relations by econometric methods and cross-spectral methods. *Econometrica*, 37 (3), 424–438.

Green, R. K. (1997) Follow the leader: how changes in residential and non-residential investment predict changes in GDP. *Real Estate Economics*, 25 (2), 253–270.

Han, S. S. and Ofori, G. (2001) Construction industry in China's regional economy, 1990–1998. *Construction Management and Economics*, 19, 189–205.

Hess, P. (2010) Determinants of the adjusted net saving rate. *International Review of Applied Economics*, 24 (5), 591–608.

Hillebrandt, P. M. (2000) *Economic Theory and the Construction Industry*. Macmillan: Basingstoke.

International Council for Research and Innovation in Building and Construction (CIB) and United Nations Environment Program (UNEP) (2002) *Agenda 21 for Sustainable Construction in Developing Countries: A Discussion Document*. CSIR Building and Construction Technology: Pretoria.

Jones, C. I. (1995) R&D-based models of economics growth. *Journal of Political Economy*, 103 (4), 759–784.

Kuznets, S. (1963) Notes on the take-off. In W. W. Rostow (ed.) *The Economics of Take-Off into Sustained Growth*, Proceedings of Conference Held by the International Economic Association. Macmillan: London.

Kuznets, S. (1968) *Towards a Theory of Economic Growth*. New York: W. W. Norton.

Lange, J. E. and Mills, D. Q. (1979) An introduction to the construction sector of the economy. In J. E. Lange and D. Q. Mills (eds) *The Construction Industry: Balance Wheel of the Economy*. Lexington Books: Lexington, MA.

Lean, S. C. (2001) Empirical tests to discern linkages between construction and other economic sectors in Singapore. *Construction Management and Economics*, 19, 253–262.

Lewis, W. A. (1954) Economic development with unlimited supplies of labour. *Manchester School of Economic and Social Studies*, 22, 139–191.

Lopes, J. (1998) The construction industry and macroeconomy in sub-Saharan Africa post 1970. *Construction Management & Economics*, 16, 637–649.

Lopes, J., Ruddock, L. and Ribeiro, F. L. (2002) Investment in construction and economic growth in developing countries. *Building Research and Information*, 30 (3), 152–159.

Lopes, J., Nunes, A. and Balsa, C. (2011) The long-run relationship between the construction sector and the national economy in Cape Verde. *International Journal of Strategic Property Management*, 15, 48–59.

Low, S. P. (1994) *The Link between Construction, Marketing and Economic Development: A Study of Africa and Asia*. UCT Research Paper Series, Paper No. 3, University of Cape Town.

Lucas, R. (1988) On the mechanics of economic development. *Journal of Monetary Economics*, 22, 3–42.

Mabogunje, A. L. (1989) *The Development Process: A Spatial Perspective*, 2nd edition. Unwin Hyman: London.

Maddison, A. (1987) Growth and slowdown in advanced capitalist economies. *Journal of Economic Literature*, 25 (2), 649–698.

Madsen, J. (2008) Semi-endogenous versus Schumpeterian growth models: testing the knowledge production function using international data. *Journal of Economic Growth*, 13, 1–6.

Mankiw, N. G., Romer, D. and Weil, D. (1992) A contribution to the empirics of economic growth. *Quarterly Journal of Economics*, 107 (2), 407–437.

Moavenzadeh, F. and Koch Rossow, J. A. (1976) *The Construction Industry in Developing Countries*. Technology Adaptation Program, Massachusetts Institute of Technology: Cambridge, MA.

Ofori, G. (1990) *The Construction Industry: Aspects of Its Economics and Management*. Singapore University Press: Singapore.

Ofori, G. (1993) Research on construction development at the crossroads. *Construction Management and Economics*, 11 (3), 175–185.

Ofori, G. (2007) Construction in developing countries (guest editorial). *Construction Management and Economics*, 25 (1), 1–6.

O'Neil, D. (1995) Education and income growth: implications for cross-country inequality. *Journal of Political Economy*, 103 (6), 1289–1301.

Organization of African Unity (OAU) (2001) *New Partnership for Africa's Development*. Framework document approved at the 37th Summit of the OAU, July.

Owen, A., Videras, J. and Lewis, D. (2009) Do all countries follow the same growth process? *Journal of Economic Growth*, 14, 265–286.

Paap, R., Franses, P. and van Dijk, D. (2005) Does Africa grow slower than Asia, Latin America and the Middle East? Evidence from new data-based classification method. *Journal of Development Economics*, 77, 553–570.

Reynolds, Lloyd G. (1985) *Economic Growth in the Third World: 1850–1980*. Yale University Press: New Haven, CT.

Romer, P. M. (1986) Increasing returns and long-run growth. *Journal of Political Economy*, 94 (5), 1002–1037.

Romer, P. M. (1989) Human capital and growth: theory and evidence. *Carnegie–Rochester Conference Series on Public Policy*, 32, North-Holland

Romer, P. M. (1990) Endogenous technological change. *Journal of Political Economy*, 98 (5), S71–S102.

Romer, P. M. (1994) The origins of endogenous growth. *Journal of Economic Perspectives*, 8 (1), 3–22

Rosenstein-Rodan, P. N. (1943) Problems of industrialisation in eastern and southern Europe. *Economic Journal*, 53, 202–211.

Rostow, W. W. (1960) *The Stages of Economic Growth: A Non-Communist Manifesto*, 1st edition. Cambridge University Press: Cambridge.

Rostow, W. W. (1963) The leading sectors and the take off. In W. W. Rostow (ed.) *The Economics of Take-Off into Sustained Growth*, Proceedings of the conference held by the International Economic Association. Macmillan Press: London.

Ruddock, L. and Lopes, J. (2006) The construction sector and economic development: the 'Bon curve'. *Construction Management and Economics*, 24 (7), 717–723.

Solow, R. M. (1956) A contribution to the theory of economic growth. *Quarterly Journal of Economics*, 70, 65–94.

Solow, R. M. (1988) Growth theory and after, Nobel Lecture delivered at Stockholm, Sweden, December 8, 1987. *American Economic Review*, 78 (3), 307–317.

Solow, R. M. (1994) Perspectives on growth theory. *Journal of Economic Perspectives*, 8 (1), 45–54.

Strassmann, P. (1970) The construction sector in economic development. *Scottish Journal of Political Economy*, 17 (3), 390–410.

Summers, R. and Heston, A. (1991) The Pen World Table (Mark 5): an expanded set of international comparisons 1950–1988. *Quarterly Journal of Economics* 106, 327–368.

Thirlwall, A. P. (1994) *Growth and Development: With Special Reference to Developing Countries*, 5th edition. Macmillan Press: London.

Todaro, M. P. (1992) *Economic Development in the Third World*, 3rd edition. Longman Group: New York.

Tse, R. Y. C. and Ganesan, S. (1997) Causal relationship between construction flows and GDP: evidence from Hong Kong. *Construction Management and Economics*, 15, 371–376.

Turin, D. A. (1966) *What Do We Mean by Building*? Inaugural lecture delivered at University College London on 14 February.

Turin, D. A. (1973) *The Construction Industry: Its Economic Significance and Its Role in Development*. UCERG: London.

UN (United Nations) (1970–) *Demographic Yearbook* (various years). Statistical Office, Department of Economic and Social Affairs: New York.

UN (1985) *International Comparisons of Prices and Purchasing Power in 1980*. United Nations: New York.

UN (2000) *United Nations Millennium Declaration*, Resolution adopted by the General Assembly – Fifty-Fifth Session, September, New York.

UN (2010a) *Yearbook of National Accounts Statistics: Main Aggregates and Detailed Tables*. Statistical Office, DESA: New York. http://www.un.org/statistics

UN (2010b) *The Millennium Development Goals Report 2010*. United Nations: New York.

UNCHS (1982) *Role and Contribution of Construction to Social Economic Growth of Developing Countries*. United Nations Centre for Human Settlements: Nairobi.

UNDP (United Nations Development Program) (1990–) *Human Development Report* (various years). United Nations Development Program: New York.

UNDP (2005) *UN Millennium Project 2005: Investing in Development – A Practical Guide to Achieve the Millennium Development Goals*. United Nations: New York.

UNIDO (1985) *The Building Materials Industry in Developing Countries: An Analytical Appraisal*. Sectoral Studies Series, No. 16, Vol. 1. UNIDO: Vienna.

Wells, J. (1986) *The Construction Industry in Developing Countries: Alternative Strategies for Development*. Croom Helm: London.

WHO/UNICEF (2010) *Progress on Sanitation and Drinking Water: 2010 Update*. World Health Organisation: Geneva.

Wong, J., Chiang, Y. and Ng, T. (2008) Construction and economic development: the case of Hong Kong. *Construction Management and Economics*, 26, 815–826.

World Bank (1984) *The Construction Industry: Issues and Strategies in Developing Countries*. World Bank: Washington, DC.

World Bank (1989) *Sub-Saharan Africa: From Crisis to Sustainable Growth*. World Bank: Washington, DC.

World Bank (1991) *World Development Report 1991*. IBRD: New York.

World Bank (1994) *Adjustment in Africa: Reforms, Results and the Road Ahead*. IBRD: New York.

World Bank (1998) *World Development Report 1998*. IBRD: New York.

World Bank (2008) *Africa Development Indicators 2008–2009*. IBRD: Washington, DC.

World Bank (2009a) *Africa's Infrastructure: A Time for Transformation*, edited by Vivien Foster and Cecilia Briceno-Garmendia. Africa Development Forum Series. World Bank: Washington, DC.

World Bank (2009b) *World Development Report 2010*. IBRD: Washington, DC.

Yiu, C. Y., Lu, X. H., Leung, M. Y. and Jin, W. X. (2004) A longitudinal analysis on the relationship between construction output and GDP in Hong Kong. *Construction Management and Economics*, 22 (4), 339–345.

4 Construction and Millennium Development Goals

George Ofori

Introduction

In a report on progress towards the attainment of the Millennium Development Goals (MDGs) which heralded a special meeting of world leaders to review progress towards the attainment of the MDGs during the first decade of the implementation of the programme, and determine courses for accelerated action in the last five years of the target period, the Secretary-General of the United Nations noted (UN, 2010, p. 3):

> This report shows . . . that the Goals are achievable when nationally owned development strategies, policies and programmes are supported by international development partners. At the same time, it is clear that improvements in the lives of the poor have been unacceptably slow, and some hard-won gains are being eroded by the climate, food and economic crises.
>
> Meeting the goals is everyone's business. Falling short would multiply the dangers of our world – from instability to epidemic diseases to environmental degradation. But achieving the goals will put us on a fast track to a world that is more stable, more just, and more secure. Billions of people are looking to the international community to realize the great vision embodied in the Millennium Declaration. Let us keep that promise.

It is pertinent to consider some relevant questions. What is the role of the construction industry in the efforts to attain the MDGs? How well has the industry played this role? What can be done to enable the construction industry to contribute more effectively to the efforts to attain the MDGs? Before addressing those questions, it is appropriate to consider the following: What are the MDGs, anyway? What has been the progress towards their attainment? What remains to be done?

Millennium Development Goals

The Millennium Declaration

The Millennium Declaration was signed in New York in September 2000 by 189 heads of state of United Nations (UN) member countries. The leaders reaffirmed the commitments of their countries to the UN charter, and outlined 'certain fundamental values . . . essential to international relations in the twenty-first century' (UN, 2000) including freedom, equality, solidarity, tolerance, respect for nature and shared responsibility. The objectives in the declaration were (i) Peace, Security and Disarmament; (ii) Development and Poverty Eradication; (iii) Protecting Our Common Environment; (iv) Human Rights, Democracy and Good Governance; (v) Protecting the Vulnerable; (vi) Meeting the Special Needs of Africa; and (vii) Strengthening the United Nations. Under objective (ii) of the declaration, it was stated (UN, 2000, section iii, paras 19 and 20):

19. We resolve further:

To halve, by the year 2015, the proportion of the world's people whose income is less than one dollar a day and the proportion of people who suffer from hunger and, by the same date, to halve the proportion of people who are unable to reach or to afford safe drinking water.

To ensure that, by the same date, children everywhere, boys and girls alike, will be able to complete a full course of primary schooling and that girls and boys will have equal access to all levels of education.

By the same date, to have reduced maternal mortality by three quarters, and under-five child mortality by two thirds, of their current levels.

To have, by then, halted, and begun to reverse the spread of HIV/AIDS, the scourge of the malaria and other major diseases that afflict humanity.

To provide special assistance to children orphaned by HIV/AIDS.

By 2020, to have achieved a significant improvement in the lives of at least 100 million slum dwellers as proposed in the 'Cities Without Slums' initiative.

20. We also resolve:

To promote gender equality and the empowerment of women as effective ways to combat poverty, hunger and disease and to stimulate development that is truly sustainable.

To develop and implement strategies that give young people everywhere a real chance to find decent and productive work.

To encourage the pharmaceutical industry to make essential drugs more widely available and affordable by all who need them in developing countries.

To develop strong partnerships with the private sector and with civil society organizations in pursuit of development and poverty eradication.

To ensure that the benefits of new technologies, especially information and communication technologies . . . are available to all.

These two paragraphs (19 and 20) of the Millennium Declaration were summarised into eight MDGs, which were, in turn, translated into 18 specific targets (see Table 4.1). There were also 48 indicators for monitoring progress in achieving the goals (http://www.developmentgoals.org). Progress towards the MDGs is now measured through 21 targets and 60 official indicators (UN, 2010). Most of the MDG targets have a deadline of 2015, and progress is gauged against 1990 as the baseline. When relevant and available, data for 2000 are also presented, to indicate changes since the Millennium Declaration was signed. In most reports on the MDGs, country data are aggregated at the sub-regional and regional levels to show overall advances over time.

The MDGs were not the first 'global development targets' to be set out; there had been a number of them in the 1990s (Fay *et al.*, 2005). For example, a series of 'development indicators' such as the 'human development index' were formulated to provide a detailed gauge of the state of socio-economic progress in countries. There have also been other composite indices (see later). However, arguably, the MDGs have formed the basis for the most comprehensive and wide-ranging set of co-ordinated actions by the largest number and widest range of actors to attain their targets. The governments of virtually all countries, and international and national organisations have committed themselves to a global partnership to achieve the MDGs. They have formulated policies and programmes, with specific targets, monitoring frameworks and assessment tools. For example, the MDG Gap Task Force, created in 2007, comprises 20

multi-lateral economic and development agencies brought together to improve jointly the monitoring of the set of international commitments and targets under 'Goal 8' of the MDGs which are considered to be essential to the global effort to realize the MDGs.

(MDG Task Force, 2010)

Since 2008, the Task Force has issued annual reports on the status of those commitments and targets. As another example, the World Bank's Africa Action Plan of 2005 (http://www.worldbank.org/afr/aap) sought to achieve development results in good governance, closing the infrastructure gap, building capable states, and more equitable distribution of the benefits of development.

Governments have taken the lead in the implementation of programmes to attain the MDGs. The private sector has also been playing a role and it is relevant here to consider this involvement. It is suggested that business, as an

Table 4.1 The Millennium Development Goals and targets

MDG	Targets
Goal 1: Eradicate extreme poverty and hunger	• Reduce by half the proportion of people living on less than a dollar a day • Reduce by half the proportion of people who suffer from hunger
Goal 2: Achieve universal primary education	• Ensure that all boys and girls complete a full course of primary education
Goal 3: Promote gender equality and empower women	• Eliminate gender disparity in primary and secondary education preferably by 2005 and at all levels by 2015
Goal 4: Reduce child mortality	• Reduce by two-thirds the mortality rate among children under five
Goal 5: Improve maternal health	• Reduce by three-quarters the maternal mortality rate
Goal 6: Combat HIV/AIDS, malaria and other diseases	• Halt and begin to reverse the spread of HIV/AIDS • Halt and begin to reverse the incidence of malaria and other major diseases
Goal 7: Ensure environmental sustainability	• Integrate the principles of sustainable development into country policies and programmes; reverse the loss of environmental resources • Reduce by half the proportion of people without sustainable access to safe drinking water • Achieve significant improvement in the lives of at least 100 million slum dwellers by 2020
Goal 8: Develop a global partnership for development	• Develop further an open trading and financial system that is rule-based, predictable and non-discriminatory, includes a commitment to good governance, development and poverty reduction – both nationally and internationally • Address the least developed countries' special needs. This includes tariff- and quota-free access for exports; enhanced debt relief for highly indebted poor countries; cancellation of official bilateral debt; and more generous official development assistance for countries committed to poverty reduction • Address the special needs of landlocked and small island developing states • Deal comprehensively with developing countries' debt problems through national and international measures to make debt sustainable in the long term • In co-operation with the developing countries, develop decent and productive work for the youth • In co-operation with pharmaceutical companies, provide access to affordable, essential drugs in the developing countries • In co-operation with the private sector, make available the benefits of new technologies – especially information and communications technologies

engine of growth and development, has a critical role to play in accelerating progress towards the MDGs through increasing investment, creating jobs, increasing skills, and developing goods, technologies and innovations which can make people's lives better (European Business Review, 2010). The private sector also plays a role as a source of capital for developing countries: the private sector accounts for over 85 per cent of investment and financial flows globally. The World Business Council for Sustainable Development (WBCSD, 2010) argues that it is in the enlightened self-interest of global business to invest in providing sustainable solutions to development challenges: (i) business cannot succeed in countries that fail – stable and prosperous societies offer better business opportunities; (ii) by developing a better understanding of, and addressing, socio-economic and environmental concerns, companies can better manage their risks and enhance their ability to operate, innovate and grow; (iii) developing inclusive business models (sustainable business solutions that expand access to goods, services and livelihood opportunities for low-income communities in commercially viable ways) will help companies to build positions in the growth markets of the future; and (iv) the transition to a more sustainable world represents a business opportunity owing to the long-term demand for related products and services needed to sustainably meet the needs of growing populations in developing countries. The WBCSD (2010) believes that the leading companies of the future will be those that align profitable business ventures with the needs of society. Therefore, business should be viewed as an enabler of social and economic progress and as a key partner in a common effort to achieve inclusive and sustainable development.

Non-governmental organisations (NGOs) at both the local and international levels have played advocacy and monitoring and reporting roles (see, for example, Oxfam International, 2010; Save the Children, 2010). (A useful list of recent reports on the MDGs can be found at http://www.guardian.co.uk/global-development/2010/sep/20/millennium-development-goals-progress-reports1.)

Funding the MDG programmes

Foreign financial assistance was seen as a key determinant of the attainment of the MDGs. It was estimated that, to attain the MDGs, developing countries must grow by 7–8 per cent per annum (World Bank, 2004), and that US$50 billion per annum in additional external funds would be required. Official aid flows were projected to rise from US$69 billion in 2003 to US$135 billion in 2006, and then to US$195 billion by 2015 (Millennium Project, 2005). The International Monetary Fund (IMF, 2010) notes that a major step towards meeting the MDGs was taken in Mexico in March 2002 when the international community pledged to provide greater and more effective financial support to lower-income countries pursuing sound policies and good governance, and the international community would also establish an enabling international economic and trade environment. At the G8 summit in 2005, the leaders of the world's eight richest countries pledged to double development aid to Africa from US$25 billion in 2004 to US$50 billion per year by 2010, and to deepen debt relief especially for countries with sound financial management and a commitment to poverty reduction.

Progress in most of the pledges of financial support has been slow. The MDG Task Force (2010) notes that the delivery of official development assistance (ODA) will fall short of the Gleneagles targets for 2010. Aid from members of the Development Assistance Committee (DAC) reached almost $120 billion in 2009; this was less than a 1 per cent increase, in real terms. The share of ODA in donor gross national income was 0.31 per cent, compared with the UN target of 0.7 per cent, which has been reached by only five countries. The perceived need among many donor countries to start fiscal consolidation soon following the stimulus packages which they implemented in response to the global economic crisis could put the flow of aid under further pressure. The gap between the provision in 2009 and the 2010 target was US$26 billion (in 2009 dollars). In 2010, Africa is expected to receive about $45 billion in ODA; this will be $16 billion less than the Gleneagles target (in 2009 prices).

The arguments on many aspects of foreign aid continue. They include: (i) whether it is necessary, and can really be effective; (ii) the volume required; (iii) who should provide it, and for what reason; (iv) the most appropriate nature it should take; (v) the prerequisites for success in its application; and (vi) the most effective mechanisms for administering it. Baulch (2006) found that most donors are not distributing their aid in a way that is consistent with the MDGs, that is they do not direct large shares of their concessionary aid to the poorest countries. The allocation of aid across countries is uneven, and has been increasingly concentrated in a few countries. The top 10 aid recipients accounted for 38 per cent of total country-allocatable ODA provided to countries in 2008 (MDG Task Force, 2010). The largest recipient of aid, Iraq, received twice the amount received by Afghanistan, which, in turn, received almost 50 per cent more than the third largest recipient, Ethiopia. DAC aid to least developed countries was just 0.09 per cent of donor gross national income ($36 billion) in 2008.

Powell and Bird (2010) examine the relationship between debt relief and other foreign aid in 42 sub-Saharan African countries using panel data for 1988–2006. They note that the international community is paying some attention to helping low-income countries achieve the MDGs. The results confirm the significance of population, the conduct of economic policy, and the need of a recipient. Debt relief schemes since 1988 all seem to have had a significant positive effect on net transfers to participating countries. The study also revealed that, for much of the period up to 2000, aggregate net aid transfers to sub-Saharan African Africa actually fell in both real and nominal terms.

From the above discussion, it is evident that the MDGs cannot be attained only from the provision of additional foreign aid. Dalgaard and Erickson (2009) address these questions: (i) How much growth should aid flows have produced in sub-Saharan Africa over the last three decades? and (ii) How much aid would be needed to attain the first MDG (MDG1) of cutting poverty in half by 2015? Their analysis indicates that expectations for aid in fostering growth and poverty reduction have been too high, and that aid may not be as effective in reducing poverty as other analyses have suggested. However, they point out that it would be a mistake to interpret their results as showing that aid is ineffective. Rather, the results indicate that the potential overall effect of aid on growth is likely to be modest. In another report, the Independent Evaluation Group of the World Bank (IEG) (2006) found that only two in five of the countries borrowing from the World Bank recorded continuous per capita income growth in 2000–2005 and only one in five over 1995–2005.

MDGs in the literature

There is much discussion in the literature on many aspects of the MDGs. First, not all researchers welcome the MDGs. Some observers point out that countries which have done well in socio-economic development in the last few decades, such as the countries in South-East Asia (Malaysia and Singapore) and East Asia (Hong Kong, South Korea and Taiwan) in the 1980s and 1990s, and more recently China, India and Brazil, did not do so by following any global development benchmarks. Some authors believe that the goals might lead to the setting of wrong priorities. Maxwell (2003) points out that the MDGs might encourage oversimplified interventions emphasising social indicators at the expense of economic growth. The IEG (2006) notes that achieving high-quality development results takes time, but pressure to show results can divert attention from the quality of results. For example, it notes that efforts to attain the MDG of ensuring universal completion of primary education have led to initiatives to increase enrolments, often at the expense of attention to learning outcomes. In Uganda, there were 94 children per classroom, and three children shared a textbook, whereas in Ghana the development programme in the education sector combined policy reforms with the provision of school buildings and teaching materials.

Second, some authors believe that the MDGs are unrealistic, and cannot be attained. Clemens and colleagues (2007) find fault with the way the MDGs were set, and assert that they are impossible to meet. They observe that indicating that the MDGs can be met merely with increased resources 'contributes to the illusion that the goals are attainable for all countries' (p. 747). They suggest that the specific MDG targets 'have set up many countries for unavoidable "failure"' (p. 747), even as they pursue good policies and make progress on some development indicators.

Third, some authors point out that the MDGs are an imposition as the developing countries did not participate in their formulation. White and Black (2004) note that the MDGs would not be effective as accountability for failing to meet them is diffuse. Some authors consider the MDGs to be unsuitable for the circumstances and needs of the countries in Africa. Easterly (2009) argues that some arbitrary choices made in defining 'success' or 'failure' as achieving or not achieving some numerical targets for the MDGs made attainment of the MDGs less likely in Africa than in other regions even when progress on these fronts on the continent was in line with or above the historical or contemporary experience of other regions. This has the effect of making African successes look like failures. He finds flaws in each of the first seven MDGs from Africa's perspective. The continent was starting from a lower base in terms of income and primary school enrolment and completion rates, and had the highest child mortality rates. Thus, it had further to go and was less likely to meet the targets than other regions. Moreover, there was a lack of data on which to base targets or attainments with regard to maternal mortality rates and prevalence of AIDS, malaria and tuberculosis. Easterly (2009) also notes that Africa was seen as relatively falling behind on reducing the percentage without access to clean water, but it would have been relatively catching up if it had been measured the conventional way of percentage with access to clean water.

Easterly (2009) is one of several authors who have observed that the implied picture that Africa is failing to meet all the first seven MDGs is not fair because the continent has made much progress in many of the social indicators. The negative picture is demoralising to African leaders and activists, and might have adverse consequences for foreign investment inflows, which would set the countries back further. Indeed, the vice-chair of the UN inter-agency committee which designed the MDGs has indicated that they were meant to apply only at the global level, not at the country or regional level (Vandemoortele, 2007), and he also highlighted the demoralising effect of labelling Africa an MDG 'failure'. Commenting on this statement, Tabatabai (2007) suggests that, if this is the intention, then the MDGs are 'not so much misunderstood as misconceived'.

Fourth, some authors argue about the need for particular goals, and also about the relationships among them. Fay and colleagues (2005) note that some observers believe that, since improvements in most indicators of development are highly correlated with increases in per capita income, MDGs 2–8 are superfluous as long as the first goal is tackled. However, studies have found relationships among many of the MDGs, appearing to indicate the relevance of each of them. For example, Abu-Ghaida and Klasen (2004) note that many empirical studies have found that gender equity in education promotes economic growth and reduced fertility, child mortality and undernutrition. Fay and colleagues (2005) observe that improvement in primary education in a particular country may depend on better transportation networks in rural areas, lower infant mortality may depend on clean water, or gender equality in school enrolment may hinge on access to piped water (facilitating school attendance by girls). They note that earlier studies (such as that by Chong and Hentschel, 2003) had found that multiple interventions in infrastructure can be shown to yield economies of delivery. In Morocco, the construction of an all-weather road in some rural communities increased girls' primary school attendance from 28 per cent to 68 per cent (World Bank, 2007a). Jalan and Ravallion (2001) also argue for the need to combine infrastructure interventions with effective public action to promote health knowledge. Thus, Fay and colleagues (2005) concluded from their study that achieving the health MDGs will require more than health and education interventions. In particular, infrastructure services play an important role. Fay and colleagues (2005) suggest that the MDGs will be useful tools only if they are not seen as narrow objectives with uni-dimensional interventions, but the multi-sectoral nature of interventions and development is stressed.

Finally, 'better' goals and targets for national development are suggested by some authors. Tabatabai (2007) proposes that 'The real yardstick for judging performance and effort is whether they have done the best they could under the circumstances'. Clemens and colleagues (2007) suggest that future development goals should (i) be country-specific and flexible; (ii) take historical performance into account; (iii) focus more on intermediate targets than outcomes; and (iv) be considered benchmarks rather than goals which are technically feasible with sufficient funds alone. Thus, the MDGs have not been the last word on indicators of development; other composite benchmarks have emerged. For example, the Multidimensional Poverty Index (MPI), developed by Oxford University, includes variables such as access to food and nutrition, sanitation and hygiene, schooling, and gender and social equality (Velloor, 2010).

In 2005, the Dutch National Committee for International Cooperation and Sustainable Development commissioned a study to develop a generic framework that measured the contributions of the private sector and civil society to the eight MDGs (WBCSD, undated). The framework was revised in 2006 to focus on measuring the impact and indirect contributions of the private sector to the MDGs. The framework includes 77 indicators consisting of questions which investigate policies, management systems and performance. The six companies assessed in the study were ABN AMRO, TNT, Philips, Heineken, BHP Billiton and Akzo Nobel. The researchers collected and analysed public and corporate information on the firms and allocated scores within the MDG framework. The study found that all six participating companies contribute positively to the MDGs and have taken specific steps towards implementing the MDGs in their operations. Some of the companies contribute directly to the MDGs by developing products and services particularly aimed at developing countries whereas others have limited opportunity to contribute to the MDGs because of the nature of their businesses.

While the argument on the merits of the MDGs continues, the need for development has never been more real and more pressing. This is the case in the poorest countries, which are in a group of 35 'Low-income Countries Under Stress' which

> are home to almost 500 million people, roughly half of whom live on less than a dollar a day. These countries face poor governance, conflict or post-conflict transitions, and a multiplicity of problems that make the achievement of development results particularly challenging.
>
> (IEG, 2006, p. 18)

The needs of these countries puts the issue of addressing the MDGs in perspective. As a final note on this point, it is pertinent to cite a statement by Mr Ban Ki-Moon, the UN Secretary-General, in his foreword to the latest MDG report (UN, 2010, p. 3):

> The Goals represent human needs and basic rights that every individual around the world should be able to enjoy – freedom from extreme poverty and hunger; quality education, productive and decent employment, good health and shelter; the right of women to give birth without risking their lives; and a world where environmental sustainability is a priority, and women and men live in equality.

Current results and need for further action

Current results

Studies indicate that efforts to reach the MDGs are falling short of the targets. The findings from the early studies were not encouraging. Sahn and Stifel (2002) assessed progress on six MDGs relating to living standards in Africa and painted a discouraging picture. Much progress has since been made. The number of people living in extreme poverty in developing countries fell from 1.8 billion in 1990 to 1.4 billion in 2005 (i.e. from 42 per cent to 25 per cent of the population), with the decline being largest in East Asia. However, the UN (2010, p. 5) noted:

> unmet commitments, inadequate resources, lack of focus and accountability, and insufficient dedication to sustainable development have created shortfalls in many areas. Some of these shortfalls were aggravated by the global food and economic and financial crises. Nevertheless, [there is] clear evidence that targeted interventions, sustained by adequate funding and political commitment, have resulted in rapid progress in some areas.

The world is on target to reduce extreme poverty by half by 2015. The 2009 Global Monitoring Report (GMR), published by the IMF and the World Bank, concludes that, although the global economic crisis slowed progress, the goal of halving extreme poverty between 1990 and 2015 remains within reach at the global level based on current growth projections. However, performance varies among regions. Greatest progress has been in Asia, especially in China and India. In China, between 1990 and 2005, the number of people living on less than US$2 a day fell by over 400 million. In India, the level of poverty fell from 36 per cent in 1993–1994 to 26 per cent in 1999–2000. It was considered that sub-Saharan Africa was unlikely to meet the goal (World Bank, 2006). The World Bank (2007b) reported that more than 314 million Africans – nearly twice as many as in 1981 – live on less than $1 a day. Thirty-four of the world's 48 poorest countries, and 24 of the 32 countries ranked lowest on the UN's Human Development Index, are in Africa. There is also much disparity, even within countries. For example, in India, whereas 14 per cent of the population in Delhi and 16 per cent in Kerala are poor, in Jharkhand and Bihar the figures are 77 and 81 per cent respectively (Velloor, 2010). It has also been found that in India, although the proportion of poor people has declined, the absolute numbers have increased. The number of poor people in India is estimated to be some 421 million, more than the 410 million in the 26 poorest African nations (Velloor, 2010).

It should be noted that progress has been made in sub-Saharan Africa. The World Bank (2007b) reports that 13 sub-Saharan African countries have attained middle-income status, with another five on course. Africa is the world's third fastest region in the pace of reforms to reduce the time and cost needed to start a business (World Bank and IFC, 2006). Most African countries remain high-cost, high-risk places to do business (World Bank and IFC, 2006). As a result, Africa receives only about 10 per cent of foreign direct investment to developing countries. As shown in Box 4.1, one of the reasons for this is that the promised enabling international trade environment has not materialised.

The Latin America and Caribbean region grew by 6.0, 4.5 and 5.0 per cent in 2004, 2005 and 2006 respectively because of higher export revenues and volumes resulting from high commodity prices and world growth (World Bank, 2007c). The countries have achieved significant progress in education. However, the region continued to face the related development challenges of increasing growth, while reducing poverty and inequality – some 106 million people (nearly 21 per cent of the population) live on less than $2 a day.

The 2009 GMR concluded that, on current trends, most human development MDGs are unlikely to be met at the global level. Deaths of children under five declined worldwide, to about 9 million in 2007, from 12.6 million in 1990, despite population growth. Nevertheless, sub-Saharan Africa and, in some cases, South Asia are likely to fall short of most targets, especially in the areas of child and maternal mortality, access to basic sanitation and reducing child malnutrition. The HIV prevalence rate has shown some decline in Africa but has risen in some other regions. Major accomplishments

Box 4.1 The global challenge

At the Millennium Summit held in 2000, world leaders agreed that strong international partnerships would be crucial to achieving the Millennium Development Goals. Tremendous progress has been made in strengthening those partnerships, especially through increased official development assistance and generous debt relief. The efforts put forth are yielding dividends, as a number of countries are now on track towards achieving several of the Goals. At the same time, . . . many other countries are falling short, and . . . as a result of the global economic crisis, larger numbers of people are facing much more difficult conditions.

The agreed deadline of 2015 is fast approaching, and there is still much to be done. Despite the renewed commitment to international cooperation, economic upheaval and uncertainty have taken a toll on progress towards achieving Goal 8, which is to strengthen the global partnership for development. Delivery of official development assistance is slowing down. The Gleneagles commitments to doubling aid to Africa by 2010 will not be met. The Doha Round of multi-lateral trade negotiations remains stalled. Debt burdens have increased, with a growing number of developing countries at high risk or in debt distress. And rising prices are hampering access to medicines, while investment in technology has weakened.

Nevertheless, economic uncertainty cannot be an excuse for slowing down our development efforts or backing away from international commitments to provide support. Quite the contrary: the uncertainty is one reason to speed up delivery on those efforts and commitments. By investing in the Millennium Development Goals, we invest in global economic growth; by focusing on the needs of the most vulnerable, we lay the foundation for a more sustainable and prosperous tomorrow . . . By fulfilling the promises that have been made to the poor, the vulnerable and the marginalized, we can build a world that is more prosperous, more just and more secure.

Source: MDG Task Force (2010)

were made in education. Enrolment in primary education in developing countries as a whole reached 88 per cent in 2007, up from 83 per cent in 2000. In sub-Saharan Africa and Southern Asia, enrolment increased by 15 percentage points and 11 percentage points respectively, from 2000 to 2007. Although the goal of universal primary school completion will be missed on a global basis, the attainment will be close. Again, the largest shortfalls are likely to be in sub-Saharan Africa and South Asia. The goal of eliminating gender disparity in primary and secondary education seems attainable by 2015, although sub-Saharan Africa is likely to fall short.

The latest MDG report (UN, 2010) presents the results outlined below.

1   Progress on poverty reduction is being made, despite setbacks due to the 2008–2009 economic downturn, and food and energy crises. The developing world, as a whole, is on track to achieve the poverty reduction target by 2015. The overall poverty rate is still expected to fall to 15 per cent by 2015 (however, it is pertinent to note that some 920 million people would still be living under the international poverty line – half the number in 1990).

2   Major advances have been made in getting children into school in many of the poorest countries.

3   Improvements in key interventions – for malaria and HIV control, and measles immunisation, for example – have reduced child deaths from 12.5 million in 1990 to 8.8 million in 2008.

4   Between 2003 and 2008, the number of people receiving anti-retroviral therapy increased tenfold – from 400,000 to 4 million (this represents 42 per cent of the 8.8 million people who needed treatment).

5   Increases in funding and a stronger commitment to control malaria have accelerated delivery of malaria interventions such as bed net protection and treating children with effective drugs.

6   The rate of deforestation, though still alarmingly high, appears to have slowed, thanks to tree-planting schemes and natural expansion of forests.

7   Increased use of improved water sources in rural areas has narrowed the large gap with urban areas, where coverage has remained at 94 per cent since 1990. However, the safety of water supplies remains a challenge and urgently needs to be addressed.

8   Mobile telephony continues to expand, and is increasingly being used for m-banking, disaster management and other applications for development. By the end of 2009, cellular subscriptions had reached the 50 per cent mark.

Thus, the UN (2010, p. 5) observes:

> Though progress has been made, it is uneven. And without a major push forward, many of the MDG targets are likely to be missed in most regions. Old and new challenges threaten to further slow progress in some areas or even undo successes achieved so far.

Further action

More work needs to be done to attain the MDGs. The UN (2010) considers the most pressing to be the following:

1 Efforts to provide productive and decent employment for all, including women and young people, must be intensified.
2 The war on hunger must be waged with more vigour.
3 Greater efforts must be made to get *all* children into school, especially those in rural areas, and eliminate inequalities in education based on gender and ethnicity, and among linguistic and religious minorities.
4 More should be done to reduce maternal mortality, especially in sub-Saharan Africa and Southern Asia, where not much progress has been made so far.
5 Much more should be done to bring improved sanitation to the 1.4 billion people who did not have access to it in 2006 (at the present rate of progress, the 2015 sanitation target will be missed).
6 Greater efforts must be made to improve the living conditions of the urban poor (slum improvements are not keeping pace with the rapid growth of cities in developing countries).
7 Greater priority must be given to preserving the natural resource base and to combat climate change.

In Box 4.1, the MDG Task Force (2010) makes the case for strengthening the global partnership, which was seen as one of the main bases of the efforts towards attaining the MDGs.

The UN (2010, p. 6) declares that the MDGs 'are still attainable'. It continues:

> The critical question today is how to transform the pace of change from what we have seen over the last decade into dramatically faster progress. The experience of these last ten years offers ample evidence of what works and has provided tools that can help us achieve the MDGs by 2015.

It notes that increased attention should be paid to those who are most vulnerable. Policies will be needed to eliminate the persistent inequalities between the rich and the poor and between those living in rural or remote areas or in slums and the better-off urban populations, and to address the needs of those disadvantaged by geographic location, sex, age, disability or ethnicity.

The former Secretary-General of the UN, Mr Kofi Annan (2010, p. A17), added:

> But achieving the MDGs is only the first step. For even if we succeed and meet all eight goals by 2015, almost a billion people will continue to live below the poverty line, hundreds of millions will remain hungry and millions will continue to die from preventable diseases or unnecessary complications.

We will certainly need to take the MDGs to the next level after the initial deadline. While there is some skepticism about the utility of naming specific goals as a basis for development strategies, I remain an advocate. After all, who can argue with and objective as simple and powerful as access to food and clean drinking water, jobs, health care and education for everyone?

Construction and MDGs

Potential of construction

The construction industry plans, designs, builds and maintains the productive facilities and items of infrastructure which enable, support and facilitate production, as well as the social facilities which enhance the quality of life. Thus, it can be argued that construction is the main vehicle through which the MDGs can be realised. The features of the industry may be used as a framework to discuss the potential of construction in these regards (see, for example, Hillebrandt, 2000; Ofori, 2000).

First, buildings and items of infrastructure are vital inputs for economic activity (both production of goods and provision of services), leading to economic growth and increased incomes (MDG1) in the short run, and national development in the long run. For example, the provision of school buildings (MDG2) and health facilities (MDG4 and MDG5) as well as the houses and items of infrastructure which enable the slum improvement objective in the Millennium Declaration to be met are all directly from construction. In delivering its products, construction is a significant sector of the economy. It contributes around 5 per cent of GDP. Some researchers put this figure much higher, with some estimates being as high as 20 per cent (Ruddock, 2007), depending on how 'construction output' is measured. At the same time, construction is responsible for about half of gross domestic capital formation (Hillebrandt, 2000), and thus is a major player in the creation of the nation's savings.

Second, construction activity has extensive linkage effects, and stimulates activities in other sectors of the economy from which the industry obtains its inputs, such as manufacturing, commerce and financial services (banks and insurance companies) and business services (lawyers and accountants). This contributes further to economic growth and development, implying that investment in construction has significant multiplier effects. Using the 2000 input–output tables for Malaysia, Chan and Lim (2007) found that, of the total input to the 'building and construction sector', 44 per cent consists of intermediate inputs, and the remaining 56 per cent of primary inputs. They found that the construction industry's Type I output multiplier (which includes direct or initial spending, indirect spending or businesses buying and selling to each other) is 1.62. Thus, to produce an additional US$1.00 of output, construction would stimulate production in other sectors of the economy by US$1.62. Construction's Type II multiplier is 2.27. This multiplier includes Type I output multiplier effects and household spending based on income earned from the direct and indirect effects (businesses buying and selling to each other) and induced effects (household spending of income earned from direct and indirect effects). Thus, to produce an additional US$1.00 of output, construction would stimulate production in all industries by US$2.27. Shuja and colleagues (2008), using the Malaysian input–output tables for 1983, 1987, 1991 and 2000, found that construction is one of the key sectors in the Malaysian economy (a key sector is one whose growth will generate or promote growth in other sectors via its technological linkages, and which has backward and forward indices greater than 1 after normalisation). The building and construction sector was ranked the highest among the sectors in 1983, but was fourth (out of 58 sectors) in 2000, as the relative importance of key sectors of an economy change over time.

Third, construction provides employment opportunities (Hillebrandt, 2006) in the form of direct employment in the industry and part-time work. The firms adopt flexible recruitment practices to attain greater control over their business operations. This provides the temporary and part-time job opportunities which, for example, farmers can take up after the planting or harvesting seasons (MDG1). Studies consistently show the construction sector accounting for between 5 and 10 per cent of total employment in many developing countries. For example, in 2008, the construction industry in Malaysia accounted for nearly 9.4 per cent of total employment (Department of Statistics Malaysia, 2009). The figure was found to be about 5 per cent in Indonesia (Suraji *et al.*, 2009). Furthermore, construction workers are effective consumers in the economy, further stimulating activities in other sectors and raising incomes generally, thus creating and maintaining jobs.

Fourth, construction activity is location specific; the physical activities on almost all constructed items must be undertaken in the locations where the items are to be placed. Thus, the employment generation potential and stimulation of the local economy can be realised wherever the construction activity actually takes place. Thus, not only are most construction jobs (apart from some of the planning and design services) impossible to outsource wholly or in large part from other countries, but also the jobs can be created in all parts of the country. The activities in construction are also varied in terms of their skills requirements; there are ample opportunities for unskilled workers on the projects. Thus, construction can be used to create full-time or supplementary jobs, and boost local incomes in the rural areas of developing countries.

Fifth, the linkage between construction and the economy is a virtuous closed loop. A growing economy and higher employment will lead to increased demand for construction (Wells, 1986). Thus, there is a cyclic relationship between construction activity and the attainment of the MDGs: construction helps to attain the MDGs and realisation of the MDGs boosts construction activity.

Table 4.2 The Millennium Development Goals and role of construction

MDGs	Contribution of construction	Indicators for construction
Goal 1: Eradicate extreme poverty and hunger	• Effective and efficient production of buildings and infrastructure • Maximum linkages of construction to other sectors of national economy to create stimulus • Generation of employment opportunities • Continuous development of industry	• Performance of industry, company or project on key indicators, such as time, cost, quality, safety • Features of construction in national input–output tables • Number of jobs created, average wages • Average corporate profits; profits on projects • Estimate of total capacity of national industry, and that of the firm
Goal 2: Achieve universal primary education	• Design and construction of suitable school buildings (in local economic, climatic contexts) • Contribution to economic growth and national development to create jobs	
Goal 3: Promote gender equality and empower women	• Creation of job opportunities for women and youth (MDG8) at all levels in construction, with close attention to working conditions on sites, pay and career progression	• Proportion of females in workforce of industry, company, project, at different levels • Average remuneration of employees of different genders
Goal 4: Reduce child mortality Goal 5: Improve maternal health	• Construction of hospitals and infrastructure • Provision of job opportunities to generate income	• Same indicators as for MDGs 1 and 2
Goal 6: Combat HIV/ AIDS, malaria and other diseases	• Effective site management to avoid health hazards • Initiatives to avoid spread of HIV/AIDS by construction workers	• Industry, company, project performance on health and safety • Industry and company policies and programmes on HIV/ AIDS
Goal 7: Ensure environmental sustainability	• Sustainable construction – life-cycle considerations of all aspects of construction • Effective management of completed buildings and infrastructure	• Industry, company, project performance on sustainable construction – waste generation • Average scores in environmental assessment of buildings • Energy performance of various types of buildings

Table 4.2 continued

MDGs	Contribution of construction	Indicators for construction
Goal 8: Develop a global partnership for development	• Construction as a partner for development • Construction as a creator of wealth and less of a burden in imported inputs • Effective logistics of construction in landlocked and small island developing states • Effective technology transfer in construction: from research to practice; from industrialised to developing countries • Partnership among industry, government, researchers • Global networks of researchers to study matters on construction and MDGs	• Same indicators as for MDGs 1 and 2

Source of MDGs: http://www.un.org.org/millenniumgoals/images/mdgs_01.gif

Table 4.2 shows the contribution which construction can make towards the attainment of the MDGs, together with indicators for measuring the progress which the industry makes.

Construction industry's performance

From the above discussion, the construction industry and its activities and processes can be a bottleneck in the effort to realise the MDGs if it is unable to play the role expected of it. It is pertinent to assess the current performance of the industries in the developing countries. This is particularly so because many studies show that there is a great need for improving the performance of the construction industries in developing countries, even beyond the essential requirements of attaining the MDGs.

First, there are many reports that, on most projects in the developing countries, the attainment of the components of the iron triangle of project performance is often elusive, and there are deficiencies such as cost and time overruns; poor quality of work leading to technical defects in completed items and, hence, poor durability; and inadequate attention to safety, health and environmental issues (see, for example, a summary in Ofori, 2007a). Some high-profile projects have failed to measure up in these regards. For example, the Commonwealth Games in Delhi in 2010 involved 71 countries, and were the largest and most expensive Commonwealth Games yet. The construction of the facilities for the games attracted widespread criticism on account of persistent and long delays and soaring construction costs. The total construction cost was estimated to be US$2 billion, double the original estimate in 2006. The Central Vigilance Committee inspected 16 venues for the games and found that every quality certificate scrutinised was either suspect or 'forged' (Ganapathy, 2010). The team reported 'major failures of concrete samples and preparation of forged testing records' and found lower standards of steel and aluminium and inferior-quality anti-corrosion coating on steel.

Several reasons are given for the poor performance in construction in developing countries. In many studies, the main root causes of the poor performance on construction projects, and the lack of capacity and capability of their construction industries, comprise factors that are outside the control of the industry (Ofori, 2009). These are considered to include the inherent features of the construction project, product and industry; the nature of its operating environment, including its inter-related political (Enshassi *et al.*, 2006), administrative, industrial, business and physical components; and external shocks such as those within the countries including economic cycles, and those from outside the countries such as immigration. The Master Builders Association of Malaysia (2006) identified the following constraints faced by the industry in Malaysia: (i) adequate and timely supply of qualified and experienced personnel; (ii) adequate and timely supply of construction materials; (iii) control of the prices of key construction materials; and (iv) the need for the reduction of red tape and administrative interference.

Some authors take a more balanced approach. For example, Suraji and Krisnandar (2008) found the following factors faced by the construction industry in Indonesia:

1 internal:
   a weakness regarding management, technological expertise, financing, lack of skilled workers;
   b lack of a strong structure for the national industry;
   c lack of synergy in terms of partnerships;
2 external:
   a inequality among suppliers and consumers;
   b lack of support from all other sectors including the financial sector;
   c lack of availability of standardised materials;
   d lack of professional and managerial training and development.

Finally, some authors focus on the construction industries themselves to find the causes of poor performance. Toor and Ogunlana (2008) found that some of the problems which the managers of construction projects in Thailand face include inadequate procurement system, lack of resources, discrepancies between design and construction, lack of project management practices, variation orders, communication lapses, cultural issues and differences in the interests of the participants.

Construction and MDGs in the literature

As discussed above, there has been some debate on the effects of initiatives to attain the MDGs, and many of these relate to the items created by the construction industry. Fay and colleagues (2005) found that better access to infrastructure (piped water, sanitation and electricity) has a large and statistically significant effect in reducing infant and child mortality and incidence of malnutrition. Ravallion (2007) disputes these findings, questioning the estimating methods adopted. He also concluded that (contrary to the findings of Fay *et al.*) there was complementarity between basic infrastructure and health care, whereby, at sufficiently high levels of initial health care, improvements in basic infrastructure reduce infant and child mortality and the incidence of malnutrition. In response, Fay and colleagues (2007: 930) highlight the complexity of the situation and call for more research. Li (2009) added to, and confirmed, the studies around the world which have shown that there is a significant correlation between children's development and the neighbourhood built environment in which they are brought up.

The employment generation potential of construction is discussed above. This potential can be most effectively realised through the adoption of appropriate procurement approaches and technologies (see, for example, McCutcheon, 2001). There are some negative aspects of this potential. For example, the International Labour Organisation (ILO) defines 'decent work' as:

> opportunities for work that is productive and delivers a fair income, security in the workplace and better social protection for families, better prospects for social development and social integration, freedom for people to express their concerns, organise and participate in the decisions that affect their lives and equality of opportunity and treatment for all women and men.
>
> (http://www.ilo.org/public/english/decent.htm,
> accessed 7 November 2009)

The ILO suggests that decent work is encapsulated in four strategic objectives: fundamental principles and rights at work and international labour standards; employment and income opportunities; social protection and social security; and social dialogue and tripartism (http://www.ilo.org/global/About_the_ILO/Mainpillars/WhatisDecentWork/lang--en/index.htm). However, construction work is widely characterised as being dirty, demanding, dangerous and low-paying (Construction 21 Steering Committee, 1999). Thus, it is not necessarily 'decent'. Studies have found that construction sites are among the most dangerous among all sectors in terms of frequency and severity of accidents, as well as fatalities. The sites and their workers contribute to the spread of mosquito-borne diseases (Vijayan and Neo, 2007) and HIV/AIDS (Meintjes *et al.*, 2007). Thus, working conditions on sites should be improved to reduce the spread of diseases among the industry's workers and the community.

The literature highlights the role of infrastructure in economic growth and development (Han and Ofori, 2001, provide a useful review). Fedderke and colleagues (2006) analysed data for South Africa from 1875 to 2001 and found that investment in infrastructure leads to long-term economic growth both directly and indirectly, the latter by increasing the marginal productivity of capital. Fedderke and Bogetić (2009) note that empirical explorations of the growth and aggregate productivity impacts of infrastructure have been characterised by ambiguous results with little robustness. They use 1970–2000 panel data for South African manufacturing and a range of 19 infrastructure measures and explore the question of infrastructure endogeneity in output equations. They conclude that controlling for the possibility of endogeneity in the infrastructure measures renders the impact of infrastructure capital not only positive, but at economically meaningful magnitudes.

The World Bank (2007d) notes that improving infrastructure in developing countries is a key factor in reducing poverty and increasing growth; it is vital to the achievement of the MDGs as it improves access to water and electricity, as well as schools, hospitals and markets. The World Bank (2007d) observes that, if Africa had attained infrastructure growth rates comparable to those in East Asia in the 1980s to 1990s, it could have achieved annual growth rates about 1.3 per cent higher than it did. Lynch and colleagues (2010) cite a study which found that landlocked countries with poor national and transit infrastructure pay 84 per cent more to export their goods than a coastal country, but this can be reduced to 33 per cent where the transport infrastructure is good. They cite the Democratic Republic of Congo as a country which has infrastructure challenges. It has been observed that food prices are kept low if there is better infrastructure (roads, railways and cold storage facilities) which can increase supply and reduce spoilage of farm products (Ghosh, 2010). Foster and Briseno (2009) present a detailed analysis of Africa's needs for infrastructure if it is to realise appreciable improvements in the quality of life of its people.

The World Business Council for Sustainable Development (WBCSD, 2005) presented some case studies of projects in several developing countries where companies had contributed to the welfare of citizens through practical initiatives which helped generate incomes for the poor. The WBCSD (2010) suggests that the appropriate legal, institutional and financial conditions required to support a transition towards a more sustainable and inclusive future include (i) promoting a fair and competitive global market that is non-discriminatory; (ii) establishing regulatory frameworks that uphold property rights, accelerate entry to the formal economy and root out corruption; (iii) providing capacity-building and general education; (iv) facilitating access to finance and investment risk mitigation instruments, in particular for SMEs; and (v) securing adequate investments into the physical infrastructure.

Box 4.2 presents an account of how important infrastructure is to the economic development of Mozambique.

Similarly, it is estimated that the lack of investment in infrastructure in the 1990s reduced long-term growth in the Latin America and Caribbean region by 1–3 percentage points, and hindered the region's ability to compete with the dynamic Asian economies (World Bank, 2007a). Eifert and colleagues (2008) observe that data from the World Bank Enterprise Surveys show that indirect costs (related to infrastructure and services) account for a relatively high share of firms' costs in poor African countries and impose a competitive burden on African firms. The difference between the indirect cost levels faced by comparable Zambian and Chinese firms is almost equivalent to the whole wage bill of the former. Kinda (2010) uses firm-level data across 77

Box 4.2 MDGs and infrastructure development in Mozambique

Maputo still has a sleepy down-at-heel charm, with its art deco houses, crumbling concrete apartment blocks and usually congestion-free streets. Yet the nostalgic 1970s feel might not last much longer. A new bridge to link the downtown area of Mozambique's capital with Catembe across the river Tembe and a new road to the South African border at Ponta de Ouro are among a clutch of projects planned by the government as it takes advantage of surging interest in the country's abundant coal, gas and mineral reserves.

These developments raise the prospect that one of Africa's poorest and aid-dependent societies may reduce and eventually eliminate its dependence on foreign assistance. 'We have to go faster', Aires Bonifacio Ali, the country's prime minister, told the *Financial Times*. Mr Ali argues that the drive to improve infrastructure – including a new road from the southern border to Cabo Delgado in the remote north – is 'crucial' if the country is to free itself from the clutch of foreign donors. 'If we have good roads and good communications it will be easier to improve our agriculture. Our dream is to connect our country from north to south.' Mr Ali is building on progress in a number of respects. Political stability has been established since the end of the civil war in 1992.

The country remains bitterly poor but it has advanced socially. 'There have been good improvements in reducing child mortality and access to schools. There has been a significant catch-up from such a low base' said one foreign diplomat.

Source: Lapper (2010)

developing countries to show that constraints related to investment climate hamper foreign investment. The results show that physical infrastructure problems, financing constraints and institutional problems discourage foreign investment. As investment climate constraints, he focused on physical and financial infrastructure problems in addition to human capital and institutional constraints.

There are many recent examples of the economic and social stimulus from investments in infrastructure. Some World Bank (2007a) projects in Peru rehabilitated 13,000 kilometres of rural roads, reducing travel time by an average of 68 per cent and increasing school enrolment by 8 per cent and visits to health centres by 55 per cent. Gunasekera and colleagues (2008) note that, when transport investments are made in relatively poor-infrastructure regions, the consequences extend beyond growth effects to some transformational changes. They estimate the direction and magnitude of some of the transformational changes induced at the firm and household level using a highway project in Sri Lanka. They conclude that, in a region with limited public capital stock, a highway improvement may potentially induce a dual structural shift: the emergence of a new social and technical environment (or a new set of economic opportunities), and a change in the pattern of relationships between the environment and social actors.

Gibson and Olivia (2010) note that, whereas access to infrastructure is identified in some studies as a factor that affects non-farm rural employment and income, less attention has been paid to the constraints imposed by poor-quality infrastructure. They analyse data from 4,000 households in rural Indonesia to show that the quality of roads and electricity affects both employment in, and income from, non-farm enterprises. The results support the view that poor infrastructure constrains rural non-farm enterprises. Moreover, there is a negative effect of poor-quality infrastructure. Thus, there would be gains from development strategies that improve both the access to and the quality of rural infrastructure.

Levels of provision

In 2010, the UN declared access to clean water and sanitation 'a human right that is essential to the enjoyment of life and all human rights' (*Straits Times*, 2010), as it estimated that, around the world, some 884 million people lack access to safe drinking water, and 2.6 billion people do not have access to basic sanitation. Some 3.5 million people die every year as a result of unsafe water and sanitation; and 443 million schooldays are lost each year as a result of these illnesses. The UN (2010) found that only about half of the developing world's population use improved sanitation, and it suggests that addressing this inequality will have a major impact on many of the MDGs. Only 40 per cent of rural populations are covered. Whereas 77 per cent of the population in the richest 20 per cent of households use improved sanitation facilities, the share is only 16 per cent of those in the poorest households.

The World Bank estimates that around 900 million rural dwellers, mainly in the poorest developing countries, are without reliable transport access (Roberts *et al.*, 2006). Only 34 per cent of the rural population in sub-Saharan Africa and 57 per cent in South Asia have access to the transport system. Moreover, many rural communities do not have year-round access to motorised transport services because of the low density of demand and/or high costs.

Addressing MDGs in construction

It has been suggested that construction activities and products should be utilised directly to realise the MDGs, especially MDG1. Engineers Against Poverty (EAP, 2006) identifies opportunities to improve the delivery of social development objectives by modifying the way in which public infrastructure projects are procured. It suggests that (i) project identification should be in line with national, local or sector plans and/or based on public consultation; (ii) the whole life cycle of the project should be considered during planning and design, and a maintenance strategy developed; (iii) social objectives should be identified at the planning stage and fed into design; (iv) funds should be set aside in the budget for the realisation of social objectives; (v) an appropriate procurement approach to deliver the specified social objectives should be chosen; (vi) the bidders' social performance and capacity to deliver social obligations should be considered; (vii) contractual obligations must be monitored and enforced through incentives and/or sanctions; and (viii) social performance audits should be conducted with the same rigour as financial audits.

The International Labour Office (ILO, 2006, p. 3) suggests that municipalities should launch investment policies and programmes with the following elements: (i) employment-intensive infrastructure development for upgrading unplanned settlements and rehabilitating facilities for people affected by disasters and conflicts; (ii) provision of social infrastructure for accessibility, water, health, education, markets, rehabilitation and preservation of national heritage; (iii) organisation and association building, negotiation and contracting capacity building for communities and informal economy operators, and support to SMEs; (iv) provision of support to local governments, community groups and the private sector in pro-poor procurement and community contracting; (vi) review of the local regulatory environment to improve their impact on job creation and the quality of jobs created; and (vii) integrated employment and environmental impact assessments of urban investment plans.

EAP has identified the features of sustainable pro-poor infrastructure. It suggests that such infrastructure (i) provides access for the poor to affordable services that meet their basic human rights and needs, reduce their vulnerability to natural disasters and allow them to participate in economic activity; (ii) supports substantive freedoms for individuals and communities to participate in decision making that affects their well-being and livelihoods; (iii) minimises the consumption of natural resources and the impact on biodiversity and natural systems; (iv) boosts the creation of employment in construction, operation and maintenance; (v) is economically and operationally sustainable in the long term; and (vi) is designed and operated through holistic consideration of social, environmental and economic costs and benefits.

Lynch and colleagues (2010) suggest that social risks from transport projects include delays or abandonment of the project; reputational damage; lack of user acceptance; decreased operational revenues; consumer boycotts; major modifications due to stakeholder pressure; exposure to legal action; and security problems. On the other hand, the social opportunities from such projects include better project outcomes through stakeholder input; streamlined approval processes; government and regulatory support; timely project completion; easier access to project finance; improved operational revenues through customer support; increased likelihood of support for subsequent projects or future expansions; value creation for proponent organisation; and enhanced contribution to sustainable development.

Ofori (2007b) considered the role of project managers in the efforts to attain the MDGs and stressed the need to expand educational and training programmes for them. He notes that the project manager has a major role to play in accelerating the progress towards improving the well-being and quality of life of the populations of the developing countries. He suggests that the education of project managers must provide the perspective and the tools to enable the graduates to play their part in the efforts to attain the national visions and goals, and thereby improve the quality of life of all the people in the community. He proposes a manifesto for project managers relating to the attainment of the MDGs. Ofori (2007a) discusses the role of researchers in the field of construction management and economics in the effort to attain the MDGs. Highlighting the need for the performance of the construction industry to be improved, he suggests that those researchers should seek effective solutions that would enable this to be achieved. Ofori (2010) discusses the built environment and programmes for attaining the MDGs. He observes that the built environment in any country determines the nature and pace of national development, and the quality of life of the people. It has a major influence on progress towards the attainment of the MDGs. The construction industry which produces this environment must be able to play its due role if it is not to be a barrier to progress in these regards. He concludes that research on how to improve the performance of the industry would be of benefit, and suggests a research agenda.

It is pertinent to note that the lack of agreement on the MDGs among researchers is mirrored in the construction industry. After a presentation on construction and the MDGs by Ofori (2007a), many participants at the conference in the UK considered the MDGs illegitimate and patronising impositions by outsiders (Collier, 2007). They did not share Ofori's (2007a) suggestion that researchers in construction have a responsibility to consider the relevance of their work to the MDGs, where appropriate. However, as suggested by Ofori (2010), researchers in developing countries do not have the luxury of being able to detach themselves from their nations' needs.

In order for the construction industry to play its role in efforts to attain the MDGs, it is necessary to assess performance on construction projects and to develop criteria for preparing strategies, action plans and tools for monitoring progress. Performance criteria for construction projects are considered in the next section.

Performance parameters for construction

There has been a debate on the best ways to assess performance on a construction project. Alarcon and Ashley (1992) suggested that the elements of success on projects are effectiveness; efficiency; quality; productivity; quality of work life; profitability; and innovation. Atkinson (1999) suggests the following success measures:

1 delivery stage: cost, time, quality, efficiency;
2 post-delivery stage:
   – the system – getting it right:
   a benefits to many stakeholders;
   b criteria from project manager, top management, client, team members;
   c resultant system;
   the benefits – getting them right:
   d impact on client, end users;
   e business success.

Lim and Mohammed (1999) highlighted the following criteria of project success: completion time; completion cost; completion quality; completion performance; completion safety; completion satisfaction; completion utility; and completion operation. Chan and Chan (2004) identified the following indicators of project performance: quality; functionality; end-user's satisfaction; client's satisfaction; design team's satisfaction; construction team's satisfaction; construction time, speed of construction, time variation; unit cost, net percentage variation over final cost; net present value; accident rate; and environmental impact assessment scores.

The often mentioned key performance indicators suggested at the industry level in the UK are also pertinent at the project level. They are (i) objective measures – construction time, speed of construction, time variation, unit cost, percentage net variation over final cost, net present value, accident rate, environmental impact assessment scores; (ii) subjective measures: quality, functionality, end-user's satisfaction, client's satisfaction, design team's satisfaction, construction team's satisfaction.

Several inter-related factors make the attainment of good results on construction projects in developing countries challenging. Ofori (2009) suggested that the management of projects in these countries is fraught with many problems because of the many practitioners from several backgrounds and different companies who need to participate in any sizeable construction project. In the developing countries, the large, complex project ends of the construction markets are dominated by foreign firms undertaking the international projects; the participants not only are more numerous but also come from several different countries. Moreover, apart from these internal stakeholders, the construction project tends to have large footprints in terms of physical, economic and social effects. Thus, they also have a large number of stakeholders including direct beneficiaries as well as those who might be affected in an adverse manner by the projects (Ofori, 2003). It is also pertinent to note that stakeholders, including the clients, end purchasers and users, have no knowledge of aspects of construction. This implies that there is a need for professionalism among the participants in the construction project, and a dedication to meet the objectives and aspirations of the stakeholders in the most value-adding manner for the benefit of all the stakeholders.

Another factor that makes the management of construction projects in developing countries challenging is that construction projects have long value chains because the industry obtains its inputs of materials, components, equipment, finance and other services from many other sectors of the economy. This implies that supply chain management, including logistics, is a key requirement. It also means that the level of development of several other sectors of the economy has an impact on the performance of construction projects. Where the construction industry lacks strong supporting sectors (in particular, materials suppliers, plant-hire facilities and financial institutions), value chain management is a difficult task.

Thus, guidance for project performance is much needed in the developing countries and some researchers have proposed possible yardsticks. For example, Ahadzie and colleagues (2008) identified 15 critical success factors for mass house-building projects in developing countries, which they found to lie in four groups: (i) environmental impact – overall environmental impact, environmental impact of individual houses, overall health and safety measures, health and safety on individual houses; (ii) customer satisfaction – overall customer/client satisfaction and that on individual houses, technology transfer, overall risk containment and that on individual houses, rate of delivery of individual houses; (iii) quality – overall quality and that of individual houses; and (iv) cost and time – overall cost and that of individual houses, overall project duration. In a study in Ghana, Gyadu-Asiedu (2009) found that practitioners and clients have different perspectives and expectations with respect to construction project performance. Practitioners consider the performance criteria to be the contractual ones of cost, time and quality; impact of the project on the environment and society; and management and execution efficiency. For the government clients, the criteria are whether the project performs satisfactorily as expected and whether the clients' needs and motivation for undertaking the project are satisfied – these include contributing to good governance, contributing to national infrastructure and addressing future infrastructure needs. However, clients also do show concern about, and interest in, cost, time and quality, and take them as given.

From the above discussion, a framework on project performance for attaining the MDGs in developing countries is now proposed.

Suggested framework

The construction industries in developing countries should make their due contribution to the efforts to attain the MDGs. This puts a different perspective on the parameters of construction projects in the developing countries. These projects should be undertaken, and completed:

1 at lowest capital and life-cycle costs, to attain value for money and enable greater volume of overall construction output to be realised from the available, usually limited, budget;
2 on time, to reduce the time that enterprises and agencies have to wait for their productive facilities, and the time that beneficiaries of projects have to wait for the items which they need to enhance the quality of their lives;
3 to the highest level of quality, to enhance the durability and longevity of the completed items, minimise maintenance and repair costs, and optimise the utility from the built items, and hence its contribution to the capital assets of the nation;
4 with focus on the health of the workers and residents in the environs of the projects;
5 with attention to the safety of the workers and neighbouring residents, as well as the eventual users in order to enhance the well-being of the workers and improve the image of the construction industry (taking points 4 and 5 together, construction should seek to provide decent jobs);
6 with attention to environmental considerations, to conserve resources (including land, water and energy) and reduce pollution, and with minimum wastage levels, to husband the materials, many of which are imported (in these countries), in order to reduce the balance of payment burden (it is suggested that among the factors that increase food prices is the loss of agricultural land as a result of construction in the process of urbanisation; Ghosh, 2010);
7 with due consideration of employment generation, through the selection of appropriate procurement policies and technologies;
8 to the satisfaction of the client, and contributing to the efficiency of productive activities intended to be undertaken in the building (and the competitiveness of the firm);

9 to the satisfaction of the beneficiaries, who should have played a role in key decisions on the project.

In general, the construction industry should seek to contribute to economic growth over the short term, and national socio-economic development over the long term.

Conclusion

Since the MDGs were launched a decade ago, there has been a debate on their appropriateness and usefulness as benchmarks of development. Several aspects of the goals have been questioned. These include their sources, their ownership, the time period set for their attainment, and the likelihood that they will be attained. However, there is a need to develop the poorer nations to improve the circumstances of their populations. The MDGs provide a framework to guide this process, and constitute an appropriate set of targets to attain. Thus, the MDGs are important. The governments of the developing countries are committed to attaining them. Most of the industrialised countries are providing them with support. The UN and international multinational development agencies have also been active in these regards. The business community and civil society (in the form of local and international NGOs) are also playing a role.

Some progress has been made towards attaining the MDGs. However, much more remains to be done, and the time is short. Indeed, it can be argued that the goals are quite modest and, even if they were attained by 2015, that would still leave significant levels of deprivation on all the criteria covered in the goals, such as poverty levels, infant mortality, gender equality, access to good jobs and suitable levels of infrastructure. It is also worth noting that, between 2007 and 2050, the population living in urban areas is projected to increase from 3.3 billion to 6.4 billion (UN, 2007).

The construction industry, which plays a key role in the process of national socio-economic development, is an important contributor to the effort towards attaining the MDGs. Thus, the industry should be able to do what is expected of it. Therefore, the capacity and capability of the construction industries in the developing countries should be enhanced to enable them to deliver a higher volume of output to meet the increased demand from initiatives to realise the MDGs, and to do so in a cost-effective and time-efficient manner, to a high quality and with overall value for money. It is necessary for an appropriate set of performance criteria for projects to be developed in order to guide the industries in planning projects and monitoring their delivery. These criteria should also aid the evaluation and auditing of the completed projects to obtain, among other things, lessons for good practice in future. The need for construction to help to attain the MDGs in the short period left for their realisation makes this project performance framework vital. To this end, the framework proposed in this chapter can be the basis for discussion and further development.

References

Abu-Ghaida, D. and Klasen, S. (2004) The costs of missing the Millennium Development Goal on gender equity. *World Development*, 32 (7), 1075–1107.

Ahadzie, D. K., Proverbs, D. G. and Olomolaiye, P. (2008) Critical success factors on mass housebuilding projects in developing countries. *International Journal of Project Management*, 26 (6), 675–687.

Alarcon, L. F. and Ashley, D. B. (1992) *Project Performance Modeling: A Methodology for Evaluating Project Execution Strategies.* Construction Industry Institute, University of Texas at Austin: Austin, TX.

Annan, K. (2010) Looking beyond UN's Millennium goals. *Straits Times*, September 20, p. A17.

Atkinson, R. (1999) Project management: cost, time and quality, two best guesses and a phenomenon, it's time to accept other success criteria. *International Journal of Project Management*, 17 (6), 337–342.

Baulch, B. (2006) Aid distribution and the MDGs. *World Development*, 34 (6), 933–950.

Chan, A. P. C. and Chan, A. P. L. (2004) Key performance indicators for measuring construction success. *Benchmarking: An International Journal*, 11 (2), 203–221.

Chan, T. K. and Lim, S. H. (2007) *A Macro-Economic Assessment of the Construction Industry in Malaysia and Economic Impact Review of the Construction Industry Master Plan (CIMP) – Draft Report.* Construction Industry Development Board (CIDB): Kuala Lumpur.

Chong, A. and Hentschel, J. (2003). *Bundling of Basic Services and Household Welfare in Developing Countries: The Case of Peru.* Background Report. World Bank, Washington, DC.

Clemens, M. A., Kenny, C. J. and Moss, T. J. (2007) The trouble with the MDGs: confronting expectations of aid and development success. *World Development*, 35 (5), 735–751.

Collier, C. (2007) Applied science. *International Construction Review*, 8. http://www.iconreview.org/node/31

Construction 21 Steering Committee (1999) *Reinventing Construction.* Ministry of Manpower: Singapore.

Dalgaard, C.-J. and Erickson, L. (2009) Reasonable expectations and the first Millennium Development Goal: how much can aid achieve? *World Development*, 37 (7), 1170–1181.

Department of Statistics Malaysia (2009) *Labour Force Survey.* Department of Statistics: Kuala Lumpur.

EAP (Engineers Against Poverty) (2006) *Modifying Infrastructure Procurement to Enhance Social Development.* Engineers Against Poverty: London.

Easterly, W. (2009) How the Millennium Development Goals are unfair to Africa. *World Development*, 37 (1), 26–35.

Eifert, B., Gelb, A. and Ramachandran, V. (2008) The cost of doing business in Africa: evidence from enterprise survey data. *World Development*, 36 (9), 1531–1546.

Enshassi, A., Hallaq, K. and Mohamed, S. (2006) Causes of contractors' business failure in developing countries: the case of Palestine. *Journal of Construction in Developing Countries*, 11 (2), 1–14.

European Business Review (2010) *A Business View on Development.* http://www.inclusivebusiness.org/2010/09/a-business-view-on-development.html.

Fay, M., Leipziger, D., Wodon, Q. and Yepes, T. (2005) Achieving child-health-related Millennium Development Goals: the role of infrastructure. *World Development*, 33 (8), 1267–1284.

Fay, M., Leipziger, D., Wodon, Q. and Yepes, T. (2007) Achieving child-health-related Millennium Development Goals: the role of infrastructure – a comment. *World Development*, 35 (5), 929–930.

Fedderke, J. W. and Bogetić, Ž. (2009) Infrastructure and growth in South Africa: direct and indirect productivity impacts of 19 infrastructure measures. *World Development*, 37 (9), 1522–1539.

Fedderke, J. W., Perkins, P. and Luiz, J. M. (2006) Infrastructural investment in long-run economic growth: South Africa 1875–2001. *World Development*, 34 (6), 1037–1059.

Foster, V. and Brisceno, C. (eds) (2009) *Africa's Infrastructure: A Time for Transformation*. World Bank: Washington, DC.

Ganapathy, N. (2010) Another scandal hits C'wealth Games. *Sunday Times* (Singapore), 1 August, p. 20.

Ghosh, N. (2010) Food prices: ups and downs. *Straits Times* (Singapore), 11 September, p. A8.

Gibson, J. and Olivia, S. (2010) The effect of infrastructure access and quality on non-farm enterprises in rural Indonesia. *World Development*, 38 (5), 717–726.

Gunasekera, K., Anderson, W. and Lakshmanan, T. R. (2008) Highway-induced development: evidence from Sri Lanka. *World Development*, 36 (11), 2371–2389.

Gyadu-Asiedu, W. (2009) *Assessing Construction Project Performance in Ghana*. Unpublished PhD thesis, Eindhoven Technical University, Eindhoven.

Han, S. S. and Ofori, G. (2001) Construction industry in China's regional economy. *Construction Management and Economics*, 19, 189–205.

Hillebrandt, P. M. (2000) *Economic Theory and the Construction Industry*, 3rd edition. Macmillan: Basingstoke.

Hillebrandt, P. M. (2006) The construction industry and its resources: a research agenda. *Journal of Construction in Developing Countries*, 11 (1), 37–52.

IEG (Independent Evaluation Group of the World Bank) (2006) *Annual Review of Development Effectiveness 2006: Getting Results*. World Bank: Washington, DC.

ILO (International Labour Office) (2006) *A Strategy for Urban Employment and Decent Work*. ILO: Geneva.

IMF (International Monetary Fund) (2010) *Factsheet: The IMF and the Millennium Development Goals*. IMF: Washington, DC.

Jalan, J. and Ravallion, M. (2001). *Does Piped Water Reduce Diarrhea for Children in Rural India?* Policy Research Working Paper No. 2664. World Bank, Washington, DC.

Kinda, T. (2010) Investment climate and FDI in developing countries: firm-level evidence. *World Development*, 38 (4), pp. 498–513.

Lapper, R. (2010) Bid to shake off dependence on aid. *Financial Times*, 22 June, Special Report, p. 1.

Li, L.H. (2009) Built environment and children's academic performance – a Hong Kong perspective. *Habitat International*, 33 (1), pp. 45–51.

Lim, K.C. and Mohammed, A.Z. (1999) Criteria of project success: an exploration and re-examination. *International Journal of Project Management*, 17 (4), pp. 243–248.

Lynch, C., Cavill, S. and Ryan-Collins, L. (2010) *Maximising the social development outcomes of roads and transport projects*. A Guidance Note prepared for the Chartered Institution of Highways and Transportation. Engineers Against Poverty, London.

McCutcheon, R. (2001) An introduction to employment creation in development and lessons learned from employment creation in construction. *Urban Forum*, 12 (3–4), 263–278.

Master Builders Association of Malaysia (2006) Malaysia's economy in 2005 and construction industry's performance. Presented at Council Meeting of Asian Constructors Federation (ACF), Hanoi, Vietnam, 30 November.

Maxwell, S (2003) Heaven or hubris: reflections on the 'new poverty agenda'. *Development Policy Review*, 21 (1), 5–25.

Meintjes, I., Bowen, P. and Root, D. (2007) HIV/AIDS in the South African construction industry: understanding the HIV/AIDS discourse for a sector-specific response. *Construction Management and Economics*, 25, 255–266.

MDG Task Force (2010) *The Global Partnership for Development at a Critical Juncture – MDG Task Force Report*. United Nations, New York.

Millennium Project (2005) *Investing in Development: A practical plan to achieve the Millennium Development Goals*. United Nations Development Programme, New York.

Ofori, G (2000) Challenges for construction industries in developing countries. Proceedings, Second International Conference of the CIB Task Group 29, Gaborone, Botswana, November, pp. 1–11.

Ofori, G. (2003) Frameworks for analysing international construction. *Construction Management and Economics*, 21 (4), 379–391.

Ofori, G. (2007a) Millennium Development Goals and construction: A research agenda. Presented at CM&E 25th Anniversary Conference, Reading, 16–18 July.

Ofori, G. (2007b) The project manager and the Millennium Development Goals. Keynote Paper presented at 10th Anniversary of Project Management Education at University of Moratuwa, Colombo, Sri Lanka, 18 August.

Ofori, G. (2009) Leadership and the Construction Industry in Developing Countries. Paper presented at CIB W107 on Construction in Developing Countries, Penang, 5–7 October.

Ofori, G. (2010) Built environment research and the Millennium Development Goals. Paper presented at West African Built Environment Research Conference, Accra, 27–29 July.

Oxfam International (2010) Halving Hunger: Still possible? http://www.oxfam.org. Accessed on 29 September 2010.

Powell, R. and Bird, G. (2010) Aid and debt relief in Africa: have they been substitutes or complements? *World Development*, 38 (3), 219–227.

Ravallion, M (2007) Achieving child-health-related Millennium Development Goals: The role of infrastructure – a comment. *World Development*, 35 (5), pp. 920–928.

Roberts, P., Shyam K.C. and Rastogi, C. (2006) Rural Access Index: A key development indicator. World Bank Transport Papers (TP-10). World Bank, Washington DC.

Ruddock, L. (2007) The economic value of construction: Achieving a better understanding. Presented at Construction and Building Research Conference of the Royal Institution of Chartered Surveyors, Georgia Institute of Technology, Atlanta, GA, USA, 6–7 September.

Sahn, D.E. and Stifel, D.C. (2002) Progress toward the Millennium Development Goals in Africa. *World Development*, 31 (1), 23–52.

Save the Children (2010) *Making It Count: Providing Education with Equity and Quality in the Run-up to 2015*. London.

Shuja, N., Yap, B. W., Lazim, M. A. and Okamoto, N. (2008) Identifying key sectors of Malaysian Economy: a comparison of unweighted and weighted approaches, *Statistics Malaysia: Journal of the Department of Statistics*, (1), 11–26.

Straits Times (2010) Water and sanitation a human right: UN. 30 July, p. A13.

Suraji, A. and Krisnandar, D. (2008) Productivity improvement of the construction industry: A case of Indonesia. Presented at 14th Asia Construct Conference, 23–24 October.

Suraji, A., Surarso, G.W. and Supriyatna (2009) *The Construction Sector of Indonesia*. Paper presented at the 15th ASIACONSTRUCT Conference, 19–21 October, Kuala Lumpur, Malaysia. http://www.asiaconstruct.com/past_conference/conference/15th/3Indonesia.pdf

Tabatabai, H. (2007) *MDG Targets: Misunderstood or Misconceived?* International Poverty Centre. One pager No. 33. April.

Toor, S.R and Ogunlana, S. O. (2008) Critical COMs of success in large-scale construction projects: Evidence from Thailand construction industry. *International Journal of Project Management*, 26 (4), 420–430.

UN (United Nations) (2000) *Resolution Adopted by the General Assembly 55/2 – United Nations Millennium Declaration*. New York.

UN (2007) World Urbanization Prospects: 2007 Revision. Available from: http://www.un.org/esa/population/publications/wup2007/2007_urban_rural_chart.pdf. Accessed on 25 January 2010.

United Nations (2010) *The Millennium Development Goals Report 2009*. New York.

Vandemoortele, J. (2007) *MDGs: Misunderstood Targets?* International Poverty Centre. One pager No. 28. January.

Velloor, R. (2010) Indians found wanting in poverty index. *Straits Times* (Singapore), 17 July, p. C1.

Vijayan, K. C. and Neo, M. (2007) Errant contractors hit with stop-work orders. *Straits Times*, May 25, p. H2.

WBCSD (World Business Council for Sustainable Development) (2005) *Business for Development: Business Solutions in Support of the Millennium Development Goals*. WBCSD: Geneva.

WBCSD (2010) *Business and Development: Challenges and Opportunities in a Rapidly Changing World*. WBCSD: Geneva.

WBCSD (undated) Measuring the development footprint of six multinational companies, http://www.wbcsd.org/plugins/DocSearch/details.asp?type=DocDet&ObjectId=MjI0NTc. Accessed on 16 October 2010.

Wells, J. (1986) *The Construction Industry in Developing Countries: Alternative Strategies for Development*. Croom Helm: London.

White, H. and Black, R. (2004) *Targeting Development: Critical Perspectives on the Millennium Development Goals*. Routledge: New York.

World Bank (2004) *Strategic Framework for Assistance to Africa: IDA and the Emerging Partnership Model, Africa Region*. World Bank: Washington, DC.

World Bank (2006) *World Development Report 2007: Development and the Next Generation*. World Bank: Washington, DC.

World Bank (2007a) *Infrastructure in Latin America and the Caribbean: Recent Developments and Key Challenges.* http://web.worldbank.org/

World Bank (2007b) *Regional Brief: Africa.* http://web.worldbank.org/

World Bank (2007c) *Regional Brief: Latin America and the Caribbean.* http://web.worldbank.org/

World Bank (2007d) *Global Economic Prospects: Managing the Next Wave of Globalization.* World Bank: Washington, DC.

World Bank and IFC (International Finance Corporation) (2006) *Doing Business 2007: How to Reform.* World Bank: Washington, DC.

Part II

Construction industry development: macro-level issues

5　Institution building in construction industries in developing countries

Rodney Milford

Background and introduction

The objective of this chapter is to investigate the role of institutions in the development of the construction industry in developing countries, with the aim of drawing out policy implications for governments in those countries.

This chapter starts by recognising that much can be learnt from the rapid expansion of the construction industries of several countries during the period of around 1850 to 1950. In this period, many institutions were established to support the development of the construction industries in countries such as India, Australia and South Africa – largely dominated by the then 'colonial models'. One such type of institution that developed during this period was the 'builder and contractor associations'. The role of these largely private-sector institutions in supporting the development of the construction industry, and the lessons learnt, are discussed in the next section of this chapter.

More recently, several countries have established formal or informal national programmes to support the objectives of construction industry development, and a range of new institutions have emerged in support of these national reform programmes. Typically, these national reform programmes have been driven by government departments or agencies with statutory powers, although some of the national reform programmes have been driven by non-statutory institutions. These reform programmes and the supporting development institutions are discussed in the third part of the chapter.

Although the second and third parts of this chapter examine two specific types of institutions, namely largely private-sector 'builder and contractor associations' and largely public-sector 'construction industry development institutions', it must be recognised that these institutions are only two of a range of interdependent institutions that have an impact on the development of the construction industry. These organisations include government ministries and departments, professional and trade associations, materials associations, and regulatory, educational and research organisations. The interdependencies of these institutions, and the possible role conflicts among them, are discussed in the fourth part of the chapter. Some concluding comments on institutional building together with policy implications are then given in the fifth and final part of the chapter, which draws out the key lessons learned.

A century of development institutions

The establishment of institutions to support the development of the construction industry is not new. In fact, the origins of the organisation of stakeholders within the construction industry can be traced back to the period 1850–1950. Many institutions that have had an impact on the development of the construction industry today can be traced back to that period.

This section considers the establishment of builder and contractor associations, and how they have supported the development of the construction industry.

The contribution of industry associations

Many builder and contractor associations began to emerge in, largely, the developing countries within the British Empire and Commonwealth during the period 1850–1950. For example, the development of builder associations is reflected by the establishment of the following national and international organisations:

- the Master Builders Association in New South Wales (MBA/NSW) in Australia around 1873 (Elder, 2007) (see Box 5.1);
- the Master Builders Association in KwaZulu-Natal in South Africa in 1901 (see Boxes 5.2 and 5.3);
- the Builders Association of India in 1941;
- the Master Builders Association Malaysia in 1954; and
- the International Federation of Asian and Western Pacific Contractors Associations in 1958.

These builder associations were established largely in a time of rapidly increasing construction activity associated with the rapid development of the economies in these countries. The increasing construction activity and the establishment of these builder and contractors associations also usually went hand in hand with the establishment of other professional and industry associations, as well as the setting up of a broad technical infrastructure (including academic institutions) within these countries.

The establishment of these early builder associations was largely a defensive strategy, in which industry role-players grouped together to defend themselves against various actions by, amongst others, clients, labour, regulators and materials suppliers. The role played by these associations has varied over time as the industries have passed through the typical cycles of

---

Box 5.1 Development of the Master Builders Association in New South Wales

- The MBA/NSW was established in 1873 with an initial focus on the need to obtain fairer conditions of contract and tendering processes, as well as just methods of resolving disputes between the builder and the proprietor or architect. Those objectives were also the driving force behind the formation of the 1873 association that achieved the introduction of better conditions of contract and a compulsory arbitration clause.
- Around 1912 a focus of the MBA/NSW was on lobbying the NSW Government over the training of apprentices, as well as on being an advocate of immigration as a quick response to labour shortages. In 1912, the MBA/NSW received nomination rights to an Advisory Committee in connection with Building Trades Classes at the Technical College.
- During the 1920's the MBA/NSW sought to create a united front among employers in dealing with industrial disputes in the building industry.
- In the post World War I period, the MBA/NSW tried to control the prices of bricks, timber and other building materials through, for example, encouraging the importation of timber. Further, the MBA/NSW had a policy which required its members to purchase supplies from members of its Builders Exchange branch.

Source: Elder (2007)

---

over-supply to under-supply, and from surpluses of key inputs to shortages of such resources, but, throughout the world, these industry associations have consistently played a significant role in advocating for, and propagating, the interests of their members largely in the areas of (i) fair contracting and tendering documents and procedures; (ii) dispute resolution mechanisms; (iii) industrial relations and wage negotiations; (iv) materials supply and pricing; and (v) skills development and training.

A significant theme that has been championed by builders' and contractors' associations over the past 100 years or so has related to the quality of work of their members, together with (in one form or the other) registration and/or licensing of builders and contractors. For example, in Australia, the MBA/NSW first introduced the concept of registration of builders in 1925. However, it was only in 1939 that Western Australia first introduced legislation for builders' registration, whereas registration of home builders was legislated in New South Wales (NSW) in 1971.

Box 5.2 Keeping the trade together

Arthur Reid warned in the 1890's that it was 'now necessary to form a master builders' or contractors' association for the purpose of keeping the trade together, establishing a business-like routine in the conduct of contracts and estimating for them and of providing a competent, practical body to whom the professions could look for advice and assistance in case of need.'

Source: http://www.masterbuilders.co.za

Box 5.3 'Militant trade unionism'

Turn of the century Durban building contractors were embattled in individual struggles against the demands of militant trade unionism. In 1901, James W Reid was convinced that only similar organisations on the part of the employers would enable business to deal on equal terms with labour. 'The unions have learned that unity is strength,' he said, 'and we will be taking a great risk if we do not learn the same lesson. If we do not come together and stay together we will all go down under the tide of constant wage demands and rising costs.'

Source: http://www.masterbuilders.co.za

Registration and licensing of builders and contractors has been viewed very differently around the world (and even between states or provinces). They have often been considered as either 'protectionist' or 'quality assurance' strategies. Many industry associations and governments have advocated self-regulation, and many builder and contractor associations have played an important role in self-regulating their members since the mid-1900s or so through, amongst other things (Elder, 2007), (i) the establishment of complaints committees with possible disciplinary action and (ii) the organisation of campaigns to advise the public on the benefits of dealing with members of the association.

Overall, on balance, the early builders' and contractors' associations made a major positive impact on the development of the construction industry, and many advances in the industry can be attributed directly or indirectly to such industry associations. Box 5.4 presents an account of the contributions made by such institutions in the area of occupational health and safety in construction.

Box 5.4 Construction health and safety: role of master builders

A recent study on construction health and safety in South Africa concluded that the Master Builders South Africa (MBSA) and the regional Master Builders Associations (MBAs) have historically provided substantive construction health and safety services to their members. Specifically, the MBSA and the MBAs strive to promote a positive occupational health and safety (OH&S) culture by:

- informing members of new OH&S legislation;
- providing OH&S advice and guidance;
- assisting contractors to improve their OH&S programmes and procedures;
- conducting site OH&S surveys and audits;
- conducting an annual national OH&S competition;
- assisting members with incident investigations and reports;
- arranging forums and workshops on informative OH&S topics; and
- coordinating OH&S training courses.

The MBSA has also produced an Occupational Health and Safety Manual for Construction Sites.

However, while the MBSA together with the MBAs strive to make a very important contribution to construction H&S, the uptake of the MBAs efforts has been questioned by the MBAs themselves.

Source: cidb (2009)

Newer institutions

While institutions such as builder and contractor associations have served the construction industry well over the past 100 years or so, and have helped shape the industry into what it is today, builder and contractor associations continue to be formed – often in response to local needs, as defence strategies against foreign companies, or to take advantage of changing circumstances, and/or to address minority issues. Such examples include the establishment of:

- the National Association of Minority Contractors (NAMC) in the USA in 1969 to address the needs and concerns of minority contractors (see Box 5.5);
- the Malay Contractors Association (*Persatuan Kontraktor Melayu Malaysia*) in 1975 to represent the interests of Malay contractors; by

Box 5.5 National Association of Minority Contractors (NAMC)

The National Association of Minority Contractors (NAMC) is a non-profit trade association that was established in 1969 to address the needs and concerns of minority contractors. While membership is open to people of all races and ethnic backgrounds, the organisation's mandate, 'Building Bridges – Crossing Barriers,' focuses on construction industry concerns common to African Americans, Asian Americans, Hispanic Americans, and Native Americans.

The mission statement of NAMC is:

- Provide education and training to minority contractors in construction.
- Promote the economic and legal interest of minority contracting firms.
- Advocate law and government actions for minority contractors.
- Bring about wider procurement and business opportunities for minority contractors.
- Reduce and remove the barriers to full equality for minority contractors.
- Build bridges between minority contractors and the entities they serve.
- Create a forum for sharing information and mutual support.

Source: http://www.namcnational.org/

2009, its membership comprised 7,000 Malay contractors operating throughout Malaysia;
- the Federation of Contractors' Associations of Nepal in 1990 after the restoration of a multi-party democratic political system and a recognition of the need to protect or develop the industry in the face of international competition; and
- the National Black Contractors and Allied Trades Forum (NABCAT) in South Africa in 1993 to mobilise and co-ordinate black construction in an organised institution.

Success criteria and limitations

The impact of institutions such as builder and contractor associations on the development of the construction industry is highly context specific, and depends on systemic relationships with a whole range of industry participants (see 'Development systems' below) as well as on economic and social cycles.

Specifically, the role of stakeholders and their impact on the development of the construction industry depends on (i) levels of construction activity and economic cycles; (ii) the legislative and regulatory environment; and (iii) the underlying technical infrastructure of the construction industry (including educational and research institutions).

At an institutional level, the success of such builder and contractor associations depends on several internal factors, including (i) a belief system in and commitment to the need for a fair, vibrant and healthy construction industry; (ii) individual leadership, vision and respect; (iii) collective action (including lobbying and bargaining) and a united front by member institutions; and (iv) disciplinary action against non-conforming member institutions.

Notwithstanding the successful contribution that such institutions have made in the past, their ability to have an impact on the development of the industry is largely limited to driving change within their own organisational boundaries or sphere of influence (i.e. largely their membership and/or suppliers) and is largely driven by self-interest. Therefore, there are very few examples around the world where industry bodies (even where they exist, such as the Construction Industry Joint Council in Singapore) have collectively driven developmental initiatives amongst themselves which are not necessarily in their immediate self-interest.

As a consequence, many governments have assumed lead responsibility for driving reform, or change, initiatives in the context of national socioeconomic objectives. Specifically, governments are more able to organise public-sector clients around change initiatives (albeit usually under the instructions of a higher authority), to drive change within their own business functions (such as procurement reform) and to drive change amongst their suppliers (largely through the procurement regime). Such national reform initiatives are discussed in the following section of this chapter.

National reform programmes

The previous section has sketched the role played by builder and contractor associations in the development of the construction industry over the past 100 years or so. The role played by these organisations has largely been based around defensive strategies looking after the interests of their specific industry members – and often their own self-interests.

As noted in the previous section, the ability of individual institutions to effect change outside their direct sphere of influence is limited, and more recently several countries have established national programmes to support the objectives of construction industry development, such as (Milford *et al.*, 2002; PSIB, 2004):

- the Rethinking Construction, Constructing Excellence, Better Public Buildings and other initiatives in the UK;

Box 5.6 Singapore: Objectives of 'Construction 21'

In the Singapore 'Construction 21' (C21) reform programme, six 'strategic thrusts' were defined:

- enhancing the professionalism of the industry;
- raising the skills level;
- improving industry practices and techniques;
- adopting an integrated approach to construction;
- developing an external wing (i.e. an export capability); and
- striving for a collective championing effort for the construction industry.

- the performance improvement programmes of the Building and Construction Authority (BCA) of Singapore (see Box 5.6);
- the Australian Construction Industry Development Agency (CIDA) and the National Building and Construction Committee (NatBACC) Action Agenda – Building for Growth;
- the Australian Construction Policy Steering Committee (CPSC) Construct NSW and other initiatives of the government of New South Wales;
- the Construction Industry Review Committee (CIRC) in Hong Kong and the initiatives of various public-sector clients, including the Hong Kong Housing Authority;
- the initiatives of the Construction Industry Development Board (CIDB) of Malaysia; and, more recently
- the Process and System Innovation in Building and Construction (PSIB) programme in the Netherlands.

A detailed assessment of many of these national reform initiatives is available in the *Inventory of International Reforms in Building and Construction* (PSIB, 2004).

Drivers for change

As noted by Courtney (2002), the 'drivers' that have led to the creation of these national initiatives show both common elements and local dimensions. The common elements are highlighted by Courtney (2002) and Milford and colleagues (2002). First, there was a recognition that construction accounts for a significant proportion of national economic activity and that the effectiveness of the sector has implications for other industries and for public services. Second, there was a perception that construction, in contrast to other industry sectors, has not improved its use of labour and its overall productivity as much as other sectors in recent decades and that as a consequence its outputs are becoming relatively more expensive. Third, there was a view that a key factor in the allegedly poor performance of construction is the number of different parties who have responsibilities within the construction process and therefore a desire to bring about a more integrated process. Finally, and overall, there was a view that construction should, by integrating its internal processes and adopting new information and production technologies, seek to become more similar to manufacturing sectors.

Other factors, or local elements, have been prominent in many countries which have created the need for a national or industry focus on the development of the industry; for example:

- In Singapore, there was a realisation that the industry was heavily dependent upon relatively unskilled operatives, many from outside the country, and a principal focus of the review was therefore skills development and means of making more effective use of labour.
- In South Africa, the main objective was to create an enabling environment for the transformation of the industry to create economic opportunities for all participants.
- In Hong Kong, some prominent defects in new construction works, such as inadequate foundations, had revealed shortcomings in quality practices and – in the extreme – corrupt practices. Institutional arrangements for both procurement and in the carrying out of the works (for example, the use of multi-layer sub-contracting) became a principal focus of the review.
- In the UK, major clients (notably utility companies which had been previously public-sector organisations) wished to achieve better value from their investments in construction and sought new relationships and procedures to realise this.
- In the UK also, the poor image of construction, as perceived by prospective employees and the stock market, caused firms to consider how new forms of operation could provide both more attractive employment conditions and higher levels of profitability.
- A desire to increase the international competitiveness of the construction sector, so that it could secure a higher proportion of business from other countries, was a factor in Singapore and Australia.
- In the Netherlands, a national research programme, 'Process and System Innovation in Building and Construction' (PSIB), was established in 2003 as a joint initiative of industry, government and research institutes to restore trust in the sector following a Parliamentary Inquiry into large-scale fraud which revealed an urgent need for improvement and reform of the Dutch building and construction sector.

It can be seen from these examples that the focus of the development of the industry is context specific, and depends on the needs of the country at a specific point in time.

Effecting change

International experience has shown that effecting change at a national or sector level is complicated, resource intensive and achievable only over a relatively long period of time (in some cases up to 10 years or even longer). An assessment of these initiatives (some of which themselves have been in existence for 10 years or longer) highlights key criteria for the success of such reform initiatives, and collectively these point towards a structured framework that can be adapted to national performance initiatives (PSIB, 2004; Hodgson and Milford, 2005), and some key elements of these initiatives are described in the following sections.

1 *Leadership is key*: Without exception, the role of leadership by individuals and/or organisations has been fundamental to the success of every one of the more successful international reform initiatives. Common forms of leadership that are observed in the reform initiatives include:
   – Leadership by *government* (either individuals or government departments) – demonstrating willingness and commitment to the reform initiatives. Examples of such leadership by government include the procurement reform initiatives being carried out in the Office of Government Commerce (OGC) in the UK, the *Better Public Buildings* initiative in the UK and the *Construction Client Charter* and *Demonstration Projects* initiatives in the UK.
   – Leadership by influential forward-thinking and progressive *private-sector* organisations, and in particular private-sector clients, is relatively common internationally. Examples include those private-sector clients which initially participated in the UK *Construction Client Charter*, and members of influential organisations such as the *World Business Council for Sustainable Development*.

2 *Objectives create the focus*: Performance improvement programmes are generally driven by high-level goals and objectives. For example, the Singapore programme to promote buildability derives from the national objective to limit the need for imported labour by improving productivity. Clarity on priority reform objectives is of the utmost importance to ensure focus, and is usually informed by policy, legislation and industry reviews. The objectives usually have to be cascaded out from higher-level objectives to more manageable lower-level objectives, which can then be arranged by order of priority. For example, the *Construct NSW* agenda set out an integrated framework of 20 strategies and 85 supporting actions to enable the government to achieve best value for money from its procurement of construction projects, to support its economic and social goals through construction procurement and to assist the industry to achieve its potential. These strategies were then grouped under eight

headings – analogous to objectives – including (i) strategic information for decision making; (ii) business ethics and practices; (iii) security of payment; (iv) management and workforce development; (v) continuous improvement; (vi) towards an ecological sustainable industry; and (vii) encouragement and recognition.

3 *Awareness creation and promotion* is fundamental to furthering the objectives of reform initiatives, so as to continually reinforce the reform message, and to broaden the awareness and understanding of the reform initiatives. There are numerous examples of successful (and unsuccessful) awareness creation and promotion activities internationally, including (i) targeted awareness creation in the popular and technical press; (ii) award systems, such as the *Considerate Contractor Scheme* and the *Prime Minister's Better Public Building Award* in the UK; (iii) forums, benchmarking clubs and demonstration projects; and (iv) periodic reporting on the state of the industry or industry reform.

4 *Information and tools*: The development and dissemination of appropriate information and tools to support the attainment of reform objectives is another key success factor and, as illustrated below, it can take various forms. However, many of the systems outlined below are in fact enforced through various instruments in many of the reform initiatives around the world, but the systems themselves provide a tool together with information to equip various stakeholders for change:
   – *codes, standards and guidelines*, both voluntary and enforced through legislation;
   – *best practices*, applicable to almost every reform initiative around the world;
   – *management systems*, together with supporting implementation tools, specifying processes to be adopted and reported on, varying from full ISO 9000 and 14000 accreditation (which is currently required of the larger consultants and contractors in Singapore) to the management systems developed to target specific issues – such as the NSW Australia OHS&R (Occupational Health Safety and Rehabilitation) Management Systems and Environmental Management Systems;
   – *accreditation and rating systems* together with supporting implementation tools, such as the Leadership in Environmental Engineering and Design (LEED) environmental design accreditation of professionals in the USA, the NSW Contractor Best Practice Accreditation System, and accreditation or rating systems for buildings – predominantly environmental and quality systems such as Green Mark in Singapore;
   – triple-bottom-line *reporting schemes* and methods, which are becoming increasingly common around the world.

5 *Capacity building* is key to several of the international reform initiatives, including:
   – formal training programmes for public-sector officials that support reform initiatives;
   – the establishment of public-sector Centres of Excellence (such as the OGC *Programme and Project Management Centres of Excellence* in the UK), whose aim is to achieve significant improvement to central government capability to deliver successful programmes and projects;
   – the sponsorship of formal and informal training programmes for private-sector participants impacted by reform initiatives.

   In addition, many of these capacity-building programmes in the public sector are supported by the development and implementation of performance management systems for public-sector officials that are aligned to the reform initiatives.

6 *Enforcement and compliance*: All reform initiatives around the world are dependent on enforcement and compliance mechanisms. Therefore, underpinning these national initiatives, a range of activities have been implemented to achieve the desired change in the industry, the most common of which include:
   – *legislation* to seek compliance with minimum acceptable standards (such as safety and health, and prompt payment of contractors and sub-contractors);
   – *procurement instruments*, which are among the most powerful instruments used in all reform initiatives for effecting change amongst suppliers, that is clients (typically government clients) specifying their requirements (aligned with the reform objectives) for other parties wishing to do business with them – often requiring compliance with codes of conduct, standards and guidelines, or the mandatory use of management systems;
   – *registration and accreditation* of contractors, designers and so on according to specified criteria for different types of activities including 'construction registers', which is typically implemented through legislation or procurement arrangements; and
   – *commitment to voluntary compliance* together with review mechanisms, to charters, codes of practice and/or conduct, management systems and reporting mechanisms.

7 *Monitoring, evaluation and review*: Regular monitoring, evaluation and review is an essential requirement for the successful implementation of any strategy, and is a key element of all international reform initiatives. It takes place at both the 'macro level' and the 'micro level'. For example, at the macro level the UK has instituted the *Construction Industry Indicators* and the *Quality of Life Indicators*, which set high-level

performance targets for the industry together with ongoing monitoring against these targets. The second example is that, at the micro level, the UK has initiated the OGC Gateway review process for acquisition programmes and procurement projects, and the CABE Design Review for buildings that will have a significant impact on their environment, while Singapore and Australia require closeout reviews of projects against certain criteria.

Developmental institutions

National reform programmes have been driven by either government departments (or agencies with statutory powers such as those in Australia, Hong Kong, Malaysia, Singapore and South Africa) or non-statutory organisations such those as in the UK and the USA.

The relative merits of statutory and non-statutory institutions in supporting the development of the construction industry is a debated issue but the need for statutory or non-statutory instruments depends on the context and on the issues to be addressed. Specifically, it is clear that many instruments that may be appropriate within a given context can function only in a statutory environment, including (i) many aspects of government procurement reform; (ii) enforceable codes of practice or codes of conduct; and (iii) registration of professionals and contractors.

The relative merits of statutory and non-statutory institutions also depend on the life cycle of the reform initiative. PSIB (2004) notes that the timescale for reform is measured in years and possibly decades and, over that time, political priorities change and political attention will fade. Therefore, the authors argue that long-term change will take place only if markets provide rewards and incentives for superior performance and have within them mechanisms that stimulate continuous improvement. In other words, it can be deduced that, over time, sustainable non-statutory instruments are likely to be necessary.

The long time required to effect change also introduces the concept that the development institutions themselves may need to be adaptive; specifically, a transition from a statutory to a non-statutory environment may be the most appropriate way to effect long-term change. In itself, this suggests that reform institutions may be well suited to being transformed from an initial public-sector-driven statutory institution to a largely private-sector institution in time.

Development systems

The previous sections have considered primarily the role of two institutions in the development of the construction industry: (i) builder and contractor associations and (ii) development institutions. However, these are but a few of the many inter-related institutions that contribute directly or indirectly to the development of the construction industry.

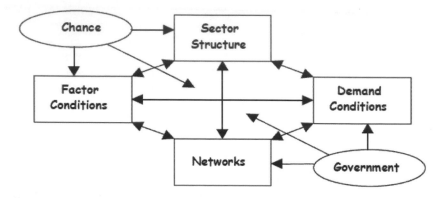

Figure 5.1 Porter's Diamond Framework.

At a high level, the dependencies between the various stakeholders and their impact on the development of the construction industry can be illustrated through frameworks and analytical tools such as Porter's Diamond Framework (see Figure 5.1), which portrays the determinants of the competitive advantage of a cluster or a nation (Porter, 1998; Milford, 2000).

Other important models for understanding the relationships and dependencies among the various role-players in the construction industry include models for national systems of innovation (Lim *et al.*, 2006, 2007).

Drawing on Porter's framework, a context for understanding the development of the construction industry can be viewed in terms of the attributes in Table 5.1 that constitute the playing field for the competitiveness of clusters on the home market, together with typical institutions that have a direct impact on these attributes.

An understanding of the attributes that have an impact on the competitiveness of a cluster and of the dependencies between the various role-players and institutions is important for policy making and strategy formulation.

A detailed analysis of the role of various institutions in the development of the construction industry is beyond the scope of this chapter. However, it is worth noting that government and public-sector bodies can play a significant role in the development of the construction industry. Their actions and initiatives can have an impact on, in particular, construction demand, labour policy and regulations, technical infrastructure (academic and research institutions, and standards and regulatory bodies) and development institutions. In practice, these functions are usually the responsibilities of different government departments, each with its own service delivery objectives. Moreover, these departments usually report to different ministries. Therefore, lack of *policy coherence* between government departments and public-sector bodies is often a major obstacle to the government when it plays a role and provides leadership in the development of the construction industry.

Another point worth highlighting with respect to the development of the

Table 5.1 Structure of a construction industry using Porter's framework

Attributes	Examples of institutions
Sector structure describing the economic behaviour of firms related to the construction industry, including rivalry and/or co-operation between local and foreign construction firms, or between established and emerging contractors, and so on	Builder and contractor associations Consulting engineering and architectural associations Materials manufacturing associations
Factor conditions or resources such as skilled labour, access to capital, equipment, raw materials and physical infrastructure	Labour organisations Training institutions Financial institutions
Demand conditions such as business conditions (availability of work and tendering competition), buying preferences, or even buying preferencing	Public-sector client bodies (which procure projects for the construction of public buildings, roads and other transport infrastructure, and water projects) Private-sector client associations
Networks of related and support industries, including the supply chain (vertical integration) and research and technical infrastructure (horizontal integration)	Academic institutions Research institutions
Government, which enhances or diminishes competition by setting market conditions, providing developmental support and so on	Standards and regulating bodies Development boards
Chance, an opportunity or threat which suddenly emerges	

construction industry is that *demand conditions* are particularly important in stimulating the development of the industry, especially in developing countries. It stands to reason that development of the industry cannot take place without sufficient demand for construction. Therefore, the role of client bodies, and public-sector client bodies in particular, in stimulating demand and in ensuring continuity of demand is of the utmost importance. Thus, development institutions and relevant client bodies have an important role in facilitating appropriate demand strategies and in facilitating the capacity to delivery such strategies that support the development of the industry.

A case study: the cidb South Africa

A case study of the South African Construction Industry Development Board (cidb) is presented in this section. This case study illustrates not only the context-specific nature of implementing interventions to support the development of the construction industry, but also synergies between various development initiatives in somewhat different contexts. Specifically, the development of the cidb in South Africa was strongly informed by the *Rethinking Construction* process in the UK, the operations and achievements of the Building and Construction Authority (BCA) in Singapore, and the development activities initiated by the CPSC of New South Wales, Australia (which in turn were informed by the *Action Agenda – Building for Growth* of the NatBACC, Australia).

Background and context

The establishment of the cidb must be seen within the context of the democratic elections in South Africa in 1994, in which political power was transferred from a white minority to the majority. Prior to 1994, the black population within South Africa was excluded from most economic opportunities as well as many social ones. These included meaningful education and skills development, as well as meaningful participation in the construction industry.

South Africa's development agenda at that time was broadly defined in the government's *Reconstruction and Development Programme* (RDP). The aim of the RDP was to dismantle apartheid social relations and address poverty and inequality. The programme identified the following key objectives:

- *meeting basic needs*: including the provision of jobs, and addressing the backlog of, and disparities in, land, housing, water, electricity, telecommunications, transport, a clean and healthy environment, nutrition, health care, social welfare and so on;
- *building the economy*: strengthening and growing the South African economy to the benefit of all participants in it, including addressing racial and gender inequalities in ownership, employment and skills;
- *democratising the state and society*: including the Constitution and Bill of Rights, government, the public sector and parastatals;
- *developing human resources*: equipping people to be involved in the decision-making process, in implementation, in new job opportunities requiring new skills, and in managing and governing South Africa's society; and
- *nation building*.

However, while the need for transformation was clear, prior to 1994 (as in many other countries) the demand for construction in South Africa was in decline (see Figure 5.2), resulting in increased competition, shedding of labour and skills, and limited recapitalisation of equipment.

Simply transferring economic opportunities in the construction industry from one section of the community to another was not a viable option, and the development of the industry depended on:

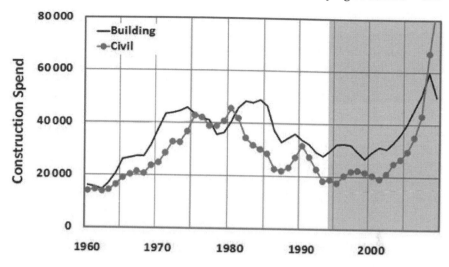

Figure 5.2 Construction spend in South Africa, 1960 to 2000 (rand million, 2000).

- *growth in infrastructure investment* – providing increased opportunities – which was, to a large extent, dependent on public-sector delivery capacity;
- *growing the capacity of the industry to meet the increased demand* – and particularly amongst the previously disadvantaged sector; and
- *improving the performance of all participants* to deliver value for money to clients and to meet the socio-economic objectives of the nation.

A consultative policy process

Towards the end of 1995, the Department of Public Works (DPW) generated a position paper for consideration by government, entitled *Establishing an Enabling Environment to Ensure that the Objectives of the RDP and Related Initiatives by Government are Realised in the Construction and Allied Industries*. In February 1996, the support of Cabinet was obtained to develop a construction industry policy and the Department of Public Works was given the mandate to lead this initiative. To ensure the alignment of policy development in construction, the government appointed a steering committee and working group comprising the Departments of Transport, Water Affairs and Forestry, Housing, Labour, State Expenditure, Trade and Industry, and Education, with Public Works as the co-ordinator.

Further research led to a discussion document, which formed the basis for a round of consultation workshops with industry stakeholders towards the end of 1996. The resulting Green Paper was released for broad comment in November 1997. At the same time, and in line with the Green

Paper proposals on the way forward, an *Inter-ministerial Task Team on Construction Industry Development* was appointed. The members of the Task Team were drawn from industry and government, and it was supported by a Department of Public Works Secretariat. The Task Team engaged intensively in consultations with industry stakeholders, and advised on the finalisation of the White Paper. Comments on the White Paper were received from more than 20 organisations and individuals. The Task Team subjected these views to a thorough analysis, which further modified the government's thinking on policy. The output of this review was presented to a Construction Industry Development Reference Group of industry stakeholders in July 1998, and gained their full support.

Thus, the White Paper was the result of a broad public policy-making process and represented a significant milestone in the development of the South African construction industry. It established an enabling framework within which the construction industry can play a more strategic role in social development and economic growth. The White Paper, published in 1999, concluded with the recommendation to establish a statutory Construction Industry Development Board (cidb) 'to advise on policy and on any existing and proposed legislation which impacts on the industry – to champion the programmes of the enabling environment as well as integrate and promote the industry at large'.

Establishing the cidb

The cidb was formally established in 2001, under the cidb Act (Act 38 of 2000). The first cidb Board was appointed by the Minister of Public Works in 2001, and comprised private- and public-sector individuals appointed based on their individual knowledge and expertise. The cidb's first CEO was appointed later in the year.

The Act gave the cidb the mandate to (see Figure 5.3):

- *provide strategic leadership* to construction industry stakeholders, developing effective partnerships for growth, reform and improvement of the construction industry;
- *promote sustainable growth* of the construction industry and the sustainable participation of the emerging sector in the industry;
- determine, establish and *promote improved performance* and best practice of public- and private-sector clients, contractors and other participants in the construction delivery process;
- *promote uniform application of policy* throughout all spheres of government and promote uniform and ethical standards, construction procurement reform, and *improved procurement and delivery* management, including a code of conduct; and
- develop systematic methods for *monitoring and regulating the performance of the industry* and its stakeholders, including the registration of projects and contractors.

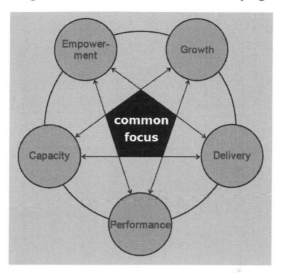

Figure 5.3 The cidb developmental framework.

Key instruments that the cidb was mandated to develop are highlighted below.

1 *cidb Register of Contractors*: The *Register of Contractors* categorises contractors according to a financial grade (from 1 to 9) and a class of works (i.e. general, civil, electrical, mechanical and special works). Only a cidb-registered contractor is allowed to undertake construction works on a public-sector contract, and it can do so only within the class of works for which it is registered, and subject to a maximum level of project value determined by the contractor's financial grade. Any organisation which carries out or attempts to carry out any construction works on a public-sector project and which is not a contractor registered by the cidb will be guilty of an offence and liable, on conviction, to a fine.

   The *Register of Contractors* was established in 2004, and the growth in the number of firms registered as general building contractors in Grades 2 to 9 is shown in Figure 5.4 (Grade 1 is the entry level, and the contractors are not included in the figure).

   Apart from being a critical element in public-sector procurement, the cidb *Register of Contractors* also provides a wealth of information on the structure of the construction industry, including the state of contractor development, and transformation of the industry (cidb, 2010). For example, the number of black-owned general building companies is shown in Figure 5.4 (where black ownership is defined here as greater than 50 per cent equity ownership).

2 *cidb Best Practice Contractor Recognition Scheme*: The cidb is also required to establish a Best Practice Contractor Recognition Scheme to:

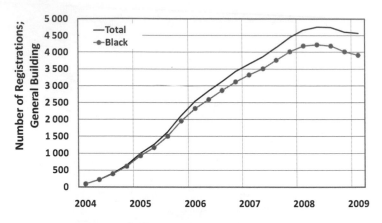

Figure 5.4 Contractors registered with cidb.

–   enable organs of state to manage risk on complex contracting strate-
    gies; and
–   promote contractor development in relation to best practice stand-
    ards and guidelines developed by the Board.

Once implemented, the cidb Best Practice Contractor Recognition
Scheme will encourage 'value-based procurement' by providing a
mechanism for clients to recognise the track record and competence of
contractors. This can then be used for:

–   assessing the suitability of contractors for registration, pre-qualifica-
    tion, selective tender lists or expressions of interest; and/or
–   adjudication for the award of a contract.

3   *cidb Register of Projects*: All public- and private-sector contracts above
    a prescribed tender value must be recorded in the *Register of Projects*.
    The Register gathers information on the nature, value and distribution
    of projects, and also provides the basis for the cidb *Best Practice Project
    Assessment Scheme* (see below).

    The cidb Register of Projects has been rolled out incrementally
    since 2004, and all public-sector tenders also have to be recorded in
    it. Compliance with the requirements for registration of projects has
    been poor, and this issue is being given focused attention by the cidb at
    present.

    cidb-registered contractors can also elect to receive notification of
    tender opportunities by electronic mail or mobile phone messages (sms).

4   *cidb Best Practice Project Assessment Scheme*: The cidb is also required
    to establish a Best Practice Project Assessment Scheme based on best
    practices identified by the Board. Once the legislation has been passed,
    all construction contracts above a prescribed tender value will then be

subject to an assessment of compliance with the best practice standards and guidelines published by the cidb.

Best practices currently being developed include requirements on public-sector contracts for skills development, enterprise development, energy efficiency and quality management plans.

5 *cidb Code of Conduct*: The Act requires the cidb to 'publish a code of conduct for all construction-related procurement and all participants involved in the procurement process'. The Code of Conduct published by the cidb requires all parties engaged in construction procurement to (i) behave equitably, honestly and transparently; (ii) discharge duties and obligations timeously and with integrity; (iii) comply with all applicable legislation and associated regulations; (iv) satisfy all relevant requirements established in procurement documents; (v) avoid conflicts of interest; and (vi) not maliciously or recklessly injure or attempt to injure the reputation of another party.

The cidb may, as appropriate, sanction those who breach the Code of Conduct – but, to date, no sanctions have been applied in terms of the Code.

6 *cidb Standard for Uniformity in Construction Procurement*: The Standard for Uniformity establishes minimum requirements within the public sector that:
-   promote cost efficiencies through the adoption of a uniform structure for procurement documents, standard component documents and generic solicitation procedures;
-   provide transparent, fair and equitable procurement methods and procedures in critical areas in the solicitation process;
-   ensure that the forms of contract that are used are fair and equitable for all the parties to a contract; and
-   enable risk, responsibilities and obligations to be clearly identified.

Compliance with the Standard for Uniformity is mandatory for public-sector organisations which solicit offers from the construction industry.

Ongoing review

In support of good governance and relevance to the industry, the cidb Act requires that the cidb facilitate an independent review of its activities every five years. The first such review took place in early 2007 (cidb, 2007). The panel, which comprised an international team of eminent practitioners and administrators, was appointed by the Minister of Public Works.

The review included (i) a *peer review*, in which the quality of the programmes of the cidb was assessed in terms of international best practice; (ii) a *stakeholder review*, reviewing the performance of the cidb in relation to the mandate and objects of the cidb; and (iii) a *management performance review*.

It is relevant to highlight some key findings of the independent review of the cidb undertaken in 2007 – and it is possible that the implications of these key finding could be applicable to reform initiatives in other countries.

Stakeholders expressed strong support for the objectives of the cidb. Generally, they recognised that the cidb had achieved much in its first five years, even though many challenges lay ahead. Some specific views obtained from the review are highlighted below:

- Some large, specialised clients considered that the cidb had not added significantly to their procurement functions.
- There were concerns from emerging contractors over aspects of the Registers and the low level of developmental activity to date.
- Some stakeholders perceived the cidb as primarily concerned with regulation and protection of the public sector.
- Local government client bodies considered the cidb to have had substantial impact on procurement processes, while other public clients were less positive.
- There was concern that the cidb should demonstrate its ability to secure compliance with its prescripts, and should be seen to be influential with government departments.
- The individual programmes of the cidb were generally considered to be of high quality, with particular praise for the procurement documentation.
- The Register of Contractors was seen as a powerful tool for structuring the industry, but there was concern at the (then) delays in registration and detailed comments on the criteria employed, which were seen to disadvantage some emerging contractors.
- The Register of Projects was less accepted, clients seeing this as an administrative imposition without clear benefits for their organisations.
- With the establishment of the Registers and procurement prescripts, there was an urgent need for more developmental action – including impacting on contractor development, skills and R&D.

All of the recommendations provided in the independent review are being addressed by the cidb.

Looking forward

It is now some 10 years since the establishment of the cidb, and the general view of stakeholders is that the cidb has made significant progress, and has achieved much. However, as identified in the independent review in 2007, there are many challenges that lie ahead, and much still remains to be done. Specific challenges (or needs) that the cidb is facing are:

- monitoring of, and enhancing, compliance with the cidb procurement prescripts;

- monitoring of compliance with, and enhancing the integrity of, the cidb Register of Projects; and
- enhancing the cidb's role in contractor development and skills development.

Concluding comments

The previous sections have examined the role of private-sector 'builder and contractor associations' and public-sector 'construction industry development institutions' in the development of the construction industry in developing countries. However, it is stressed that these institutions are only two of a range of interdependent organisations that have an impact on the development of the construction industry (they include government departments, professional associations, materials associations and regulatory, educational and research organisations).

Although both the private and public sectors have an important role to play in the development of the construction industry, governments are more able to organise public-sector clients around change initiatives, to drive change within their own business functions (such as procurement reform) and to drive change amongst their suppliers (largely through the procurement regime). Thus, the government can play a major role in the development of the construction industry, and policy recommendations that can be drawn from the previous sections are discussed below.

First, it is important that all the interdependent institutions that have an impact on the development of the construction industry function appropriately for the development of the industry. Specifically, there should be clear responsibility for co-ordination of government policy that could have an impact on the development of the construction industry, and there needs to be coherence among the various government policies.

Second, where government then plays a leading role in the development of the construction industry, such policy needs to be driven by realistic and achievable objectives. These objectives would be context specific, and could be focused around either national socio-economic objectives or (government) client-driven objectives. It is also fundamental that these objectives be sustainable. Specifically, the development of an indigenous construction industry cannot take place in the absence of sustained demand.

Various pointers that can be used as input into the development of policy have been discussed in this chapter, but it has been noted that effecting change at a national or sector level is complicated, is resource intensive, and can be achieved only over a relatively long period of time (up to 10 years or even longer). Therefore, long-term change will take place only if the markets provide rewards and incentives for superior performance and have within them mechanisms that stimulate continuous improvement. Sustainable non-statutory instruments are also likely to be necessary, and so the private sector is of fundamental importance in the sustainability of change and in the development of the construction industry. Therefore, public–private partnerships could be extremely significant in the development of the construction industry.

References

cidb (2007) *5 Year Review of the Construction Industry Development Board.* Construction Industry Development Board: Pretoria. http://www.cidb.org.za

cidb (2009) *Construction Health and Safety in South Africa: Status and Recommendations.* Construction Industry Development Board: Pretoria. http://www.cidb.org.za

cidb (2010) *cidb Quarterly Monitor: October 2010.* Construction Industry Development Board: Pretoria. http://www.cidb.org.za

Courtney, R. (2002) Private communication. Cited in Milford, R. V., Hodgson, S., Chege, L. and Courtney, R. (2002) *Construction Industry Development: A Developing Country Perspective.* CIB W107: Conference on Creating a Sustainable Construction Industry in Developing Countries. Stellenbosch, South Africa.

Elder, J. R. (2007) *The History of the Master Builders Association of NSW: The First Hundred Years.* Unpublished PhD thesis, School of Business, Faculty of Economics & Business, University of Sydney.

Hodgson, S. and Milford, R. V. (2005) *Construction Safety, Health, Environment and Quality: A Framework for Performance Improvement in South Africa.* Rethinking and Revitalizing Construction Safety, Health, Environment and Quality. 4th Triennial International Conference, Port Elizabeth, South Africa.

Lim, L. J., Ofori, G. and Park, M. (2006) Stimulating construction innovation in Singapore through the National System of Innovation. *Journal of Construction Engineering and Management,* 132 (10), 1069–1082.

Lim, J. N., Ofori, G., Linga, F. Y. Y. and Goh Bee Huaa, G. B. (2007) *Role of National Institutions in Promoting Innovation by Contractors in Singapore.* Construction Management and Economics, V 25, I 10, October.

Milford, R. V. (2000), *National Systems of Innovation with Reference to Construction in Developing Countries,* Proceedings of 2nd International CIB TG29 Conference on Construction in Developing Countries, Gaborone, November.

Milford, R. V., Hodgson, S., Chege, L. and Courtney, R. (2002) *Construction Industry Development: A Developing Country Perspective.* CIB W107: Conference on Creating a Sustainable Construction Industry in Developing Countries. Stellenbosch, South Africa.

Porter, M. (1998). *The Competitive Advantage of Nations.* Macmillan: New York.

PSIB (2004) *Inventory of International Reforms in Building and Construction.* Publication PSIB017_S_04_2341. PSIBouw, June. http://www.psib.nl

6 Championing of construction industry development programmes and initiatives

George Ofori and Patricia Lim Mui Mui

Introduction

One of the six strategic thrusts of the Construction 21 (C21) blueprint for reinventing Singapore's construction industry was the adoption of 'A Collective Championing Effort for the Construction Industry' (Construction 21 Steering Committee, 1999): a collaborative effort by all players in the value chain to radically transform the industry. The Building and Construction Authority (BCA) was identified as the championing agency to spearhead the implementation of the C21 initiatives; the Construction Industry Joint Committee (CIJC) would support the BCA in undertaking this task in the 'collective championing effort'.

The C21 study was one of a series of master plans prepared for different sectors of Singapore's economy to prepare them for the knowledge age at the turn of the millennium (see a review in Ofori, 2002). The study also followed a number of reviews and development plans for Singapore's construction industry, which can be traced back to the report of the government-appointed Committee of Inquiry into the Capacity of the Construction Industry of Singapore (1961). The committee was set up to review the capacity of the industry to undertake the huge volume of construction planned under the mass housing and industrialisation programme under the country's first national development plan. Ofori (1993a, 1994a) discusses these studies on the construction industry in Singapore. The developmental effort has mainly taken the form of government programmes; the industry generally expects increasingly more government support. Wong and Yeh (1985) noted that, as a result of a comprehensive support programme for construction enterprises in the country to enable them to deliver on the massive public housing programme, the industry had become dependent on government assistance. The situation has not changed, and Ofori and Chan (2001) observed that, although the Singapore government could provide some support, ultimately, the initiative, drive and direct action which will lead to progress in the industry must come from the construction firms and practitioners, and their umbrella organisations.

In many countries, there appear to be sufficient programmes and specific measures for improving the construction industry. The question is how to implement the initiatives successfully. For example, on the UK industry, Latham (1994, p. vii) noted: 'Previous reports on the construction industry have either been implemented incompletely, or the problems have persisted'. He observed:

> Many people have asked whether action would be taken to implement any proposals which the Final Report made. They pointed out that there had been widespread agreement on the Simon Report [1944], the report by Sir Harold Emmerson [1962] and the Banwell Report [1964] but rather less action. Some of the recommendations of these reports were originally implemented. But other problems persisted and do so to this day . . . The initiative must be taken by Government, in conjunction with major clients and other leaders of the construction process. If this opportunity is not taken, it may not re-occur for many years, and a new report may be commissioned in the year 2024 to go over the same ground again.
>
> (p. 3)

Thus, mechanisms for leading and co-ordinating the implementation of the initiatives have been among the key provisions of the blueprints and strategies which have resulted from the construction industry reform programmes in many countries. As another example, one of the six thrusts of the strategic plan for developing Hong Kong's construction industry is: 'Devising a new institutional framework to drive the implementation of the change programme for the industry' (CIRC, 2001, p. 3).

In particular, owing to such gaps in leadership and co-ordination, the developing countries have achieved limited results in the attempt to improve the capacity, capability and performance of their construction industries despite the surfeit of research findings, and dedicated effort in many nations (Ofori, 1994b, 2000). Thus, there has been a yearning for the situation to change. For example, when it was formed in 2001 after a process which had started in 1997, the Construction Industry Development Board (cidb) of South Africa was heralded as a 'new statutory body to champion construction industry development' (*Akani*, 2001). As another example, the Minister for Public Works of Zimbabwe, Mrs Theresa Makone, said her ministry would champion for the Zimbabwe Construction Industry Council bill to become an Act of Parliament (*The Herald*, 2009). This would ensure that Zimbabweans will be able to regulate the construction industry.

Research objectives

This chapter considers the championing of programmes aimed at developing construction industries, focusing on Singapore and drawing lessons for other countries. The objectives of the study on which this chapter is based were to:

- examine the need for, and the nature of, the championing of construction industry development in various countries and focus on such efforts in Singapore;
- investigate, through a field study, the championing effort during the first three years of the implementation of the most recent blueprint for developing Singapore's construction industry and determine the difficulties faced by the designated champions, the BCA and CIJC, in the collective effort to improve the performance of Singapore's construction industry;
- consider the main developments and championing efforts in the second half of the implementation of the plan; and
- recommend measures which can be taken to realise effective collaboration between the government and the construction industry in the development of the industry in Singapore, and draw lessons for the developing countries.

Past works

Championing as a concept

The business management literature refers to various forms of 'championing'. Schon (1963) is credited with being the initiator of the study of championing when studying the process of innovation in business enterprises (see, for example, Shane, 1994). However, the 'change agent' identified earlier by Burns and Stalker (1961, p. 199) is also considered to be a type of champion. Schon (1963, p. 82) defines the champion as 'a man willing to put himself on the line for an idea of doubtful success. He is willing to fail. But he is capable of using any and every means, informal sales and pressure in order to succeed'. Thus, Schon (1963) suggested that (in the process of innovation) if an idea does not have a champion it will not survive. Championing is defined by Floyd and Wooldridge (1996) as 'the persistent and persuasive communication of proposals that either provide the firm with new capabilities or allow the firm to use existing capabilities differently'. Howell and Higgins (1990, p. 40) define champions as 'individuals who emerge to take creative ideas (which they may or may not have generated) and bring them to life'.

Nepal (2003) presents a summary of definitions of championing behaviour proposed by various authors, with regard to innovation, including (i) expressing confidence in innovation, involving and motivating others to support innovation, and persisting under adversity; (ii) identifying or generating an issue or idea, packaging it as attractive and selling it to organisational decision makers; and (iii) recognising and proposing a new technical idea or procedure and pushing it for formal management approval. It has been suggested that the key behaviours common in championing include mobilising resources; persuading and influencing; pushing and negotiating; and challenging and risk-taking (Kleysen and Street, 2001). Howell and colleagues (2005) identify three traits of a champion. Such a person (i) expresses enthusiasm and confidence; (ii) persists under adversity; and (iii) gets the right people involved.

Research has considered the roles which champions play in corporations; they range from the development of new technology, through new products, to even new businesses. The theoretical construct of 'champion' has been used to characterise a variety of individuals and roles. Championing is often considered in the context of the socio-political behaviours involved in innovation (Maute and Locander, 1994; Zintz, 1997) which are essential to realising the potential of ideas and solutions. Championing has also been studied in relation to the fields of organisational studies, marketing and environmental management. In construction, Nam and Tatum (1997) stress the importance of champions in ensuring technological innovation. Nepal (2003) examined the role of the project manager as a champion of innovation on the construction project.

Shane (1994) notes that researchers have taken different approaches towards understanding the role of champions. These have included the perspectives of leadership theory; agency theory; and the documenting of championing roles observed in organisations. Shane (1994) observes that research shows that approaches to championing vary across corporate cultures and suggests that approaches might also vary across national cultures. Shane investigated and established the latter. It is pertinent to note that the champion is only one of the roles necessary for change. This puts the contribution of the champion into perspective. Moreover, championing is not a static process; role switching is an important aspect of it (Markusson, 2007).

From the discussion so far, it is clear that the consideration of champions in the literature mainly relates to individuals. Champions are seen as going beyond their formal roles in their organisations to promote the development of new things or ideas (Shane *et al.*, 1995). Champions may also be groups or units within the organisation, or organisations as a whole. Championing may also relate to the relationship between the organisation and parts of its value chain (Davey *et al.*, 1999, 2001).

Markusson (2007) suggests that approaches to conceptualising what a champion is lie on a spectrum from 'action' to 'structure' orientation. At one end of the spectrum the models see the champion as possessing certain qualities, such as enthusiasm and willingness to take risks. The champion is a hero, and everyone who does not go along with the champion's ideas becomes a villain. At the opposite end of the spectrum is a structural model of championing. Here, the organisational context stimulates the emergence of champions by creating opportunities for employees to assume this role. However, it also stresses the career and status interests of the champions, and the choice they make in seizing the opportunities provided to them by their organisations. Therefore, this model also includes an action aspect. Furthermore, self-interest is part of what motivates the champions. Heroic accounts of champions tend to be uncritical of the champion's goals, and describe them as benefiting the whole organisation although this might not necessarily be the case (Markusson, 2007). Consideration of structural factors balances the tendency of some authors on championing to focus on the aspects of voluntarism, and even bordering on heroism, thus putting the champion into perspective.

In the context of this chapter, championing relates to the collective action among the players in an industry to realise their common goal. The chapter considers the championing activities of organisations rather than individuals. In Singapore, the championing of an industry by an organisation is not new. The Singapore Tourism Board (established in 1964) has promoted the tourism industry to make it a key sector of the economy; the Economic Development Board (set up in 1961) has masterminded the development of the manufacturing sector and spearheaded the effort to attract foreign direct investment in high-technology industries and services; and the Infocomm Development Authority (formed in 1999 from a merger of the National Computer Board and Telecommunications Authority of Singapore) is giving a spur to the nation's initiatives to benefit from information and communications technologies. Each of these statutory agencies works with umbrella organisations of the businesses in the relevant clusters.

The agencies have been revitalised and restructured over time, to enable them to play their role as champions more effectively. Moreover, several additional agencies have recently been formed to champion various programmes for the nation. They include the Workplace Development Agency (which was formed in 2003 to enhance the employability and competitiveness of all in the workforce; http://app2.wda.gov.sg/web/Contents/Contents. aspx?ContId=8); the Health Promotions Board (set up in 2001 as the main driver for national health promotion and disease prevention programmes; http://www.hpb.gov.sg/about/default.aspx); and the Media Development Agency (formed in 2003 'to champion the development of a vibrant media sector in Singapore'; http://www.mda.gov.sg/AboutUs/Overview/Pages/ default.aspx).

Championing construction industry development

Many recent reviews of the construction industries in many countries stress the implementation of the proposed initiatives. Latham (1994) urged the UK government to be a best practice client encouraging continuous improvements, and deliberately use its spending power to obtain value for money and enhance the industry's productivity and competitiveness. He sketched out a championing framework. He suggests that a government department should take the lead to ensure that the recommendations in his report are implemented. A construction clients' forum should also be created to promote forward thinking on key issues. The broad championing framework should start with a Standing Strategic Group of the Construction Industry chaired by a minister and comprising the presidents and top officers of major client, contractor and consultant groupings; and an Implementation Forum with representatives from the Standing Strategic Group and the government. It should lead possibly to the formation of an agency for industry development 'to encourage the delivery of best practice and performance, and to deliver teamwork' (p. 110) or the appointment of 'an Ombudsman, to examine allegations of poor practice and issue public comments on them' (p. 110).

A few years after Latham's (1994) report, the construction industry task force led by Sir John Egan produced *Rethinking Construction*, focused on improving construction quality and efficiency in the construction industry of the UK (DOETR, 1998). It suggested that the industry, government and major clients should collaborate to radically change the way industry built. Major clients must lead by implementing projects demonstrating approaches that would improve quality and efficiency. A forum of major housing associations and major housebuilding and construction firms could be the catalyst for change, encouraging long-term partnering arrangements between clients and providers to secure consistency, continuity, innovation and value for money. Proactive support and encouragement from the government would also be essential. To implement the initiatives recommended in the report of the task force, the Strategic Forum for Construction was set up in 2001; it was a collective body of industry representatives working to deliver fundamental improvement in the industry (DTI, 2002).

In Australia, the National Building and Construction Committee (NatBACC) (1999) enabled the pooling and sharing of knowledge and resources to enhance the industry's competitiveness in the face of increasing international competition. The proposals aimed to provide leadership and facilitate change within the industry to increase productivity, research, innovation and global business opportunities. The Committee recommended the establishment of a permanent advisory mechanism to enable the industry to discuss issues directly with relevant ministers. The government should invite the industry to set up a forum to enable monitoring and dialogue between the industry and government; and both should work together to create a marketplace that rewards enterprise and encourages firms and individuals to make the most of their talents.

Box 6.1 presents some championing efforts in the construction industries in some states in the USA.

The Construction Industry Review Committee (CIRC) of Hong Kong had the vision of creating 'an integrated construction industry that is capable of continuous improvement towards excellence in a market-driven environment' (CIRC, 2001, p. 2). It noted that committed leadership and a common purpose among all stakeholders were required to advance the collective interests of the industry. To overcome the fragmentation of the industry, which had inhibited learning and sharing, the government should appoint a lead agency to ensure that the concerns of the construction industry were properly reflected and ensure better co-ordination in policy making. The industry must take ownership of the change programme to reform itself. A co-ordinating body including all industry stakeholders should be established. It would be a statutory agency generating consensus on strategic issues and

---

Box 6.1 Championing of construction industries in some states in the USA

The Construction Industry Council of Westchester and Hudson Valley Inc. in New York State, US, founded in 1978, represents the region's construction and utility contractors and materials suppliers. Its 550 member businesses (overseeing work on more than US$2 billion worth of projects) have banded together to negotiate 14 labour contracts, as well as promote their own industry and lobby for laws and construction projects. Over the next few years, the council will focus on improving the area's construction workforce and championing new projects. It has teamed up with environmental groups to reverse pollution due to increased development; together, they have lobbied for federal money to fund sound cleanup projects.

In Maryland's Construction Industry Workforce Report, it was concluded that the industry's critical workforce development challenges fall into three broad categories: (i) image and branding; (ii) alignment and connectivity of education and training elements; and (iii) pipeline development. The report noted that, to implement the blueprint for building a workforce that supports the continued growth of Maryland's construction industry, the stakeholders of the industry – businesses, educational institutions, the building trades, independent training vendors, industry associations and state agencies – must become engaged in and responsible for the implementation of the selected recommendations. Each of the initiatives which were proposed to address these challenges had a clearly identified 'champion/lead organization'.

Sources: *Westchester County Business Journal* (1998);
Construction Industry Initiative (2009)

---

communicating the industry's needs to the government. The Construction Industry Council (CIC), a statutory industry co-ordinating body, was established in February 2007 after the enactment of the Construction Industry Council Ordinance on 24 May 2006.

In Malaysia, the Construction Industry Development Board (CIDB) (2005) formulated the *Construction Industry Master Plan (2006–2015)* after discussions and consultations with leaders representing various segments of the construction industry. The master plan has the following strategic thrusts: (i) integrate the construction industry value chain to enhance productivity and efficiency; (ii) strengthen the construction industry's image; (iii) move towards the highest standards of quality, occupational safety and health and environmental practices; (iv) develop human resource capabilities and capacities in the construction industry; (v) innovate through research and development and adopt new construction methods; (vi) leverage on information and communication technology in the construction industry; and (vii) benefit from globalisation including the export of construction products and services. The CIDB has formed a committee comprising eminent practitioners and officials to monitor the implementation of the initiatives under the blueprint, as well as a special unit under a Construction Industry Master Plan Co-ordinator.

In Singapore, the government-appointed C21 Steering Committee (1999) sought to address the inefficiencies of the construction industry as the country prepared to transit to a knowledge-based economy. The C21 blueprint aims to transform the industry from a dirty, demanding and dangerous industry to a professional, productive and progressive one. It suggested that the BCA, as the champion agency, must lead the efforts towards the smooth functioning of the industry, initiate actions to address its challenges and proactively develop new areas and expertise to transform it into a knowledge industry. The CIJC could provide a platform for feedback from the industry, and rely on the executive powers of BCA to encourage best practices among practitioners (C21 Steering Committee, 1999).

Thus, the need for effective and systematic leadership, co-ordination and monitoring (or championing) of change programmes has been specifically identified in countries at various levels of development, especially those which have undertaken comprehensive reviews of their construction industries. In many developing countries, the absence of effective championing has meant that programmes for enhancing the capacity and capability, and improving the performance, of the construction industries are not effectively implemented (see, for example, Ofori, 1993a,b, 1994b). A study of the construction industry in Zambia observed that 'A number of studies have been conducted and recommendations made, but with limited success in implementation, due to absence of an institution to champion and steer the industry' (Uriyo *et al.*, 2004, p. 2). For example, whereas a National Construction Industry Policy was launched in 1995, 'There is still need for clear leadership and monitoring as regards the implementation of the Policy' (p. 23). There was also a lack of commitment in the various agencies and the government to take forward the initiatives. This had led to a situation of poor overall performance in the construction industry (Uriyo *et al.*, 2004, p. 24). The researchers hoped that, with the promulgation of the National Council for Construction Act in 2003, the council, a statutory body, would 'take up its role of championing the industry in collaboration with the Associations and thus implement the various initiatives' (p. 24).

In a review of progress in the construction industry in the UK since the publication of *Rethinking Construction*, Construction Excellence (2009, p. 5) notes:

> We encountered disappointment at the lack of progress in implementing the recommendations, and pessimism about the future outlook for change. A recurring theme from our findings is that our industry needs . . . 'a little less conversation and a lot more action please'.

That is why we focused on why the industry has yet to embrace the changes and to propose what can be done to unlock the potential that clearly exists. In our opinion, it's no longer about whether this is the right stuff to be doing, it's more about what stops us, the industry, from doing something about it.

In a survey in the UK in which more than one thousand construction practitioners participated, Construction Excellence (2009) found that some 90 per cent of respondents reported that there had been a positive impact from *Rethinking Construction*, 'but this has been limited by partial uptake. In summary there has been too little change, too narrowly adopted and at too slow a rate' (p. 8). It notes that:

> It is clear that the stated aim of genuinely embedding the spirit of changes has not been met. There is not enough evidence of a united resolve across the diverse constituencies of UK construction to achieve Egan's vision of a modern construction industry. Where there are commitments, they are superficial and expedient, not tangible and sustainable.
>
> (p. 10)

The review team found the following to be among the factors which prevent progress (p. 15): (i) lack of cohesive industry vision; (ii) few business drivers to deliver meaningful change; (iii) construction 'does not matter' as its costs have low impact on clients' business; (iv) no incentives for change as clients' business models are short-term oriented; (v) construction is seen as a commodity purchase – clients do not focus on the long-term value created; and (vi) industry culture is driven by economic forces and transactional relationships – rather than partnering.

Construction Excellence (2009) observed that in the UK: 'The construction industry as a whole suffers from a lack of champions in Government' (p. 23). It noted that the average tenure of ministers in charge of construction between 1997 and 2009 was only one year. The report of the Business and Enterprise Committee (2008), aptly titled *Construction Matters*, highlighted the need for government leadership at the strategic level and recommended that there should be a role which both government and industry accept as having overall responsibility for the construction industry, that of a Chief Construction Adviser. The Government Chief Construction Adviser (CCA) was appointed in the UK in 2009 (OGC, 2009). The role of the CCA is to work with the government and the construction industry and ensure that the UK's construction industry is equipped with the knowledge, skills and best practice to take advantage of future opportunities. The Construction Minister said: 'The Chief Construction Adviser will play a vital role in championing the sector with Government and work with industry to ensure we have a strong, sustainable construction sector in the UK.' The CCA is an independent role, but the incumbent will report to the ministries responsible for finance, and for trade and industry. The terms of reference of the role include (i) chairing the Construction Collaborative Category Board, which drives the implementation and further development of best-value government construction procurement; (ii) chairing the sustainable construction strategy delivery board to help ensure that policy regarding the industry is effectively co-ordinated; (iii) assessing the key barriers to growth in the UK's low-carbon construction sector to ensure the UK industry is well placed to serve developing needs and markets; (iv) working with the industry, through the Strategic Forum for Construction, to deliver the industry improvement agenda; (v) promoting innovation in the sector; and (vi) co-ordinating the government's response to reports featuring construction.

In the next section, championing efforts by relevant organisations in the construction industry of Singapore are discussed.

Championing efforts in Singapore

The Construction 21 Steering Committee envisaged that the industry could be leaner and more flexible, with greater professionalism, if everyone worked together, in the right spirit (Mah, 2001). Several authors including Low and Tan (1993) and Ofori (1993a) discuss the efforts to develop the construction industry in Singapore from the early 1960s. There have been two prime movers of the initiatives and leaders of their implementation. First, the Housing and Development Board (HDB), the national housing agency and the largest client of the industry at that time, implemented incentive and assistance schemes to upgrade its contractors (Wong and Yeh, 1985). Second, the Construction Industry Development Board (CIDB), which was set up in 1984, spearheaded the development of the industry (CIDB, 1994). In 1999, the CIDB was restructured into the BCA with both developmental and regulatory functions (BCA, 2000).

The next section highlights some initiatives taken by Singapore's government and the construction industry, in the first three years of the implementation of the C21 programme, which were aimed at transforming the industry and improving its performance.

Government efforts

Under the leadership of the BCA, many important changes have been realised in Singapore's construction industry. For example, in the area of procurement, since the late 1980s, design and build (D&B) has been identified in Singapore as having the potential to encourage integration in project teams, enable the contractor's experience to be fed into design, and lead to the preparation of more buildable designs. D&B accounted for 60 per cent and 50 per cent of projects in Australia and Japan respectively (Tan, 2001), but only 13 per cent and 34 per cent of building and civil engineering works respectively in Singapore. In the private sector, D&B was only 16 per cent of total projects. The enhancement of the proportion of D&B

projects was one of the targets of C21. The BCA has been promoting D&B and creating a conducive environment for its utilisation (Koo, 2001a); an inter-agency committee it led produced a standard contract form for D&B, and formulated guidelines on public-sector D&B procurement to address concerns including waste during tendering.

The BCA published guides on good practices in key trades as part of its effort to help the industry to upgrade work methods and processes by documenting practices of firms which had consistently delivered high-quality work. The BCA, professional bodies and other government agencies developed a set of National Productivity and Quality Specifications (Koo, 2001b) to help to reduce wastage, minimise rework, shorten construction time, and improve quality and productivity. This was another initiative in the C21 report. Under the Quality Mark Scheme, the BCA had dedicated a section of its website to provide information to the public on the quality of workmanship of contractors and developers, in order to generate a market force which would help to raise the quality consciousness and performance of the construction industry (http://www.bca.gov.sg).

The need for strategic application of information and communication technology (ICT) was also highlighted in the C21 report. The Construction and Real Estate Network (CORENET) aimed to re-engineer the business processes of the industry to achieve a quantum leap in turnaround time, productivity and quality (http://www.corenet.gov.sg) (Koo, 2001b). For example, a system in CORENET called e-Submission allowed some 900 consultancy firms to make electronic applications for building plan approval simultaneously to 12 government agencies (Koo, 2001b); e-Buildable Design Appraisal System (e-BDAS) enabled architects and engineers to check the buildability compliance of their designs; and e-Catalogue provided technical and marketing information on resources.

Even the largest Singapore contractors are small, both in terms of the official definition of small and medium-sized enterprises (SMEs) and in comparison with their foreign counterparts; this hinders their competitiveness on larger projects (Chen, 2001). It was considered in the C21 report to be necessary to facilitate the formation of multi-disciplinary firms that can provide a range of services. Thus C21 recommended the removal of legal impediments to such groupings such as limitations on the activities which registered architects and engineers can undertake. The BCA revamped the Contractors Registration System (CRS) in 2002 to promote upgrading and growth, requiring contractors to meet more stringent turnover, track records, financial strength and professional personnel criteria. Among the 1,550 or so registered building contractors at the time, only 2.4 per cent qualified for the top (A1) grade and 1.8 per cent for A2. The CRS reduced the number of construction companies in the highest registration category from 83 firms (in the former G8 classification) to 39 in the new A1 grouping.

According to figures published by the BCA, by 2002, some 991 firms had benefited from more than S$26 million (US$1.00 = S$1.30) of grants under various financial assistance schemes (*Framework*, 2002). However, the take-up rate was lower than expected (Balakrishnan, 2002). Some of these schemes had been remodelled to encourage the development and application of new technologies and best practices. The schemes which were on offer at that time include those outlined in Table 6.1.

Since the mid-1980s, an export strategy administered by the BCA and the then Trade Development Board has helped local construction-related firms to venture overseas. This included missions to neighbouring countries led by

Table 6.1 Financial assistance schemes for construction industry in Singapore in 2002

Scheme	Aims and criteria
BCA Construction R&D Programme	Encourages, supports construction R&D by tertiary institutions, private organisations. Funds percentage of qualifying costs
Innovation Development Scheme (IDS)	Encourages, assists organisations to engage in innovation of products, processes, applications and develop depth in innovation capabilities. Innovation should lead to improvement in productivity
Industry Productivity Fund (IPF)	Encourages companies in an industry or value chain to collaborate to implement fundamental or radical changes in strategy, operations or practices, leading to significant gains in productivity, competitiveness for industry
Investment Allowance Scheme (IAS)	For enhancing productivity of construction industry through mechanisation, automation. Approved capital expenditure on equipment or machinery eligible for investment allowance
Local Enterprise Technical Assistance Scheme (LETAS)	Aims to attain quality, productivity upgrading in industries. Reimburses cost of engaging external expert for approved short-term assignment for upgrading operations or imparting knowledge or skills
Jumpstart Construction	Special assistance programme by BCA and SPRING Singapore for IT consultancy on usage, application of CORENET. Accelerates adoption of 'ready-made' e-commerce solutions by SMEs
CRS Step-up	Assists contractors to upgrade in corporate development, readiness in IT, building up of technical capability, formulation of business strategies. Aims to accelerate pace of upgrading SMEs to meet requirements of new CRS
Local Enterprise Computerisation Programme (LECP)	Encourages SMEs to achieve higher level of competitiveness through more effective use of IT
Local Enterprise Finance Scheme (LEFS)	Fixed interest rate financing programme to encourage, assist local firms to upgrade, expand their operations

senior politicians. For example, a mission in 2002 comprised 70 businesspersons and public officials, and aimed to enable Singapore construction-related firms to network and build up contacts with their Chinese counterparts (Chong, 2002).

Industry's efforts

There had been several attempts over the years to establish one umbrella organisation for Singapore's construction industry but none of them succeeded for an appreciable period of time until the turn of the millennium (see Ofori, 1993a). The CIJC was formed in 2000. Its members are the presidents of the professional institutions and trade associations in, or relating to, the industry. The organisations are the Real Estate Developers Association (REDAS), the Singapore Institute of Architects (SIA), the Institute of Engineers Singapore (IES), the Association of Consulting Engineers Singapore (ACES), the Singapore Institute of Surveyors and Valuers (SISV), the Singapore Institute of Building Limited (SIBL), the Singapore Institute of Planners (SIP), the Society of Project Managers (SPM) and the Singapore Contractors Association Limited (SCAL). The CIJC aims to provide a platform for discussing issues facing the industry and effecting changes in how the industry operates. It aspires to be the think tank for the industry, providing good-quality feedback to, and partnering with, the government to find solutions to the industry's problems.

REDAS produced the REDASD&B standard form of contract in 2001 to establish greater certainty in the industry. The Construction Quality Assessment System (CONQUAS) was introduced by the then CIDB in 1989 to raise quality of workmanship (CIDB, 1994). Some private developers were rewarding contractors which achieved good CONQUAS scores on their projects with monetary incentives. Many contractors also set up quality management systems to ensure higher consistency in workmanship.

The Singapore Contractors Association Ltd (SCAL), formed in 1937, has more than 2,000 members. It considers itself 'the official representative of the construction industry in Singapore' (http://www.scal.com.sg/index.cfm?GPID=1). Its vision is 'To facilitate members in becoming world class builders'. It expresses a commitment to the industry in (i) ensuring that the needs of its members on industrial, contractual and technical matters are well looked after; (ii) promoting and encouraging uniformity in the custom and practice of the construction industry; (iii) improving the standards of the industry and enhancing the status of its members; (iv) promoting the development of the industry; and (v) participating in social and community services and projects. SCAL had taken actions along the parameters each of these objectives.

The private sector has made progress in the development of ICT applications in construction in Singapore. In 1999, SCAL proposed the creation of an electronic marketplace for the industry. The Electronic Procurement

Environment for the Construction Industry (EPEC) sought to set appropriate industry standards, re-engineer business processes for greater efficacy while eliminating redundancies, and raise the level of construction ICT usage and proficiency (http://www.corenet.gov.sg). The BCA and 13 major developers launched a pilot implementation of Project Website in 2000 (*Framework*, 2000) for online sharing of project information and enhancement of cross-discipline collaboration and multi-party co-ordination.

Analysis of interviews

Interview structure, interviewees and questions

Interviews were undertaken with senior practitioners and administrators who had been involved in the BCA–CIJC collective championing of the C21 initiatives. Invitations were sent by e-mail or fax, and the questions were sent to those agreeing to be interviewed in advance. The interviews were conducted over a three-week period in 2002; sessions lasted an average of 95 minutes.

The industry practitioners who were interviewed were the President, Vice-President and/or Honorary Secretary of the constituent organisations of the CIJC: IES, SCAL, SIBL, SPM and SISV. Also interviewed were a Deputy Chief Executive Officer and a Director of the BCA.

Broadly, the interviewees were asked questions on the following points:

- the nature and duration of their involvement in the CIJC and how well they were able to fulfil their roles in the CIJC;
- the effectiveness of the BCA and CIJC in their effort to improve the industry, whether other statutory boards should be involved and whether a Ministry of Construction should be formed;
- the quality of the relationship between CIJC and BCA, and the benefits and problems of the collaboration;
- key features of an effective mechanism for the collective championing effort.

CIJC in general

The first question in the interview sought to ascertain the origins and activities of the CIJC, about which little had been published, and the interviewees' level of awareness of the collective efforts (between the BCA and CIJC) towards developing the construction industry. Most of the interviewees had been involved in initiatives to develop the construction industry since the formation of the CIJC; therefore they could provide insights into the history and activities of the Committee.

The idea for the formation of the council originated within the construction industry. The rationale was to create a forum for discussing issues of common interest to the different segments of the industry, and gather the opinions of the industry to present them in a coherent whole to the government and other relevant parties. The industry players proposed the formation of a 'joint council' but the term 'council' in the title was not supported by the BCA. A senior BCA officer who was interviewed explained that 'A council is more powerful. It is not workable for the private sector to regulate the industry in Singapore'. The grouping settled on 'joint committee' in the title instead. The CIJC was formed in 1999; the formal agreement was signed in 2000.

The presidents of the major professional institutions and trade associations in Singapore's construction industry constitute the CIJC. It was agreed that each member would chair the committee for a term of one year. The CIJC is funded by the member organisations; each contributes a modest sum annually. Meetings are held once every two months to discuss issues related to the industry. Sub-committees comprising representatives of the CIJC members have been formed to spearhead and monitor the activities and attainments under the strategic thrusts and action plans in the C21 report.

The CIJC considers itself as being responsible for promoting and advancing the image of the construction industry. Its plan in these regards is to educate college students about construction. As an example of its unified lobbying efforts, in 2002 the CIJC took the initiative to submit a proposal on ways to improve the construction industry in Singapore to the Economic Review Committee which had been appointed by the government to chart strategic directions for Singapore's economy in the wake of the Asian economic and financial crisis. The first CIJC–BCA meeting was held in the year 2000. The dialogue sessions were held quarterly; they are now held three times a year. The Chief Executive Officer (CEO), the two Deputy CEOs and Directors of BCA are involved in the dialogues with the industry representatives.

Difficulties faced by CIJC

Since the members of the CIJC work on a voluntary basis, one of their main difficulties lies in finding time to deal with CIJC matters. They are unlikely to focus much on the issues of the CIJC as they have to manage their own businesses as well as their various professional and trade institutions. One interviewee pointed out that the CIJC is a grouping of the unwilling, and felt pessimistic about its prospects. Another interviewee noted:

> Every member has his own interest. More often, conflict of interests occurs. There is a lack of concerted action and direction. As a result, the progress of the [work on the] Action Plans [of Construction 21] has been slow.

There are few incentives for the CIJC members to be committed. For this reason, they do not have the drive to work hard on matters pertaining to the construction industry as a whole, including the various C21 action plans they are concerned with.

Relationship between BCA and CIJC

The CIJC members who were interviewed considered their relationship with the BCA to be good. However, when they were asked whether the CIJC's proposals and suggestions were always acted upon by the BCA, the answers were generally negative. Some of the interviewees felt the BCA does not take the CIJC's suggestions seriously. This does not bode well for the close co-operation which is necessary, and which was envisaged under the joint championing effort. One interviewee suggested that the CIJC should work in parallel with, and not under the influence and direction of, the BCA, as seemed to be the case.

Views on the government's championing efforts

Despite the extensive programme of industry development implemented by the BCA and other agencies, many of the interviewees felt that the government had not done enough to help local construction firms to upgrade; some compared Singapore unfavourably with Malaysia in these regards. Many government initiatives were discussed with the interviewees to ascertain the reasons for their generally negative views.

The practitioners shed light on some commonly held views on Singapore's construction industry. As the property market was depressed at the time of the interviews, it was not surprising that most interviewees highlighted public-sector procurement policies. One of them noted:

> Many large-sized projects are still awarded to foreign contractors; as such, the local contractors will never have the track record and expertise to compete locally, not to mention overseas.

That interviewee suggested that local contractors should be given a chance to tender for large and complex projects, especially the landmark ones. Some interviewees observed that foreign firms managed to outdo the local companies right from the pre-qualification stage of some projects as some of the pre-selection criteria ruled out the latter. One noted:

> An overhaul of the current tendering system towards fair rewarding is required. Reimburse consultants and contractors adequately for their work and then they will grow.

The new CRS was of interest to the contractors interviewed. One public officer noted that the revamping of the CRS was intended to accelerate the consolidation of the industry through mergers and acquisitions. Many local contractors were unwilling to give up the identities of their family-owned companies to become bigger. However, the businessmen interviewed were not convinced that increasing the size of a firm would necessarily improve its commercial prospects.

Most interviewees wondered whether there would be sufficient numbers of large projects for the firms remaining after the implementation of the CRS to survive, given the prevailing depression in the market. One noted that the industry had reached a mature state, saying: 'We will never go back to the heydays of construction volume of $20 million'. Another practitioner observed (on the number of firms in the highest registration category, A1):

> In fact, thirty-nine is still a big figure. Many of them are HDB grown, and depended on public housing projects to survive. With the current oversupply situation, it is inevitable [that we will] see these contractors downsizing and downgrading in due course.

An officer from a public-sector agency agreed that there were still too many large (A1) contractors. He highlighted the next stage of the registration and classification of firms:

> There is a need to differentiate the contractors. More capable contractors must get the credit and recognition. In two years' time, phase two of CRS will rate the contractors in the top four categories in terms of financial status; management and development skills; and track records and performance.

Ultimately, the public-sector clients will be able to categorise the contractors in accordance with the different requirements of their projects. The officer went on:

> The details and progress of the ratings will very much depend on the economic situation and property market performance. The new CRS has already received complaints from the contractors, especially when there are already no jobs for them. We do not want to preside over the further deterioration of the situation.

Given the prevailing market situation, most interviewees felt that the large contractors must explore opportunities overseas. One commented:

> Contractors should start looking into the neighbouring countries like Myanmar, India and Vietnam. These countries require major infrastructure. HDB can export their planning and design expertise, together with our HDB contractors to build cheap and good quality housing for them.

On statutes and regulations and their impact on the development of the construction industry, the general comment of the interviewees could be summed up as 'No one likes legislation'. The legislation on buildable design which came into effect in 2001 was highlighted by the interviewees. Most of them felt it was not significant for industry improvement. A practitioner noted that architects opposed the legislation because they felt that emphasis on buildability hinders creativity. The public agency explained that the main rationale for promoting buildable designs is that they help to improve labour efficiency and productivity. An officer from the agency commented:

> Every sector has moved on, except for the construction industry. Legislation is seen to be necessary to push the conservative industry forward.

On the architects' views, the officer went on:

> Designers must change their mindsets. They are not the ones to do the construction works. Thus, they will not understand the difficulties. Moreover, not all creativity is affected by buildability.

The practitioner interviewees felt that the government's export promotion efforts were inadequate. Communication between the government and the private sector was poor in these regards. For example, some interviewees gave negative comments on the mission to China referred to above. One of them noted:

> No detail about the mission was circulated to the private sector, and the agenda was only received one week prior to the trip.

One of the interviewees who joined the mission trip observed that the trip was meant more for the public client organisations and companies partly or wholly owned by the government, termed 'government-linked companies' (GLCs), than for the private-sector firms. An officer from a public-sector agency admitted that the trip was not well organised. It was suggested that local companies should partner with the GLCs when venturing overseas. The GLCs could provide the project funds and the companies from the private sector could provide the entrepreneurship. The officer added that GLCs would get things done, whilst the private companies know how to make profits.

The interviewees generally noted that practitioners wish for a fairer and more balanced form of contract. The practitioners revealed that the construction industry showed resentment when the government launched the Public Sector Conditions of Contract for D&B. One respondent noted that the industry was not consulted about the standard form. On the REDASD&B standard form, a quantity-surveying consultant remarked:

> It is just another standard form; the practitioners may not think that it will have great impact on the industry as a whole. We can still stick to the modified JCT form of contract for D&B projects.

Indeed, the two leading quantity-surveying consultancy firms had not adopted the REDASD&B standard form for any of their D&B projects, mainly because of what had been referred to as 'the all-in clause' on 'fitness-for-purpose' on the part of contractors.

Another initiative which is often proposed by administrators, researchers and practitioners alike is that Singapore construction firms should invest in R&D (see, for example, C21 Steering Committee, 1999). The interviewees observed that it is impractical to expect firms to have the revenue to invest in R&D when design and other consultancy commissions and construction contracts are awarded at rates of professional fees and contract prices which imply little or no profits for the service providers.

Role of BCA

Some interviewees commented that the BCA could have done more and better in its activities. One interviewee observed that the prevailing weightage of the two main aspects of the BCA's broad activities was inappropriate, and needed to be adjusted:

> BCA spends 50 per cent of its time regulating and 20 per cent in developing the industry. It should also take up the role of marketing.

Some interviewees noted that the BCA did not appear to be independent or strong. It does not have the same drive to advance the construction industry as similar agencies they knew about in Japan and Malaysia. It was suggested that the BCA should exercise its political influence to assist local firms. Many practitioner-interviewees pointed out that the government tended to treat construction as a 'sunset' industry and ignore its contribution to the nation's economic development. This was seen as a failure of the BCA to get the right message across to the policy makers. A complete mindset change in the bureaucracy was vital and the interviewees expected the BCA to lead this effort.

The BCA must find out the difficulties faced by consultants and contractors which had ventured overseas. It could also help firms to pool their resources, and support the local consultants and contractors to form consortiums and export their services. One consultant suggested that a true 'Team Singapore spirit' should be nurtured among the industry players. This could be accomplished through the collective effort of BCA and CIJC.

On an often made proposal, none of the interviewees thought that there was a need for a Ministry of Construction in Singapore. One practitioner observed that the industry was not large enough to require the formation of a ministry. Some interviewees suggested that the Ministry of National Development should require and empower the BCA to truly push the industry forward. The BCA should involve the CIJC and key players so that they could collectively champion the industry. The BCA, with the help of CIJC, should assess and monitor the reactions of the industry when new initiatives are implemented, and it should endeavour to rectify teething problems as soon as possible.

Financial assistance schemes

Studies have found that few organisations use the government's financial assistance schemes for construction firms. It is essential to ascertain and understand the reasons for the low level of utilisation. The interviewees were not surprised about the low take-up rates. They observed that the schemes are designed almost solely for contractors; focus on the needs of SMEs; and mostly aim at increasing productivity. Although some of the interviewees had applied for the grants, most of their applications had been rejected. One complained that the processing of applications for grants was slow. In general, the interviews confirmed findings by researchers (see, for example, Ofori *et al.*, 1999) that the industry is not well informed of the assistance schemes although the BCA had announced each new scheme in press releases and in its newsletter, had published pamphlets on existing schemes and provided information on them on its website.

In response to the practitioners' views, an officer of a public-sector agency explained that the BCA only administers the financial schemes, and they are part of the economy-wide package of incentives from the Economic Development Board (EDB) and SPRING Singapore. He also disclosed that an evaluation of the relevant schemes is done every two years to fine-tune them to suit the needs of the industry, if necessary.

The interviewees perceived a dichotomy of views, experiences and, most importantly, objectives between the CIJC members and the BCA relating to the same issues, despite the designation of the two organisations as forming the 'collective championing effort' for the development of the construction industry. This is a matter which should be addressed.

Remaking CIJC

The practitioner-interviewees noted that one of the strengths of the CIJC is that the constituent organisations reflect the structure of the construction industry; hence, it can obtain views from different parts of the construction value chain. However, certain segments of the industry, such as smaller contractors (which have their own grouping and are not members of SCAL), are not represented in the CIJC. Other (smaller) associations in the construction industry had approached the CIJC to be allowed to join it but the CIJC had rejected such propositions because it feared that if it became any larger it would be difficult to co-ordinate its aspirations and activities. There is a feeling among CIJC member organisations that, as they represent the larger players in the industry, which account for dominant shares of the segments of the industry, the Committee can effectively represent the entire industry. However, for the full potential of the CIJC to be attained, even practitioners who are not members of any institutions or associations must be able to get their voices heard.

The presidents of the professional institutions and trade associations who constitute the CIJC usually hold their positions for one or two years. The lack of continuity hinders the effective planning and implementation of joint industry initiatives. Some interviewees noted that many issues are not properly followed up owing to the annual rotation of the chairmanship within the CIJC. This is compounded by the term limits which the member institutions of the CIJC impose on their leadership positions. This has inhibited progress in the development of the organisation to become the think tank it was set up to be.

There was a consensus among interviewees that the CIJC has not been performing well. All of them supported the establishment of a permanent secretariat for the CIJC. However, it was realised that such a set-up would require funding. Thus, the interviewees noted that the various institutions and associations must recognise the importance of the CIJC, and the need to strengthen it, and therefore be willing to provide the necessary financial support to enable it to function effectively.

Benefits and problems of BCA–CIJC collaboration

The interviewees agreed that possible benefits from the collaboration between the BCA and CIJC include the sharing of knowledge, formulation of practical solutions for the improvement of the performance of the construction industry, and maintenance of a proper mechanism for feedback from the industry. The interviewees noted that the CIJC members must drive the joint initiatives for industry development through their executive influence within the institutions and associations they lead.

Although the CIJC offers a feedback mechanism, the concern of many interviewees was 'How willing is the government or the BCA to accept the proposals of the industry?' Many interviewees pointed out that they did not receive any feedback on the actions taken, if any, after they had submitted views and proposals to the BCA. Despite the regular dialogue sessions, the practitioner-interviewees felt that the communication between the BCA and CIJC has not been productive. One interviewee opined that the BCA should not filter the CIJC's proposals, but should channel them directly to the government. The CIJC wants a more transparent process in its dealings with the BCA, which must also undertake closer follow-up actions. One interviewee pointed out that the joint committee would be influential only when BCA listened to it, and suggested that the BCA should recognise the potential of the CIJC as a catalyst for change since it comprises the dominant players in the various segments of the industry.

On the issue of disseminating the results of discussions between the BCA and CIJC, the interviewees disclosed that most of the decisions were passed on to the members of the institutions and associations. They suggested that the BCA should use the institutions and associations as vehicles for disseminating information to the industry.

Recent developments in Singapore construction

There have been progressive changes in many of the aspects of the construction industry in Singapore which were highlighted in the C21 report as well as those which were mentioned in the interviews. The changes have mainly been the results of the efforts of the BCA, but the private sector has also realised some significant attainments. Some of these new developments are discussed in this section.

In another review of Singapore's construction industry, the Construction Working Group of the Economic Review Committee (2002) reiterated the need for radical change in the industry and recognised that changing mindsets would require a 'paradigm shift' in thinking. It noted that firm leadership from the major clients and public institutions was essential to radically change the attitude and culture of the industry. There was scope for greater value-adding activity by professional bodies and trade associations, and potential for more joint action with the government. The committee strongly urged the construction industry to explore opportunities overseas, especially in China and India; it set a target for the proportion of projects to be won overseas. Thus, continuous championing of the development of the industry remains necessary.

The BCA's current mission statement reflects an expanded scope of interest. Its mission is 'we shape a safe, high quality, sustainable and friendly built environment' (http://www.bca.gov.sg). One of its five strategic thrusts is 'We lead and transform the building and construction industry by: Enhancing skills and professionalism; Improving design and construction capabilities; Developing niche expertise; and Promoting export of construction related services' (http://www.bca.gov.sg). In recent years, the BCA has (i) provided early information on aggregate levels of supply, demand and output to enable the industry to make relevant investment decisions; (ii) provided information and advice to government to manage levels of construction activity by withdrawing non-urgent planned projects when there is danger of overheating, and launching more projects when levels of activity fall and pump priming is required – this also enables the public sector to take advantage of lower prices in the construction market; (iii) revised the procurement guidelines (for the engagement of both consultants and contractors) to reduce the reliance on price and enhance the weightage of quality (including safety); (iv) radically restructured the contractor registration system into a licensing and rating system which is enhancing professionalism; and (v) prepared many more guides on good practice on areas such as appropriate detailing, working methods and environmentally sound technologies.

The use of ICT is now pervasive in the construction industry in Singapore. At the national level, it is used in the submission of plans and designs; for checking on compliance of designs with regulations; for finding out contracts advertised; for submitting bids; and for finding out the results of tenders. The public–private collaboration under BuildSmart Singapore provides leadership in ICT development and application in construction. In construction companies, ICT is used for business, technical and administrative purposes. It is also used in all aspects of the management of construction projects. There have also been positive developments on the research front; the government has provided a S$50 million fund for R&D, which is administered by the BCA, and has been used to undertake several major research projects leading to the development of key technologies.

The main programmes of the BCA in recent years are now discussed. Since 2005, environmentally responsible building has been one of the focal activities of the BCA. It launched the Green Mark scheme in 2005. The objectives are to promote environmental sustainability in the construction industry and raise awareness among developers, owners and professionals of the environmental impact of their projects; recognise building owners and developers who adopt practices that are environmentally conscious and socially responsible; and identify best practices in the development, design, construction, management and operation of buildings. The scheme is applicable to both new and existing buildings. New buildings are assessed under Energy Efficiency; Water Efficiency; Site and Project Development and Management; Indoor Environmental Quality and Environmental Protection; and Innovation. On existing buildings, Site and Project Development and Management is replaced with Building Management and Operations. To ensure that buildings given the Green Mark are well maintained during their operation, they are assessed once every two years. Buildings can be awarded Certified, Gold, GoldPlus or Platinum ratings, corresponding to energy efficiency improvement of 10–15 per cent, 15–25 per cent, 25–30 per cent or more than 30 per cent respectively. Different assessment benchmarks have been developed for office interiors, infrastructure and park design and development.

Other initiatives of the BCA relating to environmentally responsible building include the passing of legislation to make the attainment of Green Mark Certified a mandatory requirement for all new buildings. New government buildings and those built in strategic locations must attain even higher standards of performance. The government will also introduce schemes to encourage 80 per cent of the existing building stock in Singapore to achieve the Green Mark Certified rating. The second initiative is the provision of incentives for building owners and design consultants for attaining various levels of the Green Mark. The third is the setting of an example by spearheading the design and construction of, and investing directly in, a high-performing building, the 'Zero Energy Building at BCA Academy' to showcase appropriate environmentally sound technologies. The fourth initiative is capacity building through the offering of educational programmes and training courses at various levels, including a post-graduate degree programme.

The *Second Green Building Masterplan* (BCA, 2010a) outlined the following strategic thrusts: (i) public sector taking the lead; (ii) spurring the private sector; (iii) furthering the development of green building technology; (iv) building industry capabilities through training; (v) profiling Singapore and raising awareness; and (vi) imposing minimum standards.

It is pertinent to discuss the incentive schemes for environmentally responsible building. In December 2006, the BCA launched the S$20 million Green Mark Incentive Scheme to encourage developers to seek to attain higher Green Mark ratings for their developments (http://www.scal.com.sg/index.cfm?GPID=449). In May 2008, the scheme was enhanced to also provide incentives to architects and mechanical and electrical engineering consultants. By January 2010, the scheme had lapsed as the fund had been fully committed. However, the Green Mark Gross Floor Area Bonus is still available for new developments. Another incentive scheme is the S$15 million Sustainable Construction Capability Development Fund, which was set up by the BCA (http://www.scal.com.sg/index.cfm?GPID=458) in June 2010 to develop capabilities of the industry in delivering sustainable materials and adopting sustainable construction methods. The fund focuses on developing capabilities in recycling of waste arising from the demolition of buildings and in the use of recycled materials for construction. Firms which wish to test-bed relevant technologies or materials can also apply for support under the fund.

Another current major focal activity of the BCA has been productivity development. This has been in response to the government's programme to enhance the productivity of the economy of Singapore generally. The 'strong' measures introduced include enhancing the quality of the workforce and fine-tuning the foreign worker policies, including adjustments to the foreign worker levy and the number of foreign workers who may be employed for a given value of projects (under the Man-Year Entitlements scheme). The BCA will also review its Construction Registration of Tradesmen (CoreTrade) scheme to build up a larger core of skilled and experienced workers. The government set up a S$250 million fund 'to steer the construction sector towards higher productivity and build capability' (BCA, 2010b). The incentives under the fund cover the following: (i) workforce development – co-funding attendance of training programmes and providing scholarships; (ii) technology adoption – providing funding support to encourage the adoption of technologies and acquisition of equipment; and (iii) capability development – giving financial support for promising firms to build up their capabilities in complex civil engineering projects to take advantage of developments in the construction market.

Among the longest-lasting incentive schemes administered by the BCA is the economy-wide Investment Allowance Scheme (IAS), which provides a tax allowance for the purchase of new equipment. In construction, it has been used to guide firms towards the acquisition of plant and equipment which would help them to attain particular desired objectives of the government;

the enhancement of productivity has been a major aim. Thus, from January 2005, the following conditions apply: (i) the equipment must be new, and contribute to increasing project or company productivity; (ii) the equipment must bring at least 20 per cent improvement to the project or work trade; (iii) the project in which the item of equipment, machinery or tool will be utilised must fall in at least one of the categories of construction ICT, buildability or miscellaneous, such as quality, environment and safety; (iv) the equipment must represent a new technology compared with the current norm; and (v) preferably, the equipment should support projects to enable them to meet the requirements of the Code of Practice on Buildable Design.

The third pertinent current major focus of the BCA is assisting Singapore construction and construction-related firms to export their services. The board's Export Digest portal provides information on potential overseas construction opportunities and market intelligence. This includes country and market information; project leads; announcements on forthcoming activities and events; articles on overseas news; relevant government incentive schemes; and contact persons for various country markets. The BCA maintains networks in East Asia, South-East Asia, South Asia and the Middle East. It also participates in project facilitation through doing preliminary feasibility studies; assisting in forming consortia for projects; sourcing for funds; and liaising with government agencies in host countries. The Export Link Services assists member firms to explore overseas business opportunities, providing project referrals and leads, marketing of companies, and market intelligence. Finally, member companies can benefit from customised services including project facilitation; BCA serves as a link between overseas clients and member companies, helps members to form value chain consortia, and provides networking opportunities.

The private sector has also been active in implementing initiatives for developing the construction industry in Singapore. This has mainly been evident in programmes and actions of professional institutions and trade associations. For example, the manifesto of the SIA, released in 2007, is a strategy for developing key aspects of the architectural profession in Singapore. It had the following points (SIA, 2010):

1   Singapore – architecturally the most exciting city in the world:

   a   Singapore must be an architectural capital to compete with the world;
   b   architects must take the lead to push the frontiers of design;
   c   local architects must take the leading role to shape Singapore;
   d   developers must encourage creative architecture;
   e   industries need to upgrade skills to support good architecture;
   f   government should continue to review guidelines to encourage good architecture; and
   g   Singapore citizens and residents must embrace good architecture.

2   Sustainable and environment-friendly island:

a   Singapore should be a model of sustainability and environment-friendliness;
b   the successful public housing programme should be turned into a model of a sustainable community;
c   urgent research on alternative construction methods and materials is required;
d   end-users and the general public should be educated on the importance of a sustainable environment; and
e   Singapore should be a nation that appreciates and cares for its environment.

3   Socially responsible profession and an enlightened society:

a   design for a greying population;
b   design for a safer environment;
c   foster community spirit through revamp of the public environment;
d   maintain our architectural heritage;
e   recognise and reward good architects and industry players to achieve the vision of an excellent city; and
f   grow the next generation in an excellent environment.

4   Globalise:

a   expand overseas;
b   the Singapore Brand; and
c   globalise with a heart.

The SIA has been following the points in its manifesto in developing the architectural profession over the past few years. For example, it has recently prepared a blueprint for green design.

The clients' organisation, REDAS, has also been active in providing leadership on particular subjects. For example, the third edition of the 'REDAS Design and Build Conditions of Contract' was launched in 2010. The main changes relate to the introduction of an Option Module (with Employer's Architectural Design), which was included in response to requests by developers for REDAS to produce a set of contract conditions for the Partial Design and Build Contract Model whereby the employer retains part of the design responsibility for the works.

On its part, SCAL has three subsidiaries which have played a role in the development of the construction industry in Singapore (http://www.scal.com.sg/index.cfm?GPID=118). The first is SC2 Pte Ltd, which was founded in 1993 as a non-profit subsidiary of SCAL. Its main purpose is to provide professional safety audit on occupational safety and health. The second subsidiary is SCAL Resources Pte Ltd, established in 1997 to assist practitioners in the construction industry with their manpower and staffing needs; supply quality workers for the industry; and manage the welfare of migrant workers within the local construction industry. The company is involved in the development and management of foreign worker dormitories. Finally, the SCAL Academy seeks to plan, design and deliver training courses to address the construction industry's training needs and contribute towards the development of the management and supervisory personnel. SCAL also administers the Singapore List of Trade Subcontractors (SLOTS), which registers sub-contractors in the various trade categories. Box 6.2 provides information on the Local Enterprise and Association Development Scheme (LEAD), in which SCAL participated. It provides an example of collaborative initiatives which an institution in construction has undertaken with a partner outside the CIJC and BCA construction umbrella.

One of the five strategic thrusts of the BCA is: 'We forge effective partnerships with the stakeholders and the community to achieve our vision' (http://www.bca.gov.sg). Thus, in addition to the BCA's co-operation with the CIJC, it teams up with individual professional institutions or trade associations to pursue particular objectives. A manifestation of such collaboration is the formation of the Singapore Green Building Council in 2010. Another is the BCA–REDAS series of seminars, such as the annual updates on the construction industry, and the seminar on quality and productivity in 2010.

Conclusion and recommendations

The collective championing effort in Singapore's construction industry seeks to address the mutual interests of all key players in radically changing the industry. Although there have been many significant changes which have led to the advancement of the construction industry across many broad fronts and the industry has been well developed since the idea of the collective championing was proposed, the majority of the initiatives have been implemented by the government, and mostly with legislation, without the active involvement of the industry. Thus, the effectiveness of the collective effort is low.

Despite its statutory mandate, administrative responsibility and identification in C21 as the championing agency, from the interviews, the BCA should leverage upon its political influence and deepen the scope of its efforts for the benefit of the industry.

The CIJC aspires to be the think tank of the construction industry in Singapore and hopes to create closer collaboration among its member organisations. This would lead to less adversarialism and conflict on construction projects, and greater effectiveness in the CIJC's role in the championing effort. However, each CIJC member has its own interests, and the joint forum is not satisfactorily collective. Moreover, the CIJC is not well known within the industry and its potential role is not recognised. The leadership

Box 6.2 Local Enterprise and Association Development Scheme

The Local Enterprise and Association Development Scheme (LEAD) is a multi-agency initiative led by SPRING Singapore (the enterprise development agency responsible for helping Singapore enterprises grow) and International Enterprise Singapore (the agency spearheading the development of Singapore's external economy). LEAD 'aims to enhance industry and enterprise competitiveness by partnering industry associations willing to drive industry development initiatives to improve overall capabilities of local enterprises in their sectors'. In 2007, five more industry associations including SCAL joined the LEAD programme. The five new LEAD projects would generate over S$800 million in revenue (S$368m overseas), and create more than 4,000 new jobs over three years. (By October 2010, 22 LEAD associations had embarked on over 20 capability-building projects, benefiting 1,800 enterprises, which were expected to generate more than S$1.6 billion of value-added and S$4 billion of revenue upon completion.)

Mr Chong Lit Cheong, CEO of International Enterprise Singapore, noted that:

> LEAD has enabled IE Singapore to heighten the visibility and competitiveness of Singapore companies in the global market through a deeper engagement and partnership with leading industry associations. Through LEAD, forward-looking associations . . . have taken a leadership role to raise the profile of the local industries in the international scene. Today, our Singapore furniture manufacturers have set up commercial presence in more than 16 countries globally.

The Minister for Trade and Industry, Mr Lim Hng Kiang, also observed:

> Our local industry associations play a vital role in helping their respective industries enjoy and sustain such robust growth . . . Being champions of their industries, [they] have a keen understanding of the capabilities, needs and potential of their industry players. They are in the best position to conceptualise strategies and spearhead efforts to strengthen the capabilities and competitiveness of their industries.

SCAL's LEAD proposal was entitled 'Capabilities Development of Singapore Construction Industry'. The vision was 'To build up safety competencies, industry trade practices and export capabilities of the contractors so as to achieve higher efficiency and enhance competitiveness.' The proposal focuses on three key initiatives:

- Improve the Standard of Safety Practices for SMEs and Local Enterprises – SCAL would conduct safety training, certification and auditing, targeting SLOTS firms;
- Research and Innovation: Development of Construction Standards for SMEs and Local Enterprises – SCAL would develop 10 good industry practices, disseminate them to 1,800 companies, and conduct 10 seminars on them;
- Increase Export Capabilities of Contractor and Local Enterprises – SCAL would set up a facilitation centre to provide advisory services to support the internationalisation needs of its members; engage consultants to gather market information and produce market reports; organise nine mission trips and nine study trips to explore targeted markets; conduct nine seminars to share with members; engage a consultant to help in the branding and marketing of the construction industry; and produce marketing collaterals.

Sources: http://www.scal.com.sg, accessed in 2008; http://www.spring.gov.sg/et/2007_11/index10.html; http://www.iesingapore.gov.sg/wps/wcm/connect/ie/my+portal/main/press+room/press+releases/2010/next+phase+of+local+enterprise+and+association+development+programme+focuses+on+empowering+associations+through+capability+upgrading?pageDesign=IE+Default+Search+Presentation+Template

and influence of the professional institutions and trade associations (some of which are well regarded) should be utilised to serve as the link between the CIJC and its membership, to ascertain views on policies, and to disseminate government, CIJC and institutional initiatives.

Whereas the individual professional institutions and trade associations have implemented significant initiatives which have contributed to the development of the construction industry in Singapore, so far there has not been any identifiable major CIJC initiative for developing the industry. Thus, there is little evidence of its merit as a partner in the championing effort. The potential of the CIJC should be appreciated and accepted by the government and the industry players. A permanent secretariat must be formed for the CIJC; and the chairperson of the CIJC should have a longer term of office, say four years. Moreover, instead of a rotation of the position, the chairperson must be elected by merit and should possess the required expertise and experience to be the leader among the leaders.

In Singapore, the government, through the BCA, has effectively championed the development of the construction industry. It has been in consultation with the industry and has provided information on the objectives and rationale of its key initiatives. However, much more could have been attained if there had been the collective championing by the BCA and CIJC which had been envisaged. In this way, the maturity of the industry could be effectively fostered, and progress would not rely on legislation and mandatory provisions. There are still persistent problems in the construction industry in Singapore. These include a poor image, an inability to attract talent and poor safety performance.

The collective championing effort has even greater potential. The BCA and CIJC should jointly promote and improve the image of the construction industry. Efforts should be made to change the public's perception of construction. Landmark buildings and items of infrastructure could be used to publicise the high technology adopted on the projects.

A long-term partnership and trust must be nurtured and sustained between the BCA and the CIJC in order to improve the effectiveness of the implementation of initiatives to upgrade the industry and improve its performance. A truly collective championing effort between these two organisations is necessary to generate fundamental changes and accomplish the vision of 'reinventing construction' which formed the title of the C21 report. This requires more than statements to that effect in strategic plans. A change in mindset among all the parties involved is necessary.

Championing has been done in different ways in various countries. In the UK, an individual, the CCA, plays the role. In Hong Kong, Malaysia, South Africa and a number of countries, the champion is a statutory agency. In India and Indonesia, the championing is provided by an organisation established with government's leadership (and, in Indonesia, by law), but comprising both the public and private sectors. In the United States, championing at the state level is done by the private sector.

The experience of Singapore shows that championing is critical if the strategies for improving the construction industries of all countries, and especially those in developing countries, are to be effectively implemented. Active leadership by the government is important. This requires education and awareness, and appreciation of the needs of the construction industry on the part of the government officials. The construction industry would benefit from a single voice. Collective public–private sector championing would be most useful. In each country, it would be worthwhile to make conscious efforts to develop this.

References

Akani (2001) New statutory body to champion construction industry development. Launch issue. http://www.buildnet.co.za/akani/2001/july/02.html

Balakrishnan, V. (2002) Speech delivered at the BCA Awards ceremony, 25 April. http://www.bca.gov.sg/newsroom

Banwell, H. (1964) *Report of the Committee on the Placement and Management of Contracts for Building and Civil Engineering Work*. HMSO: London.

BCA (Building and Construction Authority) (2000) *Building Up: Inaugural Annual Report 2000*. BCA: Singapore.

BCA (2010a) *Second Green Building Masterplan*. BCA: Singapore.

BCA (2010b) Measures to raise productivity and build capability in the construction sector. Media release, 3 March.

Burns, T. and Stalker, G. M. (1961) *The Management of Innovation*. Tavistock: London.

Business and Enterprise Committee (2008) *Construction Matters*, Ninth Report of Session 2007–08. The Stationery Office: London.

Chen, C. N. (2001) Speech delivered at the Construction and Property Prospects Seminar, 11 January. http://www.bca.gov.sg/newsroom

Chong, V. (2002) Mah Bow Tan to lead construction team to China. *Business Times* (Singapore), 23 May. http://business-times.asia1.com.sg

Committee of Inquiry into the Capacity of the Construction Industry of Singapore (1961) *Final Report*. State of Singapore: Singapore.

Constructing Excellence (2009) *Never Waste a Good Crisis*. Constructing Excellence: London.

CIDB (Construction Industry Development Board) (1994) *10th Anniversary Commemorative Publication*. CIDB: Singapore.

CIDB (2005) *Construction Industry Master Plan (2006–2015)*. CIDB: Kuala Lumpur.

Construction Industry Initiative (2009) *Maryland's Construction Industry Workforce Report*. Maryland Governor's Workforce Investment Board. http://www.gwib. maryland.gov/news/constenforum/constructionlayout.pdf

CIRC (Construction Industry Review Committee) (2001) *Construct for Excellence*. Environment, Transport and Works Bureau: Hong Kong.

Construction 21 Steering Committee (1999) *Construction 21: Re-inventing Construction*. Ministry of Manpower: Singapore.

Davey, C. L., Powell, J. E. and Powell, J. A. (1999) Exemplary case studies in action learning. Unpublished. University of Salford: Salford.

Davey, C. L., Lowe, D. J. and Duff, A. R. (2001) Generating opportunities for SMEs to develop partnerships and improve performance. *Building Research & Information*, 29 (1), 1–11.

DOETR (Department of the Environment, Transport and the Regions) (1998) *Rethinking Construction: Report of the Construction Task Force*. HMSO: London.

DTI (Department of Trade and Industry) (2002) *About Rethinking Construction*. DTI: London. http://www.detr.gov.uk

Economic Review Committee (2002) *Construction Working Group Report: Executive Summary – Economic Review Committee Subcommittee on Domestic Enterprises*. Ministry of Trade: Singapore.

Emmerson, H. (1962) *Survey of Problems before the Construction Industries*. HMSO: London.

Floyd, S. and Wooldridge, B. (1996) *The Strategic Middle Manager: How to Create and Sustain Competitive Advantage*. Jossey-Bass Publisher: San Francisco.

Framework (2000) Buildable design: bringing the benefit of flat plate design to the home. September–October. http://www.bca.gov.sg

Framework (2002) The construction industry gets a boost with new initiatives. May–June. http://www.bca.gov.sg

The Herald (2009) Construction industry must play its role in economic turnaround. 19 May. http://allafrica.com/stories/200905190662.html

Howell, J. and Higgins, C. (1990) Champions of change: identifying, understanding, and supporting champions of technological change. *Organisational Dynamics*, 19 (1), 40–55.

Howell, J. M., Shea, C. M. and Higgins, C. A. (2005) Champions of product innovations: defining, developing and validating a measure of champion behavior. *Journal of Business Venturing*, 20, 641–661.

Kleysen, R. F. and Street, C. T. (2001) Toward a multi-dimensional measure of individual innovative behaviour. *Journal of Intellectual Capital*, 2 (3), 284–296.

Koo, T. K. (2001a) Speech delivered at the Design and Build: New Achievements Conference, 9 May. http://www.bca.gov.sg/newsroom

Koo, T. K. (2001b) Speech delivered at the *Seminar on Good Industry Practices*, 9 February. http://www.bca.gov.sg/newsroom

Latham, M. (1994) *Constructing the Team*. HMSO: London.

Low, S. P. and Tan, S. P. (1993) *The Relationship between Construction, Marketing and Economic Development*. Royal Institution of Chartered Surveyors: London.

Mah, B. T. (2001) Speech delivered at Baucon Asia 2001, 20 November. http://www.bca.gov.sg/newsroom

Markusson, N. (2007) Bringing structure to the championing of environmental improvements in innovation work. Paper presented at Science and Technology in Society: An International Multidisciplinary Graduate Student Conference, Washington DC, 31 March to 1 April. http://www.cspo.org/igscdocs/Nils%20Markusson.pdf

Maute, M. and Locander, W. (1994) Innovation as a socio-political process: an empirical analysis of influence behaviour among new product managers. *Journal of Business Research*, 30, 161–174.

Nam, C. H. and Tatum, C. B. (1997) Leaders and champions for construction innovation. *Construction Management and Economics*, 15, 259–270.

NatBACC (National Building and Construction Committee) (1999) *Building for Growth: Building and Construction Industries Action Agenda*. Department of Industry, Science and Resources: Canberra.

Nepal, M. P. (2003) *The Role of the Project Manager as a Champion of Construction Innovation*. Unpublished MSc dissertation, National University of Singapore.

OGC (Office of Government Commerce) (2009) Government names its first Chief Construction Adviser. Press release, 24 November. http://www.ogc.gov.uk/About_OGC_news_9747.asp

Ofori, G. (1993a) *Managing Construction Industry Development: Lessons from Singapore's experience*. Singapore University Press: Singapore.

Ofori, G. (1993b) Research on construction industry development at the crossroads. *Construction Management and Economics*, 11, 175–185.

Ofori, G. (1994a) Formulating a long-term strategy for the construction industry of Singapore. *Construction Management and Economics*, 12, 213–217.

Ofori, G. (1994b) Practice of construction industry development at the crossroads. *Habitat International*, 18 (2), 41–56.

Ofori, G. (2000) Challenges of construction industries in developing countries: lessons from various countries. In A. B. Ngowi and J. Ssegawa (eds) *Challenges*

Facing the Construction Industry in Developing Countries, Proceedings, Second International Conference of CIB Task Group 29, 15–17 November. National Construction Industry Council: Gaborone.

Ofori, G. (2002) Singapore construction: moving towards a knowledge-based industry. *Building Research and Information*, 30 (6), 401–412.

Ofori, G. and Chan, S. L. (2001) Factors influencing development of construction enterprises in Singapore. *Construction Management and Economics*, 19, 145–154.

Ofori, G., Leong, C. and Teo, P. (1999) *Influence of Foreign Contractors on Development of Singapore Construction Companies.* Centre for Building Performance and Construction, National University of Singapore: Singapore.

Schon, D. A. (1963) Champions for radical new inventions. *Harvard Business Review*, March–April, 41, 77–86.

Shane, S. (1994) Cultural values and the championing process. *Entrepreneurship: Theory and Practice*, 18 (4), 1–17.

Shane, S., Venkataraman, S. and Macmillan, I. (1995) Cultural differences in innovation championing strategies. *Journal of Management*, 21, 931–952.

Simon, E. (1944) *The Placing and Management of Building Contracts: Report of the Central Council for Works and Buildings.* HMSO: London.

Singapore Institute of Architects (SIA) (2010) *Manifesto 2007.* http://www.sia.org.sg

Tan, S. L. (2001) Speech delivered at the 3rd International Conference on Construction Project Management, 29 March. http://www.bca.gov.sg/newsroom

Uriyo, A. G., Mwila, J. and Jensen, L. (2004) *Development of Contractor Registration Scheme with a Focus on Small Scale Civil Works Contractors*, Final Report Prepared for National Council for Construction, Zambia.

Westchester County Business Journal (1998) Construction Industry Council marks 20th year. 2 November. http://findarticles.com/p/articles/mi_qa5278/is_199811/ai_n24338598/

Wong, A. and Yeh, S. H. K. (1985) *Housing a Nation: 25 years of Public Housing in Singapore.* Maruzen Asia: Singapore.

Zintz, A. C. (1997) Championing and managing diversity at Ortho Biotech Inc. *National Productivity Review*, 16, 21–28.

7 Informal construction activity in developing countries

Jill Wells

Introduction

Studies dating from the mid-1990s describe a rapid expansion of 'informal' construction activity in many parts of the developing world. This has been accompanied by increasing diversity in the interpretation of the concept, leading to considerable confusion. Too often when talking about the informal sector people find that they are talking about different things.

This chapter is divided into five sections. In the first section, the origins of the 'informal sector' concept in the early 1970s and the way it has since evolved are briefly outlined. It is now widely recognised that the essence of informality is the absence of regulation. In the second section, four major areas of regulation that affect the construction industry are described. The third and fourth sections examine the evidence of an expansion of informality in each of the four areas or spheres. These sections draw heavily on the author's own research in East Africa, supplemented by a review of the literature which indicates that the developments described – although varying in degree – are pretty universal across the developing world. The final section summarises the main threads of the discussion and suggests some key issues for research and analysis in the coming years.

The informal sector concept

The concept of the 'informal sector' originated in studies of the urban economy in developing countries in the early 1970s. The term 'informal sector' was used to describe that part of the urban economy where those not in regular employment were somehow able to make a living. Hence, the informal sector comprised the survival activities of the urban poor working in marginal or peripheral segments of the urban economy (ILO, 2002).

The term was originally defined with reference to the characteristics of employment (Hart, 1973). Thus, there were formal sector jobs in which the terms and conditions of employment were regulated and workers received some protection from the law; and there were informal jobs in which workers were mostly self-employed, and were subject to little regulation or protection.

A slightly different approach was taken by the mission of the International Labour Office (ILO) to Kenya (ILO, 1972), which distinguished between the formal and informal sectors on the basis of the characteristics of enterprises, rather than employment situations. Thus, formal sector enterprises were generally large, heavily capitalised and regulated by the state; whereas informal enterprises were very small, unregistered and operating with little capital in unregulated markets. It is this definition of the informal sector as a collection of small enterprises that has been more influential over the years, probably because it is the definition used in statistical surveys. However, whether defined according to the nature of employment or the characteristics of the enterprise, the informal sector was essentially seen as an arena where the state did not reach, a place where economic activity was conducted in the absence of regulation.

The informal sector was also originally seen as a phenomenon that was temporary and likely to disappear as countries developed. Of course, this has not happened. In many low-income countries, particularly the countries of sub-Saharan Africa, rapid urbanisation in the absence of significant economic growth has greatly increased the number of people without access to regular jobs in the formal sector. Migrants to the city are increasingly forced to eke out a living in unregulated occupations on the margins of the urban economy. The majority of them cannot afford to rent a place to live in the planned and regulated parts of town but crowd into rapidly growing high-density areas, 'squatter settlements' or 'slums'.

At the same time, formal sector enterprises throughout the world (in both developing and developed countries and in many sectors of the economy) have been shedding labour on a massive scale in the past few decades, often retaining only a small core of regular workers and relying for the bulk of their workforce on a large periphery of temporary and casual workers, often supplied through intermediaries (ILO, 2001). The intermediaries (subcontractors and labour agents) who are the new suppliers of labour find it easier to avoid registering workers and complying with various aspects of labour legislation than the contractors who previously employed them. Hence, the outsourcing of labour is generally associated with deregulation of labour markets. The result in many countries, including those that have experienced significant levels of economic growth, has been a great expansion in the number of 'informal employees', who may be found working in both formal and informal enterprises. This is the case in many parts of Latin America and South-East Asia, where the number and proportion of people working informally has, in fact, risen in line with gross domestic product (GDP) (OECD, 2009).

In response to these developments, new terms such as 'informal employment', 'informal labour', 'informal economy' and 'informal housing' have come into use. Revisiting the concept and reality of the 'informal sector' some 30 years since the term was first used, the ILO (2002) noted the changes taking place and suggested that the term 'informal sector' with its implication of a dichotomy between two distinct parts or sectors of the economy is increasingly inappropriate. It proposed that the use of the term be confined to the total collection of informal enterprises and that the term 'informal economy' be used to refer to the 'conceptual whole of informality', covering both production relations and employment relations. This redefinition is welcome and the new approach will be adopted in the current chapter.

However, the attempt to bury the 'informal sector' has not been entirely successful and the term is still in frequent use, especially in relation to construction. The following sections attempt to disentangle the various threads of informality in the construction sector of developing countries.

Four spheres of regulation

As discussed above, informal economic activity has been defined over the years by various criteria relating to enterprises and/or employment – notably the size or scale of activities, type of technology, form of ownership or organisation, and employment status. However, it is increasingly recognised that the key essence of 'informality' is the absence of regulation – the failure to comply with all, or some, of the regulations in the body of national or local legislation (Wells, 2007). Hence, it is not the intrinsic characteristics of activities that define informality, but rather the boundaries of state regulation. Informal economic activity can then be defined as activity which lies outside the institutions of state regulation (Harriss-White, 2010).

A key question that immediately arises is which regulations are not complied with, and by whom? Construction activity is subject to a number of different spheres of regulation, comprising an even larger number of regulations. In this chapter the focus will be on four spheres.

First there is regulation of the terms and conditions of employment for the construction workers, as set out in labour legislation, health and safety rules and so on. In many developing countries (particularly those in sub-Saharan Africa) labour legislation has never affected more than a tiny minority of workers. Moreover, even where they have been effective in the past (for example, in South Africa and many countries of Latin America) these regulations are increasingly flouted, as enterprises attempt to avoid the high on-costs associated with registering workers. They do this by engaging workers on a casual basis or by outsourcing their labour supply. Today, informal employees may be found working in both formal and informal enterprises. Together with the self-employed and family labour they make up 'the informal sector of the construction labour force' or 'informal labour' (Breman, 1996; Pais, 2002).

The second aspect of regulation refers to the enterprises that are involved in production, notably contractors, subcontractors and material suppliers. In most countries, every enterprise must have a licence to engage in economic activity. Enterprises involved in construction are generally subject to additional regulation by national- or local-level authorities. There seems to be a consensus in the literature that those enterprises that fail to comply with the regulations may be classed as informal. According to ILO definitions, the totality of informal enterprises then makes up the 'informal sector of enterprises'.

The third area of regulation comprises the rules, regulations and agreed procedures for the organisation of the construction process. The various steps include the hiring of architects and engineers to design the product, formal procedures for the appointment of construction contractors and sub-contractors and the use of written contracts between the parties involved (consultants, contractors, sub-contractors and suppliers) to allocate responsibilities and apportion risks. However, there is an alternative way of organising production, with fewer participants, more flexible arrangements and less dependence on contracts that are enforceable by law. This has been called the 'informal construction process' (UNCHS, 1986) or the 'informal construction system' (CIB Task Group 29, 1998).

The fourth area of regulation refers to the product. In most countries, planning permits are generally required for new construction, to show that the land is designated for building and legally owned. Moreover, building plans have to be submitted and approved before construction begins and the final structure inspected to ensure that it conforms to the building regulations in force. However, in many cities throughout the developing world, these regulations are only partially applied or not applied at all. Buildings constructed without planning permission are sometimes referred to as 'informal buildings' or, more often, 'informal housing', as this is the sub-sector in which lack of approved plans is most common. A collection of buildings or houses without planning permission then becomes an 'informal settlement'.

These four separate spheres of regulation often overlap, so that deregulation in one sphere (such as labour) is closely associated with deregulation and an expansion of informality in another. For example, unregistered enterprises may be found employing labour informally, engaged in an informal production process to deliver products that fail to comply with planning and building regulations. However, informality in one area does not necessarily imply informality in the others. It is equally likely that formally registered enterprises are employing labour informally and delivering products for which they do not have planning permission through an informal system (Wells, 2007).

It should also be noted that each regulatory sphere comprises a bundle of laws, rules and procedures which may be very large and complex and which may be followed to a greater or lesser extent (Wells, 2007). If it is assumed that observance of all of the rules, regulations and procedures indicates 'formality', and non-observance 'informality', then it is equally clear that there are many possible stages in between. An enterprise, employment situation, production process or even product may partially comply with rules, regulations and established procedures. It may also comply with some aspects of these and not with others. Thus, there is no hard and fast dividing line between formal and informal, but rather a gradation.

Evidence drawn from recent research in developing countries, of the growth of unregulated or 'informal' activity in each of the four areas is examined in the following sections.

Informal employment in construction

The extent of the informal employment of labour in the construction industry, in both developed and developing countries, has been documented in a study by the ILO (2001). In many developing countries, the practice of recruiting labour on a casual basis, usually through intermediaries, is long established. For example, at the end of the 1990s, 74 per cent of construction workers in Malaysia were employed on a casual basis through sub-contractors or *kepala* (Abdul-Aziz, 2001). Casual workers were estimated at 85 per cent of the construction workforce in the Philippines (Yuson, 2001), 66 per cent in Mexico (Connolly, 2001) and 77 per cent in the Republic of Korea (Cho, 2004). In Egypt, an estimated 90 per cent of construction workers are either hired on a casual basis or self-employed (Assaad, 1993).

However, there is also evidence from many countries of a movement in the past three decades towards even greater casualisation of the labour force. Standing and Tokman (1991) argue that casualisation has been a direct outcome of the general process of deregulation that has affected many parts of the developed and developing world. Increased competition has also played a part. The process has not been confined to the construction industry, but the shift to the use of casual and sub-contracted labour has been particularly marked in construction.

In a study of the 'informal economy' in India, Anand (2001) describes the growth in the number of casual daily-wage workers available at street corner labour markets or offered by labour agents. This development is particularly important in construction, in which the number of such workers has grown exponentially in recent years as the 'organised' (formal) sector has sub-contracted various functions, as well as outsourcing components and services. An intensive study of building labour and enterprises in two small towns in India shows the extent and diversity of sub-contracting in the Indian building industry, where it is not unusual to find all of the trades contracted out to individuals and groups of workers (van der Loop, 1992). These developments are reflected in the fact that construction employment in the organised (formal) sector increased by only 0.37 per cent between 1973 and 1987, while that in the unorganised sector rose by 9.73 per cent (Breman, 1996).

An attempt to measure the extent and growth of casualisation across industrial sectors in India on the basis of data collected from the National Sample Survey found that construction is one of two sectors in which casual workers predominate (agriculture is the other) and in which the share of casual labour increased in the 10 years between 1983 and 1993. In 1993, 64 per cent of men and 96 per cent of women working in urban construction were employed on a casual basis. The comparable figures for 1983 were 58 per cent and 89 per cent (Pais, 2002). The author argues that casual workers are only a sub-set of the larger group of workers who constitute the 'informal sector' of the labour force. If self-employed workers (who offer services to households in the 'domestic sector') are also included then 89 per cent of men and 97 per cent of women working in construction in 1993 could be considered as 'informal labour'.

China has not escaped the trend. A reform programme launched in 1984 called 'Separation of management from field operations' led to the shedding of labour by state-owned construction companies (Lu and Fox, 2001). The majority of construction field workers are now employed on a temporary basis by urban collectives. As a result, the proportion of temporary employees in the Chinese construction industry rose from 28 per cent in 1980 to 65 per cent in 1999. Together with the self-employed they make up 72 per cent of the total construction workforce (Lu and Fox, 2001).

There is similar evidence from Latin America. For example, data from the national household survey in Brazil show that the proportion of unregistered and self-employed workers in the construction workforce rose from 57 per cent in 1981 to 75 per cent in 1999 (ILO, 2001). In this case (and in other Latin American countries) evidence suggests that the employers adopt this approach to escape from the high 'on-costs' of registering workers, which are not considered proportional to the benefits delivered.

Data for African countries are more difficult to find and until recently largely anecdotal. In a series of papers based on research in Kenya and Tanzania, Wells (1998, 1999, 2001), Mlinga and Wells (2002) and Wells and Wall (2003) note a number of developments in the employment practices in the construction sector, following the adoption of liberal economic policies in sub-Saharan Africa. First, increased competition, declining workloads and/or restrictive employment regulations have led registered contractors here, as elsewhere in the world, to shed their permanent labour forces and replace them with workers recruited for short periods on a casual basis. Casual and temporary workers recruited directly by the principal employer receive no protection from the law.

More recent research in Tanzania confirmed these findings. A study of labour practices on 11 large construction sites commissioned by the ILO in 2003 found that the proportion of casual employees (hired and paid on a daily basis) was generally over 70 per cent and on some sites as high as 95 per cent (ILO, 2005). Subsequent analysis of data from the Labour Force Surveys of 1990/91 and 2000/01 revealed significant changes in employment status among the construction workforce over the 10-year period, with a dramatic fall in the proportion in paid employment, from 77 per cent to 37 per cent, and a corresponding increase in the proportion who were self-employed (Wells, 2009).

It is clear from the above that, if all of the construction workers who are employed on a casual basis, without regular contracts or any form of social protection, are included in the expanded concept of the informal economy, then a large part of the construction workforce worldwide would fall within the boundaries. Informality is now the norm, rather than the exception, in the construction industry throughout much of the developing world. By this definition, the 'informal economy' in construction is no longer marginal.

This phenomenon is not just confined to construction. Recent research by the Organisation for Economic Co-operation and Development (OECD) (2009) confirms that levels of informal employment vary widely across continents, reaching the highest share in sub-Saharan Africa, where close to 80 per cent of all employed people (outside agriculture) work informally. Africa is followed closely by South-East Asia, with over 70 per cent, and Latin America, with informal employment being close to 60 per cent of all employed persons.

Informal enterprises in construction

There is evidence from many countries that the outsourcing of the supply of labour and other services and products has opened up new opportunities for small enterprises and workers in the role of sub-contractors and labour suppliers to the formal sector. For example, the increase in the proportion of unregistered and self-employed workers in the construction workforce in Brazil in the 1980s and 1990s was accompanied by a big expansion in the number of unregistered employers, most of whom were believed to be labour brokers or *gatos* (ILO, 2001). Generally, labour suppliers are not registered, taxed or regulated by the state.

One of the most notable developments observed in Kenya in the 1990s was the growth of specialised enterprises offering labour for common tasks such as concreting or block laying (Ngare, 1998). Similar developments are reported from Tanzania (Mlinga, 1998) and from South Africa (English, 2002; Cattell, 1994). Many, if not most, of these enterprises are not registered with any government body and contracts with them are informal.

The increasing popularity of the informal building process (explained below) has also expanded the market for small producers and suppliers of building materials – the 'informal economy' of building materials supply. Clients using the informal process to develop their buildings usually buy materials in small quantities, as and when they have the funds. Unregistered, small producers are able to meet the demand for small quantities, whereas they cannot supply the large orders from the registered contractors in the more formal system. The development is well documented in Zambia by Mashamba (1997), who was probably the first to draw attention to the effects upon the construction sector of structural adjustment programmes and the withdrawal of the state from many spheres of economic activity. He describes a great expansion of the informal construction sector, particularly small enterprises, in building materials production. Similar developments in Kenya and Tanzania are described by Wells and Wall (2003), and in Tanzania by Jason (2005).

Another study in Tanzania investigated the relationships between registered (formal) and unregistered (informal) contractors in four urban settings (Mlinga and Wells, 2002). The research was founded on interviews with over 600 enterprises, one-third of which were registered with the Contractors Registration Board (CRB) and two-thirds unregistered. Strong linkages were found between the two groups of enterprises through sub-contracting, most of which takes place on a 'labour-only' basis. Small unregistered enterprises are actually supplying much of the labour required by the registered firms. Linkages were also found in terms of building materials and equipment, with more than 80 per cent of enterprises purchasing materials and supplies from unregistered enterprises. As well as supplying labour to the registered contractors, the unregistered enterprises found a market working directly for private house owners (generally building or repairing a part of a house on a labour-only basis) while the registered enterprises were working for the public sector clients and on large private-sector jobs. Hence, the enterprises worked in different sections of the market, although they sometimes competed for work in the private sector when public-sector demand was slack. Perhaps the most interesting finding from this research was the considerable overlap in terms of both size and activities between the two groups of enterprises, with some of the unregistered enterprises having greater access to resources and markets than some of the smaller registered firms. In fact, some of the more established unregistered contractors were found to possess the minimum requirements (in terms of capital and equipment) to register with the CRB but had chosen not to do so. The main reason given was that they could operate perfectly well without registration because they operate only in the private sector, which does not insist on using registered contractors (Mlinga and Wells, 2002).

Further insights into informal construction enterprises were gleaned from participatory action research over a two-year period among 38 groups of 'informal construction workers' in Dar es Salaam (Tanzania) (Jason, 2005). The members of the groups had come together spontaneously for the purpose of creating their own employment and income by selling their labour or by producing and/or selling construction materials (mostly crushed stone, timber or metal goods). They were referred to as 'informal' because none of the groups was registered or covered in any official statistics and very few had any capital. Most groups were structured and operated more like workers' co-operatives than privately owned enterprises. Their clients were individuals building their own houses, but job opportunities were found to be very limited, which meant workers staying for long periods without work and income. Even when they had work the pay was very little. The project experimented with a number of activities to try to improve the situation, including helping them to form an umbrella organisation to fight for official recognition, but the main problem remains the lack of demand due to the general poverty of the communities in which they live and work and for whom they provide a service (Jason, 2008).

Informal methods of delivery

The discussion above shows that the shift in construction employment from formal to informal in part reflects an increase in sub-contracting by formally registered enterprises and a new role for the informal enterprises as suppliers of labour. However, research in Kenya around the turn of the century (Wells, 2001) concluded that something more was happening. There are several ways of organising the delivery of a construction project. The conventional way in much of the developing world is to appoint professionals (architects or engineers) to plan and design the project. The formally approved plans are then put out to tender, with the construction firm which wins the tender engaged by the client, on a legally enforceable written contract, to construct the project for an agreed price. In Kenya, these formal procedures were still being adhered to by the public sector. However, the volume of demand for construction by the public sector was declining following the introduction of structural adjustment programmes, and an increasing number of private-sector clients were choosing to bypass formal procedures and contracts with registered enterprises, in favour of engaging directly the services of informal enterprises supplying labour. Building materials were purchased by the owner (opening up a market for informal enterprises which could only supply materials in small quantities) while building plans were provided by architects, engineers, planners or technicians, who might or might not be registered, for a small fee (Wells, 2001).

Wells (2001) concluded that a new production system is emerging in African cities, one that is characterised by a much closer nexus between building owners and building workers. These more informal arrangements offer many advantages to clients, notably lower costs and the opportunity to build in stages, as and when they have the funds. Costs are lower, not necessarily because of lower wages (in fact there is evidence that earnings may be higher in informal enterprises) but because the client absorbs more of the risks and responsibilities of construction. The fees paid for design and engineering are also at a more realistic level and there are no contractors' profits (Wells, 2001). Although this method of delivery does not offer the guarantees and safeguards of the formal system, it can be argued that, with increased sub-contracting and a decline in regulation, those safeguards may in any case be more illusory than real (Wells, 2001).

Informal or unconventional modes of housing provision in developing countries are, of course, nothing new. They flourish in developing countries primarily because of the inability of low-income groups to purchase professionally designed and constructed housing produced through the conventional process (Gilbert, 1990; Keivani and Werna, 2001). In fact, a search of the construction literature found that the most common use of the term 'informal sector' (Wells, 1998) was in relation to this particular type of activity or section of the market: traditional building in the rural areas, or what is seen as an extension of this activity into the urban areas in the form of individual low-cost house building, maintenance and repair – that part of the construction market that Hindle (1997) has called the 'domestic sector'.

For example, the World Bank (1984) refers to the informal construction sector in the context of low-cost house-building activities. This is seen as mostly self-help activity by self-employed and family labour, assisted by individual jobbers and builders, and small enough to escape legal regulation and statistical enumeration. The activity is labour intensive and a seed-bed of skills. Writing in similar vein, Chana (1981) argues that informal construction groups provide almost all construction in the rural areas of developing countries and 50 per cent of housing in the urban areas. Although the contribution of the 'informal sector' (understood as individuals and groups of workers) to the provision of housing for low-income groups is acknowledged in these studies, the housing produced is seen only as 'illegal temporary shelters' (Chana, 1981) and the activities are seen as marginal (World Bank, 1984).

This perception of the poor quality and 'marginality' of housing produced by the informal sector began to change over the years, as the inability of governments to provide affordable housing for the expanding urban population became increasingly apparent. The United Nations Centre for Human Settlements (UNCHS) (1986) emphasised the important contribution to urban housing of what it called the 'informal construction process'. The process involved unpaid family labour, self-employed builders, casual labour and 'informal sector' entrepreneurs and was responsible for a large percentage of the addition to the housing stock in many countries.

However, recent research has led to the recognition that this method of organising the building process, without the use of contractors or formal contracts, is no longer restricted to individual clients or to low-cost housing. In many countries, this is now the main method of organising the production process for all types of housing as well as many other types of building. The remarkable development in Nairobi (Kenya) of large, multi-storey buildings (comprising shops, hotels, flats and so on), constructed without the services of contractors, architects or engineers, has been well documented (Wells, 2001; Mitullah and Wachira, 2003). The same is true in Dar es Salaam (Tanzania), where the central business district of Kariakoo has been completely transformed with multi-storey private developments constructed largely by informal (unregistered) enterprises and with little input from the professionals.

It seems that, in many developing countries, only publicly funded buildings are now procured through conventional channels. According to Lizarralde and Root (2008), this is the reason for the relatively high cost and narrow reach of so-called 'low-cost' housing in South Africa and its failure to meet the real needs of the urban poor. The authors see the 'informal sector' (undefined but understood to embrace the informal delivery system) of housing as the only means to meet the needs of the poor, and the failure of public housing to embrace this method as responsible for the limited success of the programmes.

In recognition of the growing importance of the informal delivery method, the participants at the first meeting of Task Group 29 of the International Council for Research and Innovation in Building and Construction (CIB) ('Construction in Developing Countries') in Arusha, Tanzania, in 1998 concluded that 'We should also recognise that there is an Informal Construction System'. It defined this system as follows:

> a method of organising the production of buildings without the involvement of contractors. It is characterised by a close relationship between building owner and building workers. The building owner supplies materials and the building process is often incremental.
>
> (CIB Task Group 29, 1998, p. xii)

The 'informal construction system' refers to the organisation of the production process. In this context, the term 'informal' describes the relationships between the participants – the informality of the contracts – as much as the participants themselves. The informal system may involve participants from the formal sector (for example architects supplying plans) as well as the informal sector, although it is more likely to involve the latter.

Informal products

A number of authors (UNCHS, 1986; Syagga and Malombe, 1995; UNCHS and ILO, 1995) have used the term 'informal housing' or 'informal construction', or even 'informal sector housing', to refer to houses constructed without planning permission, irrespective of how they were delivered or of the status of those involved in the delivery. For example, UNCHS (1986) referred to 'informal housing', which was defined as all dwellings constructed without building permits. It noted that, in Egypt, 77 per cent of dwelling units built between 1966 and 1976 were not formally registered and therefore considered 'informal'. However, this did not necessarily mean that they were built by unregistered enterprises or using an informal delivery sector, as some dwellings without proper permits ('informal housing') were, in fact, found to have been constructed by registered contractors (UNCHS, 1986).

There are also many examples of projects with planning permission that have been constructed through informal methods. Many of the developments in central Nairobi described above actually have planning permission and building permits. It may be concluded that there is no necessary convergence between informality in the building delivery system or its participants and informality in terms of the regulation of the building product. However, it is still likely that buildings procured through the informal process will be constructed to standards and using methods and materials which the building regulations would not allow. This is particularly apparent in the case of housing for the urban poor.

The use of the term 'informal sector' to refer to the status of the products of construction activity can give rise to strange anomalies, such as informal extensions added to formal housing, leading to a situation in which 'part of the building is in the formal sector and part in the informal' (UNCHS and ILO, 1995, p. 14). Because of such anomalies (also because participants were more concerned with the process of construction rather than with planning procedures) the CIB Task Group 29, meeting in Arusha (Tanzania) in 1998, rejected the view of the 'informal sector' as construction activity that takes place outside planning or building regulations.

Nevertheless, there is evidence that, in many developing countries, such activity has increased and is likely to increase further in the future, because planning departments are overstretched and the poor are unable to afford (or the state to provide) good and safe housing for rapidly increasing urban populations.

The recognition that informal housing and informal settlements are here to stay has led to arguments that the failure of governments to enforce their own regulations is because the standards required are unrealistic, and to calls for the regulations to be relaxed. For example, Lizarralde and Root (2008) argue that state-sponsored housing policies and construction practices are inefficient because they systematically exclude the informal construction sector, which has been the only source of delivery of housing for the urban poor. Although the term 'informal' is not defined, close reading suggests that the authors are referring to *both* the involvement of informal enterprises and a different product: the use of building materials that are not normally permitted in urban housing. Similarly, Keivani and Werna (2001, p. 194) argue that 'unconventional or informal modes of housing provision in developing countries primarily exist due to the inability of the low income groups to purchase high quality and professionally designed housing produced through the conventional sector'. Therefore, the state has to tolerate and accommodate a certain degree of illegality and irregularity. They call, in particular, for the elimination of regulatory barriers to entry into the industry (regulation of enterprises) and removal of constraints to the development and use of local building materials (regulation of products).

Summary and conclusion

In this chapter, informality has been defined to mean the absence of regulation. Informal activity is that which lies outside the boundaries of state regulation. It has been shown that the construction industry is subject to many types of regulation, and four have been highlighted: (i) regulation of the terms and conditions of employment; (ii) regulation of the enterprises involved; (iii) regulation of the production process and the relationships between the participants; and (iv) regulation of the product itself. Informality in one area does not necessarily imply informality in the others. However, the four areas of regulation do often overlap, so that deregulation in one is closely associated with deregulation and an expansion of informality in another.

It is clear from the evidence assembled here that there has been a significant increase in informality in the construction industry in the past three decades across much of the developing world. However, this is not unique to construction. It is now widely recognised that in the majority of developing countries much economic activity has always fallen outside the scope of state regulation, and that the larger part of the economy in most developing countries is today outside the regulatory control of the state (Harriss-White, 2010). In the employment sphere, the OECD (2009) has recognised that informality is now the norm. Informal employment accounts for up to 90 per cent of the total employment in many countries in sub-Saharan Africa and South Asia and up to 65 per cent in the rest of Asia and Latin America. Even in the developed countries, 25–40 per cent of the employed population are in informal jobs, in either formal or informal enterprises. Thus, the trends which have been detected in the construction sector are only mirroring those affecting economies as a whole.

The factors driving this phenomenon are complex. They include the withdrawal of the state in the context of structural adjustment policies of the 1980s and 1990s, increased competition in world markets (including labour markets) associated with 'globalisation', rising population (especially the urban population), increasing poverty and the need for cheaper methods of working and products to serve the lower-income groups. Many of these forces will only intensify in coming years, so one cannot expect a return to formality any day soon.

Policy responses to the problem of informality (if indeed it is a problem) must take account of the fact that, whereas some (individuals and enterprises) may choose to operate informally, many others have no alternative. Attempting to 'formalise the informal' may be a sensible objective in some cases. However, an argument might also be made for 'informalising' the formal: reforming economic policies and institutions to include the informal economy as a legitimate part of the whole, a target of economic policies and incentives and a stakeholder in policy making and in rule-setting institutions (Chen, 2009). Where construction is concerned, this line of argument could be extended to include reforming policies that require clients (including the state) to procure buildings (including housing) only from registered enterprises using formal delivery systems and incorporating high-quality materials that put the products beyond the reach of the majority of the population. However, that will have to be the subject of another chapter.

References

Abdul-Aziz, Abdul-Rashid (2001) *Site Operatives in Malaysia: Examining the Foreign–Local Asymmetry*. Unpublished report for the International Labour Office.

Anand, Harjit (2001) *The Nature of the Informal Economy and Three Sectoral Studies*. Unpublished report for the International Labour Office/UNDP.

Assaad, Ragui (1993) Formal and informal institutions in the labour market, with applications to the construction industry in Egypt. *World Development*, 21 (6), 925–939.

Breman, J. (1996) *Footloose Labour: Working in India's Informal Sector*. Cambridge University Press: Cambridge.

Cattell, K. S. (1994) *Small Black Builders in South Africa: Problems and Prospects*, Research Paper Series Number 2. Department of Construction Economics and Management, University of Cape Town: Cape Town.

Chana, T. S. (1981) *The Informal Construction Sector*. Paper prepared for an ad hoc Expert Group Meeting on Development of the Indigenous Construction Sector, UNCHS, Nairobi.

Chen, Marty (2009) *The Informal is Normal*. Presentation at an OECD seminar, 8 April. http://siteresources.worldbank.org/INTEMPSHAGRO/Resources/Marty_Chen_OECD_Is_Informal_Normal.ppt

Cho, J. W. (2004) *Korean Construction Workers' Job Centre, Seoul, South Korea*. Paper for In-focus programme on socio-economic security, International Labour Office, Geneva.

CIB Task Group 29 (1998) TG29 on Construction in Developing Countries – Definitions. In *Managing Construction Industry Development in Developing Countries: Report on the First Meeting of the CIB Task Group 29*, CIB Publication No. 229. Arusha, Tanzania, 21–23 September. CIB: Amsterdam.

Connolly, Priscilla (2001) *Recent Trends in the Mexican Construction Industry and Outlook for the 21st Century: Its Image, Employment Prospects and Skill Requirements*. Unpublished report for the ILO.

English, Jane (2002) *The Construction Labour Force in South Africa: A Study of Informal Labour in the Western Cape*, Working Paper 188. Sectoral Activities Department, International Labour Office: Geneva.

Gilbert, A. G. (1990) The costs and benefits of illegality and irregularity in the supply of land. In P. Baross and J. van der Linden (eds) *The Transformation of Land Supply Systems in Third World Cities*. Avebury: Aldershot.

Hart, K. (1973) Informal income opportunities and urban employment in Ghana. *Journal of Modern African Studies*, 11 (1), 61–89.

Harriss-White, Barbara (2010) Work and well-being in informal economies: the regulative roles of institutions of identity and the state. *World Development*, 38 (2), 170–183.

Hindle, R. D. (1997) The structure of construction markets and their effect on the size and distribution of construction firms. In *Proceedings, First International Conference on Construction Industry Development, School of Building and Real Estate, National University of Singapore*.

ILO (1972) *Employment, Incomes and Equality: A Strategy for Increasing Productive Employment in Kenya*. International Labour Office: Geneva.

ILO (2001) *The Construction Industry in the 21st Century: Its Image, Employment Prospects and Skill Requirements*. International Labour Office: Geneva.

ILO (2002) *Decent Work and the Informal Economy*. International Labour Conference, 90th session, discussion paper.

ILO (2005) *Baseline Study of Labour Practices on Large Construction Sites in Tanzania*. Working paper no. 225. Sectoral Activities Department, International Labour Office: Geneva.

Jason, Arthur (2005) *Informal Construction Workers in Dar es Salaam, Tanzania*. Working paper no. 226. Sectoral Activities Department, International Labour Office: Geneva.

Jason, Arthur (2008) Organising informal workers in the urban economy: the case of the construction industry in Dar es Salaam, Tanzania. *Habitat International*, 32 (2), 192–202.

Keivani, Ramin and Werna, Edmundo (2001) Refocusing the housing debate in developing countries from a pluralist perspective. *Habitat International*, 25, 191–208.

Lizarralde, Gonzalo and Root, David (2008) The informal construction sector and the inefficiency of low cost housing markets. *Construction Management and Economics*, 26, 103–113.

van der Loop, Theo (1992) *Industrial Dynamics and Fragmented Labour Markets: Construction Firms and Labourers in India*. Netherlands Geographical Studies 139. Royal Dutch Geographical Society: Utrecht.

Lu, Youjie and Paul, W. Fox (2001) *The Construction Industry in the 21st Century: Its Image, Employment Prospects and Skill Requirements: Case Study from China*. Unpublished report for the ILO.

Mashamba, S. (1997) *The Construction Industry in Zambia: Opportunities and Constraints under Structural Adjustment Programmes and Enabling Shelter Strategy*. Unpublished PhD dissertation, University of Newcastle.

Mitullah, Winnie and Wachira, Isabella Njeri (2003) *Informal Labour in the Construction Industry in Kenya: A Case Study of Nairobi*. Working Paper 204. Sectoral Activities Department, International Labour Office: Geneva.

Mlinga, R. S. (1998) Significance and development of the informal construction sector in Tanzania. In *Proceedings, First Meeting of TG29, Managing Construction Industry Development in Developing Countries*, Arusha, Tanzania, 21–23 September.

Mlinga, R. S. and Wells, Jill (2002) Collaboration between formal and informal enterprises in the construction sector in Tanzania. *Habitat International*, 26 (2), 269–280.

Ngare, Jedidah Muthoni (1998) Problems facing the informal construction sector in Kenya. In *Proceedings, First Meeting of TG29, Managing Construction Industry Development in Developing Countries*, Arusha, Tanzania, 21–23 September.

OECD (2009) *Is Informal Normal? Towards More and Better Jobs*. Policy brief, Organisation for Economic Co-operation and Development Observer, March.

Pais, Jesim (2002) Casualisation of urban labour force: analysis of recent trends in manufacturing. *Economic and Political Weekly*, 16 February.

Standing, G. and Tokman, V. (1991) *Towards Structural Adjustment: Labour Market Issues in Structural Adjustment*. International Labour Organisation: Geneva.

Syagga, P. M. and Malombe, J. (1995) *Development of Informal Housing in Kenya: Case Studies of Kisumu and Nakuru Towns*. Housing and Building Research Institute, University of Nairobi: Nairobi.

UNCHS (1986) *Supporting the Informal Sector in Low-Income Settlements*. United National Centre for Human Settlements (now UN-HABITAT): Nairobi.

UNCHS and ILO (1995) *Shelter Provision and Employment Generation*. United Nations Centre for Human Settlements and International Labour Organisation: Geneva.

Wells, Jill (1998) The informal sector and the construction industry. In *Proceedings, First Meeting of TG29, Managing Construction Industry Development in Developing Countries*, Arusha, Tanzania, 21–23 September.

Wells, Jill (1999) The construction industry in low income countries: an agenda for research. In *Proceedings, 2nd International Conference on Construction Industry Development and First Conference of TG29 on Construction in Developing Countries*, University of Singapore.

Wells, Jill (2001) Construction and capital formation in less developed economies: unravelling the informal sector in an African city. *Construction Management and Economics*, 19, 267–274.

Wells, Jill (2007) Informality in the construction sector in developing countries. *Construction Management and Economics*, 25, 87–93.

Wells, Jill (2009) Dar es Salaam. In Roderick Lawrence and Edmundo Werna (eds) *Labour Conditions for Construction: Building Cities, Decent Work and the Role of Local Authorities*. Wiley-Blackwell: Chichester.

Wells, Jill and Wall, David (2003) The expansion of employment opportunities in the building construction sector in the context of structural adjustment: some evidence from Kenya and Tanzania. *Habitat International*, 27, 325–337.

World Bank (1984) *The Construction Industry: Issues and Strategies in Developing Countries*. World Bank: Washington, DC.

Yuson, Albert S. (2001) *The Philippines Construction Industry in the 21st Century: Is There a Globalisation of the Local Construction Industry?* Report for the International Labour Office and for the International Federation of Building and Wood Workers (now Building Workers International).

Part III

Construction development issues: key areas

8 Construction technology development and innovation

Emilia van Egmond

Introduction

This chapter outlines theoretical views and empirical findings regarding technology development and innovation in construction in developing countries. The main issue is: how to *manage technology development and innovation in construction* in developing countries, particularly in the context of the global effort towards attaining the Millennium Development Goals (MDGs) by 2015.

The discussion draws from, and builds on, the works of the main authors on evolutionary economics, technological innovation studies (Nelson and Winter, 1982; Hodgson, 1993; Metcalfe, 1995; Cohen *et al.*, 1996; Dosi *et al.*, 2000a,b) and organisational studies (March, 1988; Cyert and March, 1992; March and Simon, 1993).

The chapter ends with a consideration of policy implications for technology development and innovation in construction towards sustainable development in developing countries.

Background

Construction problems and challenges

The construction industry has a large impact on the development of national economies and on the quality of life of peoples (Ofori, 2000; Egmond-deWilde de Ligny and Erkelens, 2008). Yet the construction industry everywhere faces problems and challenges. These include globalisation, which has increased competition, and urbanisation, which has resulted in a large number of people living at the bottom of the pyramid in shanty towns, especially in the developing countries (Prahalad, 2004).

The construction industry accounts for some 40 per cent of the world's annual consumption of natural resources (Roodman and Lenssen, 1995), and contributes significantly to greenhouse gas emissions as well as the generation of waste in the form of construction and demolition waste (Macozoma, 2002). It is estimated that, currently, 80 per cent of the energy consumed during the whole life of a building is used during its operation (or service life), whilst the remaining 20 per cent is used for the production of materials and in construction and demolition works. The industry's emission from construction processes constitutes around 5 per cent of global anthropogenic carbon dioxide emissions (IEA, 2009).

In developing countries, rapid economic development and industrialisation have taken a heavy toll on the environment. Case studies indicate that the greenhouse gas emissions from developing countries are likely to surpass those from more developed countries before 2050. Although *Agenda 21 for Sustainable Construction in Developing Countries* was launched in 2002, sustainable building is still a relatively new concept in developing countries. Faced with survival issues, the emphasis in terms of policy has been on poverty alleviation and, in a few cases, the provision of housing to lower-income households, with little concern for long-term environmental impacts of building programmes.

Technology development and innovation for sustainable development?

Technology development and innovations (TD&I) have had a major impact on world development. Research in manufacturing has confirmed that enterprises which are able to use innovation to improve their processes or to differentiate their products or services perform better than their competitors in terms of market share and profitability. However, innovations do not always result in the expected effects and are not always appropriate or widely accepted and adopted in the relevant industries. There is now growing public concern and scepticism about the nature of innovation and technologies, and their actual impact on sustainable development.

The situation outlined above makes the pursuit of improvements in construction in developing countries particularly challenging, and calls for new *modi operandi* in the construction industry. The impact of construction on the environment is closely linked to the development and application of technologies, knowledge and innovation in design and construction. New technologies and materials, and new designs which ensure better resource consumption can substantially enhance the performance of the construction industry whilst they take into consideration the whole life cycle and end of life of buildings as well as the quality and safety of the built environment while ensuring customer and user satisfaction.

The above discussion reveals the relevance of investigating issues of TD&I in the construction industry in developing countries.

Core concepts

The core concepts are now discussed in order to provide a common understanding of the meanings of the key terms used in this chapter and, thereby, facilitate the discussion on TD&I in construction in developing countries.

Technology

'Technology' is a conjunction of two Greek words which were already in use in antiquity. The first word is *techne*, which means the skill, *know-how* or craft needed to make something; and the second word is *logos*, which refers to the rationality (*ratio*) of nature, but nowadays is usually interpreted as knowledge, which, in this context, stands for *know-why, -when and -by whom*.

Technology relates to production processes, in which the production inputs (natural resources and/or intermediate products) are transformed into the desired production output (product or service) by means of technology. In such a *production system perspective*, a distinction is made between product and process technology.

Product technology is the knowledge and skills *embodied in the output* of the production process required by the society in which it is used, and which uses it. These knowledge and skills become apparent in the attributes of product technologies. *Process technology* is the knowledge and skills embodied in the 'transformer' applied to transform the inputs in production processes into the required goods and services (Stewart, 1979; Stewart and James, 1982; Egmond, 1999). Thus, *construction product technology* refers to the knowledge and skills embodied in the building, the building system, elements and parts. This becomes apparent when one considers the complex nature of the technological attributes of the building, the building system, elements and parts, as indicated in Figure 8.1.

Construction process technologies are composed of a combination of inextricably inter-related components used during the realisation of a building project. They comprise the technoware (equipment, tools and machines), humanware (manpower), infoware (documented facts) and orgaware (organisational framework). This is depicted in Figure 8.2.

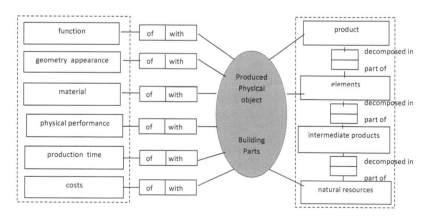

Figure 8.1 Building product technology components and attributes. Source: Egmond (1999).

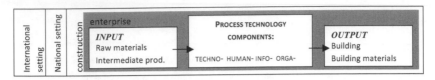

Figure 8.2 Process technologies. Source: Egmond (1999).

Technological development

Technological development refers to changes in the total range and charac-teristics of the existing stock of technologies that is used and produced in an industry in a country. Therefore, technological development refers to changes in the skills and knowledge embodied in:

*   the stock of products and services, which becomes apparent in changes in the product attributes; or
*   the stock of process technologies (transformers); thus, changes in the attributes of technology components – techno-, human-, info- and orga-ware – used in the transformation process.

Changes in product technologies often occur together with those in the pro-duction process and in the organisation (Dosi *et al.*, 2005).

The concept of technological development is often mixed up with other concepts such as technology changes, inventions or innovations. Thus, it is important to note that *technology change* simply refers to a change in the attributes of the product or process technologies.

Innovation

Innovation refers to the total cycle from *invention*, that is the development of new knowledge and technologies (products and production processes), to the *diffusion, adoption and application* of the new knowledge and technologies.

Product innovations are seen as successfully developed, introduced, dif-fused and used product technologies. Similarly, *process innovations* can be defined as successfully developed, introduced, diffused and used production process technologies (Rosegger, 1996). Innovation involves various activi-ties and interventions by actors in an innovation system such as the search for existing new products and processes; the development and generation of technological products and processes; the execution of fundamental research; the transfer and implementation of the results in the market; and the adaptation and improvement of (new) products and processes after their implementation, utilisation and evaluation.

Innovation system

An *innovation system* is defined as a network of inter-related individuals, organisations and enterprises that share a common field of knowledge and interest regarding innovation (Malerba, 1999). The actors can be found at international, national, sector, company or project levels.

Diffusion

Diffusion is the rate at which (new) technologies are *adopted* and *applied* in companies or institutions, or by people, causing the technologies to spread in society. This spread is accomplished through human interactions and communications among members of an innovation system (Rogers, 1995).

Technology flows

The essence of *technology flows* is that technologies, either material or immaterial (i.e. forms of knowledge and information), shift from a source to a recipient. New technologies and knowledge are not necessarily generated by in-house research and development (R&D) but also can be acquired through (i) simple acquisition of the technology and knowledge from elsewhere in the country or abroad or (ii) acquisition of the parts of the needed technology and knowledge that are missing and combination of these with the further in-house development and production process. Generally, technology flow involves the sharing of knowledge and technologies (*learning*) among actors in an innovation system such as R&D laboratories, industry, universities, state and local governments, and third-party intermediaries. International technology flow is defined as 'Any form of transmittal of technology over space, across the borders, between institutions or within the same institution' (Dunning, 1981).

Technology flows are seen as important mechanisms to foster learning, capability building and innovation. They also stimulate technological development and the formation and growth of entrepreneurial start-ups, thereby creating new jobs and providing the basis for economic growth and development. There are many aspects that are related to the mechanisms of technology flows (and these are also, in turn, inter-related to one another). These aspects include types of technology and knowledge, tacitness, asset specificity, prior experience, complexity, partners and partner protectiveness, control, cultural distance and organisational distance. Many of these aspects are discussed in relevant parts of this chapter.

Modes of technology flows

Modes of technology flows refers to *mechanisms* or *channels* through which technologies and knowledge flow from source to recipient. Numerous approaches have been used to classify technology flows, by using criteria such as vertical and horizontal; formal (market mediated) and informal (non-market mediated); embodied and disembodied. Other categorisations consider the degree of packaging; whether the flow is direct or indirect; the form of the institution (intra-firm, integration/investment, pure market, sales and intermediate forms). The various categorisations illuminate the different aspects of the technology flow processes. Radosevic (1999) distinguished two main ways by which technology flows take place: (i) conventional channels based on the type of contractual agreement such as foreign direct investment (FDI), licensing, joint ventures, franchising, marketing contracts, technical services contracts, turnkey contracts and international sub-contracting; and (ii) non-conventional channels such as reverse engineering and reverse brain-drain.

Theoretical views on technology development and innovation

Emergence of theoretical views on TD&I

Several paradigms and models have been developed to analyse and describe the TD&I processes. Capital accumulation (in terms of advancement of technologies and innovation) was considered to be the engine for improved competitiveness of enterprises and economic growth in Western countries in the traditional growth theories.

In neo-classical economic theories, the basic assumption was that a motivated *profit-maximising, cost-minimising and output-maximising* entrepreneur has to make choices among various production technologies in an environment of perfect competition (Rosenberg, 1976). The difficulties encountered in addressing issues of technological change are due to the particular nature of the technology.

A key issue that was investigated by means of the international production theory in an attempt to find an explanation for the differences in industry performance and economic growth was the role of trade in technology and knowledge flows, and the effects on production performance, enterprise development and economic growth. Dunning (1981) suggested that ownership, internalisation and locational advantages may favour exports, FDI and other forms of contractual technology transfer (such as licensing, technical assistance agreements and management contracts). The imperfect international technology market with monopolistic and oligopolistic tendencies, characterised by a dominance of the industrialised countries and their transnational corporations as suppliers of technology, was to be blamed for unsuccessful technology transfers and thus stagnation in economic growth. Policy responses to these findings included the regulation of technology imports, mainly in an attempt to increase the bargaining position of recipient firms and to diminish the social costs of importing technology. In some cases, these regulations on technology and knowledge flows were implemented jointly with policies aimed at import substitution (Dunning, 1981; Stewart and James, 1982).

Based on TD&I cases in manufacturing, extensions of the neo-classical economic theory (such as the new growth theory), as well as alternative approaches, have emerged over the past decade. These theoretical approaches attempt to explain and interpret the relationship between TD&I, knowledge, enterprise performance and socio-economic development. The theoretical basis for these views is in *evolutionary economics* (Nelson and Winter, 1982; Lall 1992; Hodgson, 1993; Metcalfe, 1995; Malherba, 1999; Nelson and Pack, 1999; Stiglitz, 1999; Dosi and Winter, 2002; Nelson and Winter, 2002), *technological innovation studies* (Nelson and Winter, 1982; Cohen *et al.*, 1996; Teece *et al.*, 1997; Dosi *et al.*, 2000a,b), as well as *organisational and management studies* (Cyert and March, 1992; March and Simon, 1993; March, 1988; Tidd *et al.*, 2005). Capability building through knowledge accumulation and learning in innovation systems is often put at the centre of TD&I processes.

Technology development and innovation cycles

Product life cycle theories emerged on the basis of evidence from manufacturing which indicated that newly developed products rise early in their commercial life and decline as soon as competitors start 'imitating' the products (Abernathy and Utterback, 1978). Process innovation takes place when the enterprise prepares to make the item on a larger scale. To keep its competitive position, an enterprise starts a new cycle based on the existing one (incremental technology development), and incorporates the new one developed from R&D or through acquisition from elsewhere. The rate of diffusion is related to the age of the technology (see Figure 8.3).

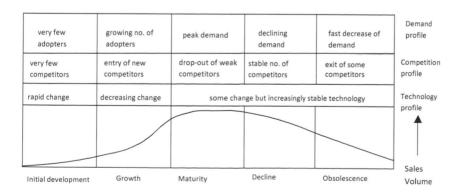

Figure 8.3 Product life cycle. Source: Egmond (2006), based on Rogers (1995).

Tidd and colleagues (2005) delineate the TD&I process into three phases:

- Phase 1 – *Search,* detect signals in the environment about potential for change.
- Phase 2 – *Select* and acquire knowledge and skills from among various market and technological opportunities that match the overall strategy of the firm.
- Phase 3 – *Implement,* by turning potential ideas and knowledge into newly prepared market-ready products or services, process changes and so on.

It is only when the target market makes the decision to adopt the innovation that the whole innovation process is completed.

In reality, TD&I is more complex and not 'chain/sequential'; the different phases overlap. Thus, innovation is a 'parallel' or 'integrated' process with upstream (with suppliers) and downstream linkages (with users), and alliances with other firms (Kline and Rosenberg, 1986). Figure 8.4 illustrates interactions in the process of technology development and innovation.

Innovation system and actor network

As opposed to neo-classical thinking, it is assumed in evolutionary thinking that innovations are not an outcome of *ex ante* economic reasoning, but of processes of evolution of social needs and economic competition, technology and the status of knowledge, government intervention and the like, analogous to biological evolution. It takes place in an *innovation system* (Nelson and Winter, 1982; Hodgson, 1993; Metcalfe, 1995). Innovation systems encompass 'interconnected institutions that create, store and transfer the knowledge, skills and artefacts which define new technologies' (Metcalfe, 1995). This also matches the views of the Social Construction of Technology (SCOT) theory, an actor-centred theory which suggests that innovation processes can produce different outcomes depending on the social circumstances

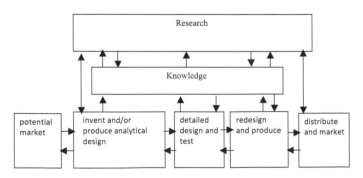

Figure 8.4 Interactions in the process of technology development and innovation.

in which they take place (Bijker *et al.*, 1987). Similarly, Rogers (1995) and Wenger (1998) suggest that TD&I is accomplished through human interactions and communications between members of a *community of practice*, which gives rise to particular trajectories that are sustained by industrial interests vested in it, assumptions about user needs and the costs of making a system change (March, 1991). Malerba (1999) adds clients to the *actor network* in innovation systems. The actor network is illustrated in Figure 8.5.

Technological capability

A notion that emerged from the studies on TD&I during the last decade of the twentieth century suggests that the core element for TD&I in an industry is *technological capability*. This is the total stock of resources, including knowledge and skills, that can be found at different levels in innovation systems and which can be categorised into three functional classes: *investment capability* (which is needed to identify, hold negotiations on, purchase and manage suitable and feasible technologies and knowledge); *production capability*; and *innovation capability* (that needed to create, diffuse, adopt and implement new knowledge and technological solutions) (Lall, 1992).

Studies have found that the capability to innovate and quickly adopt new technologies is strongly correlated with successful production and trade performance (Dosi *et al.*, 1990).

Knowledge and learning

More recently, *knowledge* has been recognised as a key component of the capabilities for TD&I competitiveness in the new world economy (Stiglitz, 1999). It is argued that new technologies are based on both *know-how* (the

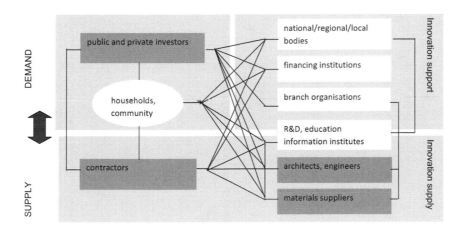

Figure 8.5 Outline of an actor network in the construction innovation system.

craftsmanship and skills required to carry out a job), and knowledge (*know-why*), and if the new technology is not supported by specific knowledge then it will collapse sooner or later. Thus, the role of knowledge is outstanding, whereas knowledge and techniques are needed in combination. The evolution in theoretical approaches has meant that greater attention is paid to the processes of the mastery of technology and knowledge, thus *learning* or upgrading of *technological know-how, knowledge and skills formation*. Learning processes have become the centre of analysis and debate especially around issues of upgrading in developing countries (Lee and Tunzelmann, 2004; Lundvall *et al.*, 2006).

The process of embodying knowledge in people (*learning*), products and processes (*application*) is costly in time and resources since knowledge is, to varying degrees, *tacit*: embedded in norms, rules, principles, routines, techniques or skills, which are learnt and which are (after some time) unconsciously applied in doing, acting or creating (Polanyi, 1967; Nelson and Winter, 1982, Stiglitz, 1999). The 'translation' of tacit knowledge into fully explicit codified statements (information) is barely possible, since it is hard to explicitly delineate or codify certain procedures or behavioural patterns. This process involves personal interactions, observation and practical experience in specific contexts (Callon, 1995). However, the non-codified or tacit knowledge stock, as well as the ways in which it is managed, is considered to be particularly important for a company's competitiveness.

Learning (technological capability building) involves various mechanisms by which knowledge and skills flow from a source to a recipient (Freeman, 1982; Bell, 1984; Pavitt, 1999). Learning takes place by doing, using, reverse engineering, imitation, searching (formalised activities such as R&D) or adoption. It can also occur by interacting with upstream or downstream sources of knowledge (suppliers or users) or in co-operation with other firms; by absorption of new developments in science and technology; and by finding out about what and how competitors and other firms within the industry or in other fields are doing (intra- and inter-industry spillovers). Learning mechanisms often overlap. For example, learning by searching may take place jointly with learning from advances in science and technology. In the developing countries, learning is partly based on (i) production experience; (ii) a deliberate process of investing in knowledge creation (formal training); and (iii) importing ready-made knowledge from industrialised countries (Lall 1992, 2003).

Technology flows as learning mechanisms

Pack (1999) distinguished two main forms of these mechanisms:

1 *Free knowledge and technology flows*. In this mode, knowledge is obtained (i) as a *byproduct* of market transactions or (ii) through *maintaining and reverse engineering* of imported new equipment or locally

produced equipment (see Figure 8.6). However, the potential gains from these free flows depend on the negotiation on the agreement and the absorption capabilities of the recipient.

2 *Technology and knowledge flows through market transactions.* Most technology and knowledge flows take place through 'commercial contracts' (enterprise-to-enterprise agreements) such as (i) trade contracts (on exports of capital goods and intermediate goods), consultancy and turnkey projects; (ii) exchange of know-how contracts (R&D collaboration, sales of patents, licences, trademarks, management and consultancy agreements); and (iii) private investments such as FDI, wholly owned subsidiaries, joint ventures and other forms of collaboration between firms (Katz, 1987; Lall, 1987; Pack, 1999). Figure 8.7 illustrates the flows of technology and knowledge.

Commercial transactions contribute to capability building but they also require capabilities which are not always available in the firms (Abbott, 1985; Lall, 1985, 1987, 1992; Katz, 1987; Carillo, 1994; Kumaraswamy, 1995; Pack, 1999; Egmond and Kumaraswamy, 2003). A real technology and knowledge flow can be considered to have taken place only when the recipient is able to use and adapt technology components on a self-reliant and sustainable basis, which may imply that, in the end, the recipient might become a competitor of the supplier of the technology (Ssegawa *et al.*, 2001).

Control on technology flows

Technology and knowledge components – such as *machines and equipment*; *documents* (such as blueprints, designs of products, manufacturing instructions, process flow charts and equipment specifications); and *personnel* in charge of specific operations or to train the labour in the recipient enterprise – are offered in commercial transactions as *part of a contractual 'package' deal* with attached conditions. The conditions determine whether components are all *included* in a package deal (such as a licence agreement including instructions, documentation or assistance on the use of the patented technology) or whether components are missing and have to be acquired separately from another source. Acquiring a complete ready-to-use package means that risks can be minimal and quality is secured; this is beneficial for the recipient.

Figure 8.8 shows the degree of control and cost of various forms of

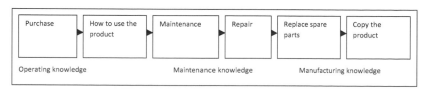

Figure 8.6 Free technology flows.

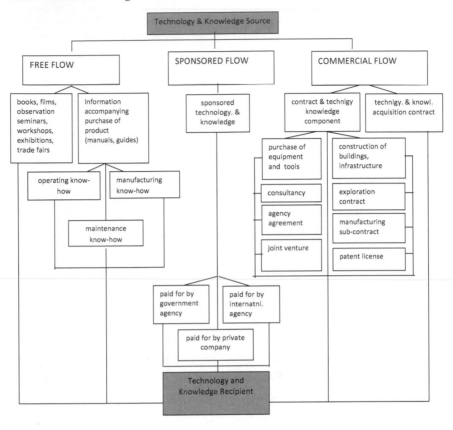

Figure 8.7 Flows of technology and knowledge.

transactions in technology and knowledge. Conditions are a form of *control* imposed by the supplying as well as by the recipient firms, and they vary from limited ones to a full and complex set of restrictions or controls, such as on the use of the technology, markets for the end-products of the technologies, and suppliers of necessary raw materials or spare parts (Stewart, 1979; Dunning, 1981). Conditions can enable the supplying firm to establish a quasi-monopoly over the components of previously non-monopolised knowledge. From this perspective, knowledge can be classified as (i) *public non-excludable* (this means that no one can be effectively excluded from using the good) *and non-rivalrous good* (knowledge that is generally non-monopolised, partially uncodified, freely and widely available, and unprotected) and (ii) *private good*, which is firm-specific monopolised knowledge, often protected by patents, and held to be trade secrets (Stiglitz, 1999). Only immaterial ('disembodied') knowledge, information, ideas, concepts and other abstract objects of thought are purely non-rivalrous.

Packaging can have many negative effects for the recipients and is associated with discontinuities in knowledge flows within, and between,

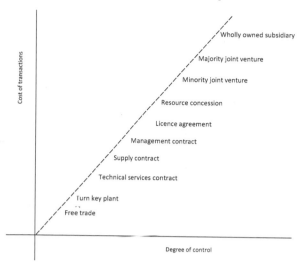

Figure 8.8 Degree of control and cost of transactions.

organisations, and restraints on organisational learning as well as on the development of an organisational memory (Stiglitz, 1999; Dubois and Gadde, 2002). Unfavourable experiences with package deals boosted the application of the unpackaging concept, which means breaking down a technology into several components and purchasing every component separately, if possible from different suppliers. This may bring down costs and limit the conditions of the transactions. It will also open opportunities for recipients to increase their technological capabilities (learning by doing), maintain control over technology flows and decrease dependence on the supplier. However, unpackaging turned out to be rather expensive, time-consuming, vulnerable to failures, especially in those cases where governmental support was not available, and usually involved 'old' technologies. Success highly depends on the technological capability of the recipient firm. Moreover, the market does not always allow unpackaged deals, for example when the technologies form the core elements which the supplying firm needs to maintain its competitive position.

Public-sector organisations tend to make use of package deals (for example, turnkey projects for public goods such as a town hall), generally through tendering, in order to obtain adequate, affordable technology under reasonable terms. Other forms of package deals also take place (examples of these are public–private partnerships for various types of projects).

Routines and technological regime

From the above discussion, it can be concluded that prevailing practices determine, to a large extent, how and whether learning through technology and knowledge flows takes place. Correspondingly the innovation theories indicate that *routines* and *technological regimes* in innovation systems are important factors that determine TD&I (Nelson and Winter, 1982). The argument is that, in production and decision making, firms rely on routines, which may explain the competitiveness and growth rates of firms and industries (Teece *et al.*, 1997). Routines are standard procedures or regularly followed courses of action that are built up over time and they exist in different social environments such as industry, science, politics and markets (Geels, 2004).

Routines are based on *technological regimes*, which are coherent collections of rules, regulations, collective memories (experience, and explicit and tacit knowledge), conventions, consensual expectations, assumptions or thinking shared by stakeholders in an innovation system. These determine their behaviour, characterise professional practice and guide innovations (Nelson and Winter, 1982; Coriat and Dosi, 1998; Dosi *et al.*, 2000c). Technological regimes are sector-specific and, to a large extent, they are typical for individual firms. This makes it difficult for other firms to imitate them (Pavitt, 1984; Patel and Pavitt, 1997; Breschi *et al.*, 2000; Marsili, 2001). The determining impact of a technological regime is shown by the evidence of the emergence of new technologies which are successful owing to their fit in the prevailing regime. Similarly, Malerba (2002) indicated that the existence of an innovation system, in which there are strong inter-relationships among the stakeholders, who share a common field of knowledge and interest (a technological regime), will facilitate TD&I.

Strategic niche management

A technological niche combines with other existing technologies and knowledge in a particular domain to form a *technological network*, within an innovation system (the organisational environment in which innovation takes place). Any of the technologies exists, or is developed, alongside other technologies, whilst serving a particular limited domain of applications. Nodes of the technological network are technologies, each of which has specific embodied knowledge and skills, which provides the basis upon which another related technology is created. The position of a technology in the network is its niche in the domain.

The new technology may be promising but under-supplied in the market because of high uncertainty or high up-front costs, or because its technical and social benefits are not sufficiently valued in the market. The properties of the technology determine whether it fits the *technological regime* in the innovation system; for example, the expected rate of return on investment determines its market potential. Different actors in the innovation system (policy makers, the regulatory agency, local authorities, a citizens group, private companies, an industry organisation or a special interest group) may act as the *niche manager* to take actions to promote, and to bring about a change in the regime, and stimulate the diffusion of a new technology (Schot and Rip, 1996).

Innovation intervention and innovation promotion agents

A number of *innovation intervention and promotion agents* can be distinguished; these agents may intervene in all the processes of the development, diffusion and utilisation of new technologies (see Table 8.1).

Contextual factors

Innovation system theories stress that innovation is the result of interactive learning (technology and knowledge exchange) taking place among individuals and organisations located in a specific national, regional or sectoral system (Lundvall *et al.*, 2006). This leads to the notion that an innovation system is basically composed of three building blocks: (i) a network of actors interacting in a specific economic area; (ii) a set of competitive technologies, products, processes and services which fulfil specific needs in markets; and (iii) a particular technological regime (Carlsson and Stankiewicz, 1991; Breschi and Malerba, 1997). It is also clear that the innovation system is embedded in a wider macro-level social context.

The technological regime amongst the actors in the innovation system supports TD&I and continuous learning to increase knowledge, skills and

Table 8.1 Technology intervention and promotion agents

Technology intervention and promotion agent	Form and field of technology promotion	Resulting in
National government	Formulation and implementation of technology policies	Favourable climate by means of subsidies, incentives, tax holidays and so on to enhance TD&I
Educational and training institutions; consultancy companies	Training and education	Human resources development (humanware)
Documentation and information centres, libraries, statistical organisations, patent and registration offices, museums; consultancy companies	Storing and lending of documented facts Training and informing	Improved knowledge and insight (infoware + humanware)
Testing, certification and standardisation laboratories	Upgrading and standardisation of technologies	Improved and standardised technoware
R&D and financing organisations; design, engineering and management consultants	Assistance for setting up of management and organisational structures	Orgaware

Source: Egmond (2006).

capabilities. The actor network is about linkages both with other firms and with other institutions and public organisations in the innovation system, which are increasingly important as sources of knowledge (these include universities and research institutes) (Hippel, 1988; Lundvall, 1993; Lall and Pietrobelli, 2005; Lundvall *et al.*, 2006).

At the national and international levels, the *technology climate* is considered to be important as a framework which facilitates the creation of a favourable climate for TD&I, competitive production and societal development (ESCAP 1989, 1994). The *technology climate* refers to the national and international setting (in particular, the local political, legal and economic situation as well as the features of the global market) in which TD&I occurs.

Theoretical framework

The discussion so far has highlighted the need for continuous innovation to meet the ongoing changing need for sustainable development in countries, thereby indicating that various parameters such as contextual, innovation systems, organisational and TD&I performance properties must be considered simultaneously and holistically.

The national and international setting frame the opportunities for, and challenges facing, TD&I. The *major indicators* of the *national setting* – classified in terms of social system characteristics and geographic physical characteristics of the natural and man-made environment – that are generally used include (i) the level of socio-economic development; (ii) the stock of science and technology personnel and R&D expenditure; (iii) the science and technology scene in the production system; (iv) the science and technology in academic institutions; (v) the efforts and advances in selected areas of specialisation; (vi) the macro-level commitment to science and technology for development; and (vii) the status of the physical infrastructure and support facilities. Indicators for the *international setting* are the features of the international technology market. However, whether or not developing countries can take advantage of the globalisation process, particularly to increase their technological capabilities through technology flows from the ever growing rich pool of global knowledge, depends on their relative positions in the global technology market.

Figure 8.9 shows the theoretical framework for TD&I flows.

Construction technology and innovation in developing countries

Construction technologies and innovation

Historically, building construction has heavily depended on the craftsmanship of the labour force and has been rather labour intensive. Over the years, construction process technologies changed, and the processes

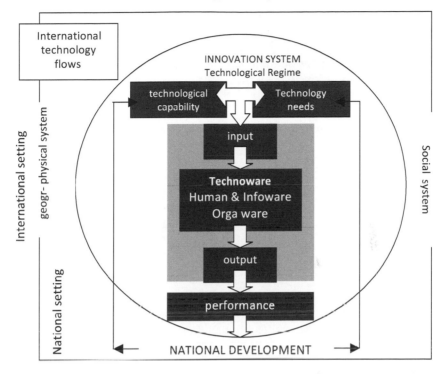

Figure 8.9 Theoretical framework for TD&I flows.

developed towards a more industrialised way of production in many countries. Technological development was enhanced by a rapid increase in the demand for construction owing to population growth and industrialisation. The increasing mechanisation, rationalisation, systematisation, standardisation, automation and flexibility of construction implied (a) the movement of building construction activities from the site to the factory and (b) the standardisation of industrially produced building components and construction technologies. Thus, construction has become an assembling process of components which facilitates the on-site construction activities and thereby can result in a better price–performance ratio as well as in the reduction of waste generation and of energy and material inputs. The building products have also changed.

What has happened in the construction industry in the course of time is that combinations of innovative solutions based on accumulated advances in technology and knowledge were adopted in attempts to move from the largely craft-based ways of construction to a systematic process whereby resources are utilised efficiently. In fact, a *convergence of technologies and knowledge* from different countries, sectors and disciplines has taken place (Egmond, 2009).

There are many examples of innovation in construction, such as the development, distribution and application of new building materials (for example, new composite materials); new design and engineering solutions (such as new piling and foundation systems); new equipment and tools; information and communications technology (ICT) for design (three-dimensional computer-aided design [CAD] and building information modelling [BIM] tools); construction and building automation; and new business models (for example, new forms of collaboration between project partners such as those used for public–private partnerships, and integrated supply chain management).

The innovations in the construction industry are mostly incremental. At present, construction enterprises are increasingly being forced to optimise their price–quality ratio in order to enhance their competitiveness. Thus, the firms need to innovate (Ofori, 1990; Ofori and Chan, 2000).

Sources of new technologies

New technologies, design and engineering solutions, among others, are developed by firms in the construction industry as well as by research and educational institutes and continuously on each construction project on the work site. Most of these new technologies are not recorded, with the result that the construction industry has an image of low innovation. Officially, many inventions are considered to have originated from manufacturing industries related to construction and making profit from their application, given scale economies (Miozzo and Dewick, 2004; Manley, 2001).

Even in the developed countries, only a few large construction firms, which have the resources or incentives, carry out R&D (Walker *et al.*, 2003). Customers may act as stimulators of innovation. In some cases, client demands may be the only driving force behind innovation in project-based firms.

State of the art of technology development and innovation in construction

The construction industry has made progress in TD&I. However, the industry is still generally traditional and lags behind most of the other sectors of the economy, especially manufacturing (Moavenzadeh, 1978; Ofori, 1990; Nam and Tatum, 1997).

There are incentives for TD&I. For example, the uniqueness of each construction project provides impetus for technology development since construction operations are rather transparent, are easy to copy, and provide opportunities for job-site training, which means that, in construction, there is extensive scope for technology flows from other projects and industries (Tatum, 1991). Yet there are various inherent difficulties, barriers and restrictions in construction that have hampered and complicated TD&I

and, particularly, the diffusion of new technologies (Abbott, 1985; Carillo, 1994; Ofori, 1996; Slaughter, 1998, 2000; Egmond, 1999, 2004, 2006, 2009; Dulaimi *et al.*, 2002; Kumaraswamy and Shrestha, 2002; Miozzo and Dewick, 2002; Egmond, and Kumaraswamy, 2003; Blayse and Manley, 2004; Egmond and Oostra, 2008). For example, many authors highlight deficits in the innovation system and the characteristics of the technological regime and routines (professional practices) in construction that tend to be detrimental to the fast development, acceptance and adoption of new technologies.

In developing countries, innovation is particularly difficult. The majority of these countries are marginal producers of new technologies and knowledge. Local innovations (if any) are based on traditional knowledge. Innovations of global significance are less frequent, and tend to be more common in emerging countries than in low-income ones. However, some of the emerging countries have gained advanced scientific and technological knowledge which can be exploited for the introduction of radical new innovations. These include those concerning solar energy systems (China) and new building materials that are cost-effective and eco-friendly, such as the utilisation of industrial and agricultural waste, for example the production of building blocks with fly-ash (India). In Latin America, R&D takes place on renewable materials such as straw panels and bamboo and innovative recycling of construction waste. However, most of these technologies are still in the experimental or demonstration stage and have not yet been widely diffused and used in construction (Egmond and Erkelens, 2007).

The bulk of innovations in developing countries entail new technologies and knowledge which have been taken up from somewhere else, adapted to fit the local requirements and conditions. After all, there is a lot of knowledge and technology available worldwide; developing countries can draw on these and develop the adopted technologies further by adapting them to their needs and capabilities. They can then use these as a source of competitiveness at less cost and in a shorter time than local R&D. However, this approach to the acquisition of novel technologies faces some barriers, besides deficiencies in the innovation system, and this complicates the technology and knowledge flows into many developing countries (Radosevic, 1999).

For developing countries there is another important potential source of TD&I in the form of the 30 per cent of the R&D professionals living in the member countries of the Organisation for Economic Co-operation and Development (OECD) who originally came from developing countries. They offer a significant opportunity to upgrade the knowledge bases in their home countries. Aubert (2004) indicates that there are examples of R&D professionals returning to their home countries (such as China, India, Israel and Mexico) who have brought into these countries significant volumes of human-embodied technologies. Alternatively, technology flows and capability building by expatriates take place through (i) the flow of FDI by them into their home countries (such as China); (ii) their acting as innovation brokers for their home countries (for example, India); and (iii) sending remittances (which is the largest source of external funding in many developing countries after FDI).

Developing countries in the global technology market

During the last decades of the twentieth century, globalisation intensified market linkages, increased trade and technology flows, reduced government intervention, and fostered production integration and networking. This enhanced economic performance in many parts of the world was expected to enable the developing countries to benefit from the technology and knowledge resources on the international market. Yet the participation of developing countries has been limited owing to the particularities of the global market.

The character of the global technology market can be outlined as follows: (i) it is extremely imperfect and has a diffused character; (ii) there are monopolistic and oligopolistic tendencies and asymmetrical power relationships between technology recipients and suppliers; (iii) transactions take place between a small number of countries, whereas transnational corporations take dominant positions, and developing countries rather weak and dependent ones (Stewart, 1979; Dunning, 1981; Lall, 2003).

The imperfect character of the market is a consequence of government interventions in the form of trade regulations, subsidies, taxes and legal systems of protection (including patents). An appropriate overall regulatory framework, a predictable technology-supporting policy regime, and good governance with few restrictions and regulations regarding intellectual property rights, as well as a proper balance between laws and regulations, foster FDI and innovation. In some countries, administrative procedures are superior to laws; this means that complicated procedures form a constraint on technology flows.

Monopoly or oligopoly positions give technology suppliers a bargaining advantage. Besides legal protection systems, the capabilities and bargaining position of technology recipients considerably influences the price to be paid for technologies (Stewart, 1979; Lall, 2003). The price might be determined by crude bargaining prior to competition. The situation is different in the construction industry, where the technologies used on most projects are still rather traditional. Here, the recipient of the technology has the potential to have stronger bargaining power. The cost and quality of the technology are determined by the searching, selection and negotiation capability of the recipient. The (quasi-)monopolistic position which the technologically advanced position of the large global companies gives them in the technology market enables them to prevent their local partners from becoming their competitors.

A large variety of actors play a role in the technology market. Each of these actors has its own motives and capabilities, and these are often in conflict. Moreover, each of them has to face certain risks. *Private enterprises and individuals* operating within the framework of a national political and economic setting generally have a short-term focus (for example, they engage in cost saving by cutting down expenditure on production inputs such as R&D) and, as they have to face potential conflicts of interest, they have to balance their motives and risks with recognition of the conditions in their operating environment (Stewart and James, 1982).

Global FDI and export statistics indicate the dominance of transnational corporations from industrialised countries. These firms, which have subsidiaries in one or more countries outside their homes, often in developing countries, play a major role in the global economy in terms of foreign investment, thereby bringing into the recipient countries information, knowledge, technology and R&D capability, management and organisational capability, and employment. The majority of transnational corporations are manufacturing firms, although the number of service companies is increasing. There are also some transnational construction companies such as those from Germany, France, Sweden and the United States. Up to 70 per cent of the construction opportunities in developing countries during the last few decades of the twentieth century (measured by size of contracts) were undertaken by these firms; these were primarily infrastructure projects (UNCTAD, 2000a). There is also an increasing number of Chinese companies with overseas operations in construction.

FDI has become the largest source of external financing in many developing countries. Thus, it is a crucial instrument and a powerful source of knowledge, capability building and trade expansion. FDI into developing countries is still focused on short-term, labour-intensive manufacturing, whilst FDI in high-technology activities such as in the services sectors is limited.

Cross-border technology flows through FDI continued to rise until the global crisis in 2008; and they have occurred in both developed and developing countries. The developed countries used to take the lead in outward FDI. However, some emerging countries such as China and India are increasingly playing an important role in these regards. The global economic and financial crisis in 2008 initially affected developed economies more severely than the developing nations, but the decline of inward investment is now also visible in developing countries (UNCTAD, 2009), and it is having an impact on the countries.

The majority of technology transactions still take place among developed countries. The majority of developing countries are 'recipients' of technologies with a relatively weak purchasing power and investment capacity, a weak R&D and absorption capacity and physical infrastructure deficiencies. However, the contribution of developing countries to world trade has increased considerably since the turn of the century; they now account for up to 36 per cent of total world exports. It is pertinent to note that the technological content of some of these exports is quite high. For example, China was the largest high-technology exporter (mainly of products with high R&D intensity, such as electronics). In 2008, China's exports of such technologies were about six times those of member countries of the European Union (EU) such as the UK and the Netherlands (UNCTAD, 2009). However, China appears to be an exception; other emerging countries such as India and Brazil export only 10 per cent of the high-technology export value of the EU countries.

Cross border south–south technology flows

South–south flows of technology have grown far faster than north–south flows. The share of exports from the south to other developing countries was 45 per cent of total exports from the south to the world in 2005 (UNCTAD, 2009). The major sources of south–south flows are the emerging countries such as Brazil, China and India.

Chinese investments in sub-Saharan Africa (of which 46 per cent were in manufacturing, 40 per cent in services and 9 per cent in resource-related industries) has increased significantly. The figure of $120 million during 1998–2002 was six times higher than that during the period 1990–1997; 20 per cent of it went to South Africa. Taiwan was also a major FDI source in Africa. During the 1980s, the main destination was South Africa but currently it invests throughout sub-Saharan Africa (World Bank, 2004). South Africa itself is also a source of south–south investments in other African countries, with about 600 projects in 2003, 15 per cent of which were in mining, 20 per cent in manufacturing, and the rest in agriculture and services such as utilities, hospitality and tourism, construction, IT and banking (Gelb and Black, 2004).

India is also a fast-growing source of investment source in Africa, again particularly in South Africa. It has focused on sectors such as services (mainly IT and banking), as well as manufacturing. Besides, India is undertaking initiatives to expand economic ties with other emerging countries such as Brazil and Argentina as well as to establish a tripartite relationship among Brazil, India and South Africa to facilitate growth in trade in services such as IT, construction and R&D. The north–south and south–south inward FDI in India are both market-seeking investments attracted to sectors with large domestic markets, and those with high export intensity and low import intensity.

Recently, China has considerably increased its trade and investment in South Asia. Factors contributing to this include China's business-friendly model, liberal trade policies, substantially reduced qualitative and quantitative barriers, improved political relations between China and the countries in this region, and increased investment in infrastructure for the purpose of facilitating trade (Bera and Gupta, 2009).

There is less south–south technology flow into Africa from Latin America beyond Brazilian firms entering Angola and Mozambique.

The investments and trade of the emerging countries in other developing countries are either *market-seeking*, driven by the liberalisation of regulations and lowering of entry barriers in the host countries, or *resource-seeking*, in a quest for natural resources as well as cheap labour, driven by trade motives, such as to facilitate exports to their home countries or third markets, for instance the United States (Gelb and Black, 2004).

Market-seeking investments increase the scope and quality of goods and services that are available to domestic firms and households as inputs into production and consumption and may stimulate domestic business, through promoting forward and backward linkages and effects on competition amongst suppliers in the domestic market. Resource-seeking investment (especially by producers seeking cheap labour) may have positive effects on employment and exports as well as on establishing or improving the production infrastructure.

Yeung (1994) stated that south–south investments are more beneficial to the host countries than those from the north, since south–south market-seeking investments will produce goods and services that are more appropriate to the host countries than those from the industrialised countries, whereas the 'technology gap' between domestic firms and those from other developing countries is smaller. This last factor may enhance the pace and volume of knowledge flows and learning. For example, the business models used by firms from the emerging countries are more informal (for example, in terms of governance arrangements, distribution of goods and services and security issues) than those applied by the industrialised countries. Spillover effects through south–south FDI might also occur faster since investors will make use of distribution and business network models that promote backward and forward linkages within the host country, and therefore support the development of local enterprises (Gelb and Black, 2004). However, this depends on the absorptive capacity of the host country.

In resource-seeking investments by foreign firms, especially by Asian ones, large numbers of expatriate managers and supervisors are often brought into the host country. An unforeseen effect that supports economic activities with spillover effects to the host countries is that many of these expatriates leave their original employers and stay in the host country to set up their own firms since prospects might be more limited for them in their home countries.

South–south technology and knowledge flows through investments or trade will provide the host countries with greater bargaining power in their relations with transnational corporations, especially those from the industrialised countries. South–south transfers enlarge and diversify the options for host countries to select the sources from which they can acquire the technologies and knowledge they need (Beausang, 2003; UNCTAD, 2006).

Construction technology exports

In all regions, the dominant providers of technologies and services in the international architectural and engineering services (AES) sector and in the construction industry were firms from developed countries. They are reported to be responsible for 32–50 per cent of international revenues for work carried out abroad. Most firms working on the international markets offer packages composed of a wide range of services in architecture, engineering and construction. In total, developing country firms accounted for only about 4.5 per cent of the total foreign billings in the AES sector in 1998. The highest ranked AES firm from the developing countries in that year was an Egyptian firm. The next were four Korean firms and one Chinese firm (providing both engineering and construction services). Among the major southern construction technology exporters in 2006 were companies from China, Egypt, Malaysia, Singapore and Thailand (UNCTAD, 2009).

The southern transnational corporations are potential providers of south–south knowledge and technology transfer (Lall, 1996; Narula and Marin, 2005) to local firms to upgrade their technological capabilities (Chaminade and Vang, 2008). However, the transnational corporations have little incentive to interact with the local firms given their lack of resources (Dunning and Narula, 2004). The activities carried out by the local firms in host countries are often in lower value-adding segments of the value chains. It is also pertinent to note that the technology and knowledge spillover effects of the foreign affiliates of transnational corporations in the host countries depend on the type of contract.

Imports of construction technologies

Imports of construction technologies into developing countries have been important for building and enforcing the technology capability of the firms in the local construction industry. Asia was the largest destination (29 per cent) for international contracts, followed by Africa (9 per cent) and Latin America (9 per cent) (UNCTAD, 2000a). Companies from developing countries have increasingly entered into ad hoc co-operation agreements with their counterparts from developed countries, focused around specific projects. Sub-contracting has proved to be an entry point to the international market for small and medium-sized construction firms from developing countries and also the transition economies. Recently, formal long-term agreements between firms from industrialised and developing countries have also been established, which involve knowledge sharing during both the design and execution phases. As discussed above, companies from the emerging countries such as China, India and Brazil are increasingly playing an important role in technology imports in developing countries.

Chinese as major players in south–south construction technology flows

Chinese contractors are playing an increasingly significant role in south–south construction technology flows, which appears to be due to the particular features of their foreign operations. A survey carried out by the CRGP (2001–2006) found that Chinese construction firms often start their operations in African countries by setting up a representative office (19 per cent) or a subsidiary (19 per cent). In 15 per cent of the cases, the entry mode is in the form of a strategic alliance; a sole venture company constituted another 15 per cent. Other entry modes were a local agent (12 per cent), joint venture (10 per cent), a sole venture project (3 per cent), licensing (3 per cent), build–operate–transfer (BOT)/equity project (3 per cent) and joint venture project (1 per cent) (Javernick *et al.*, 2007).

Chinese contractors win approximately 50 per cent of their projects in Africa through competitive international bidding on projects financed by traditional aid agencies such as the African Development Bank and World Bank. The Chinese firms bid for projects financed by local governments only when they consider the country to be politically and economically stable, thereby avoiding any uncertainties over payment.

Statistics show that the highest number of Chinese construction companies is to be found in oil and gas-producing countries such as in Algeria, Angola and Nigeria, where contractors expect to benefit from having a foothold in the host country (Javernick *et al.*, 2007). Nevertheless, resource-seeking bidding and investment by the construction firms might be risky since the profit from natural resources in the host countries is largely uncertain. The study by Javernick and colleagues (2007) found that the Chinese contractors are provided with debt support in their entry into these markets by (Chinese) financial institutions such as the Bank of China, China Exim Bank, China Construction Bank and China Agricultural Bank, and most of their insurance coverage is also provided by Chinese corporations.

Javernick and colleagues (2007) found that at least 48 per cent of the labour in the Chinese firms operating overseas is from China, 51 per cent from Africa and 1 per cent from other countries. Most of the management staff (91 per cent), technicians and site engineers as well as at least 30 per cent of the skilled labour force are Chinese. The plant and equipment are also from China. An argument used by the Chinese contractors to justify their reliance on plant and equipment from their home country is that the labour force is familiar with working with them. Moreover, the prices are lower than for plant and equipment from industrialised countries. After the completion of a project the plant and equipment are generally left behind in the host country, most often by selling them.

Construction innovation system in developing countries

Deficiencies in the construction innovation system

A major problem that is often mentioned with respect to the *innovation system of the construction industry* is its rather fragmented character, considering the variety of actors involved in any project, including government agencies, materials suppliers, designers, general contractors, specialist contractors, the workforce, professional institutions and trade associations, private capital providers, clients, owners, vendors and distributors, testing services companies, educational institutions and certification bodies. These actors are expected to form a network of closely related actors, the members of which jointly and individually contribute to a successful development, as well as, in this context, the effective adaptation, application and diffusion of new technologies.

The relationships in construction are weak and project bound, and have a temporary nature, whereas *long-term linkages* among the various actors are critical for innovation, efficiency in terms of speed and costs at which projects and technological inventions on them are realised, handed over and used by the actors, and the efficacy with which clients' requirements are met.

To a large extent, the construction innovation system in developing countries is similar to that in the developed ones. There are only few large firms with resources that carry out R&D, and a majority of small- and medium-scale firms with negligible assets. In developing countries, the construction innovation system is even more characterised by low levels of interaction among the firms, as well as with other types of organisations such as universities, technology service providers and government agencies. An idiosyncratic difference is the large number of micro-enterprises operating in the informal economy which do the work based on traditional knowledge.

The mainstream of construction is in the residential sector where, in most countries, the work is carried out in informal networks dominated by micro-enterprises. In the traditional construction processes used by these firms, the production of much of the building materials, elements and components takes place on the building site and thus is integrated into the stage of actual realisation of the planned building. The lack of alignment among the parties working side by side on construction projects translates into dysfunctional teams, poor levels of co-operation and lost opportunities for the optimum utilisation of resources (Egmond, 1999). The large projects in developing countries are carried out by a handful of foreign firms with a considerable amount of sub-contracting and other forms of collaboration between foreign and local firms, while many of the construction technologies used are imported (Drewer, 1997; UNCTAD, 2000b; ILO, 2001). Thus, innovation in this segment of the construction markets in developing countries is basically in the hands of the foreign companies. The temporary and unbalanced relations between the foreign and local firms do not contribute to capability building for local technology development (Wells, 2001; Bertelsen and Müller, 2003).

The many small- and medium-scale contractors generally have very few or no assets. Much of the plant and equipment is outsourced. The major asset of construction firms is intangible knowledge and experience for managing and undertaking the project, which is embodied in the labour force. It makes them rather vulnerable in the construction market and, thus, it does not stimulate innovation in the companies. There is even less reason for investment in innovation by contractors as every project starts all over again with limited replication and extended usage of technologies from previous projects, which means that learning effects are largely non-existent (Pries and Janszen, 1995; Kumaraswamy and Shrestha, 2002; Thomassen, 2003; Bresnen *et al.*, 2004). On the other hand, in the industrialised countries several factors, such as the shrinking and more expensive labour and the social pressure for sustainable construction, push contractors to innovate.

Governments and international organisations may play a key role by providing support from national and international power networks to overcome bureaucratic or institutional barriers. However, there is a general *lack of co-ordination and information* in developing countries regarding the formulation, execution and evaluation of technology development and the dissemination of the results to beneficiaries. This forms a real constraint to innovation in construction.

Technological regime in construction in developing countries

Many authors have noted that continuous learning and knowledge assimilation in the firms, and TD&I in the industry, are hampered by the way in which the construction industry is organised, the operational practices of the actors in construction, and the views and expectations of the actors with regard to innovation (Kadefors, 1995; Pries and Janszen, 1995; Slaughter, 1998; Dubois and Gadde, 2002; Miozzo and Dewick, 2004).

On a construction project, miscommunication and contra-productivity arise from the *varying collaborations* in temporary project alliances; the large number of participants with independent and complex tasks, targets and motivations who work together; and the pressure of deadlines and budget constraints under which work is carried out. This means that TD&I does not take place (Pries and Janszen, 1995; Barlow, 2000).

The weak relationships in the construction industry form a constraint to learning, creation of an organisational memory, diffusion of the knowledge, and capability building. Dubois and Gadde (2002) highlight the rather tight collaboration among partners in a project, which generally brings about various innovative solutions to problems which occur. However, the knowledge gained on projects is generally not secured in the firms and also not further diffused in the industry. Capability building through the diffusion of the technologies and knowledge developed on a project is even more problematic, especially in the construction contracting business since much of the technology and knowledge from project experiences is tacit and not codified, but embodied in people.

The traditional tendering system and form of collaboration adopted on construction projects also limit TD&I in more respects. The rather detailed specifications in tender documents means that there is often no opportunity and no incentive for contractors to propose and apply innovative solutions.

This situation augments the tendency towards conservatism in the industry, and the reluctance to pursue or promote technological change which is evident in the construction industry. An important factor is the very low level of profits and the various risks of possible failure, damage or losses during the execution of the project (Thomassen, 2003).

A phenomenon of construction that makes it different from other sectors of the economy is that in construction the old technologies generally remain in existence, and are used next to the newly developed technologies and materials, whereas in most other industries the older technologies become obsolete and are often completely replaced by the new ones, once these new technologies have been introduced.

It may sound paradoxical but innovation is also considered to have a negative impact on project performance. The argument is that even the smallest alteration in the project, no matter which aspect it concerns, requires discussion and negotiation, and thus extra time and expenditure (Slaughter, 1998). Given the low profits from them, and the time pressure associated with them, there is no time for ex-post evaluation of the projects or for documenting the achievements and challenges on the project (Keegan and Turner, 2002). Thus, on a recurring basis, different solutions are developed for identical situations on different projects, even when many of the participants are the same (Barlow, 2000). This means that there is little organisational learning and capability building. The opportunities to apply successful solutions on other projects are limited by the temporary nature of project alliances and relatively weak level of communications in the industry, which reduces the incentives to innovate and enhances the tendency to keep applying well-known traditional technologies (Dubois and Gadde, 2002).

Construction management is still largely based on the attainment of pre-determined time, cost and quality criteria. This short-term perspective in construction, in which the project managers do not see the need for innovation, also makes it difficult to achieve the longer-term benefits of innovation, learning and capability building (Gyadu-Asiedu *et al.*, 2007).

Conservative clients may also limit innovation by their negative views and attitudes towards it. They are particularly concerned with potential extra costs and risks of failure of the novel technologies (Briscoe *et al.*, 2004). Since there is the expectation that structures should be durable, clients tend to have a preference for proven materials and technologies, which in turn limits the incentives for manufacturers to develop new products, and for builders to change their practices (Pries and Janszen, 1995; Miozzo and Dewick, 2004).

Building codes, regulations and standards, as well as the variety of contractual agreements, may also hamper TD&I in construction, although there is the potential for them to have a stimulating impact (Nam and Tatum, 1997; Ofori, 1990; Slaughter, 1998; Barlow, 1999).

The new structure of the construction industry (in terms of organisation and geographical diversification) that evolved in the last decades of the twentieth century impinges upon the contractor's risk exposure and leverage capacity, especially in developing countries. Large construction firms moved into international markets, through mergers and acquisitions, thereby often mainly working as service companies concentrating on management and co-ordination functions, finding clients and marketing their products and services. These changes in the structure of the construction industry were induced by the increased size, complexity and technical sophistication of construction projects and processes. Together with strong international competition, this stimulated construction firms to form consortia and to engage in partnership arrangements and strategic alliances in design, engineering, tendering for, and execution of, the projects. The large international construction firms have practically withdrawn from the physical work on the construction site. This was made possible through the outsourcing of labour by engaging sub-contractors. The workers are employed on temporary and casual terms by sub-contractors and intermediaries, which are generally small, and sometimes very small, firms serving niche markets (ILO, 2001).

To a certain extent, this movement offers opportunities to firms in developing countries since the new organisational structure enables them to collaborate with the large foreign firms and acquire some of the novel technologies and knowledge that are used on the projects. However, foreign firms tend to adopt strategies which do not support host countries' efforts to develop their industries. Abbott (1985) and Carrillo (1994) observe that foreign firms are not keen to effectively transfer their technology and knowledge since they believe that this means they would be nurturing their future competitors.

Ofori (1996) underpinned the above observations by indicating that one should take into account the differences in the interests and objectives of the various construction project participants at different levels: firm, sectoral and national. These have an impact on the extent to which technology and knowledge are actually accepted and implemented. Malherba (2004) suggested that there is an increased willingness on the part of firms to share the benefits of innovation with their partners and an alleviation of the fear of opportunism in networks where the interests of the members are common, which then stimulates diffusion.

Research capacity in developing countries

Generally, the *research* community and capability in developing countries are limited. The bulk of the R&D expenditure in construction takes place in the industrialised countries; this give them an advantageous position with respect to TD&I. The developing countries lack financial and human resources, as well as the demand for R&D. In 1998, R&D spending as a percentage of GDP by the 29 OECD countries was greater than the total economic output of the world's 61 poorest countries. For this reason, the focus of the international research is often dominated by the developed countries, and this is not always in the interest of the developing countries (Aubert, 2004).

In the emerging countries, R&D capacity is growing. This can be seen in the number of patent applications. The country with the largest number of applications was Japan, followed by the United States, China and Korea (WIPO, 2010). There is some evidence that intellectual property protection becomes important at a certain stage of development: when a country has attained upper-middle-income status.

In most developing countries, the scientific and technological infrastructure is rather weak. Aubert (2004) remarks that research institutes often operate in isolation from industry and other stakeholders. The work of the universities and research institutes is only poorly related to the needs of, and opportunities in, society. The growth in R&D in emerging countries can be related to the presence of strong universities (such as the Indian Institute of Technology) and/or research centres.

Local innovations are based on traditional knowledge. It is difficult to document and protect these as there is no proof or written record of ownership. This makes it impossible to place a patent claim on such knowledge. Moreover, there is a lack of R&D support services in terms of standards, quality control and regulations although the relevant public-sector institutions tend to be numerous, including those supporting the promotion of enterprise development, exports of services and inflow of FDI. In most developing countries, there is no functioning legal framework for the protection of indigenous, local and traditional knowledge. As a result, traditional knowledge and innovation in developing countries are often kept secret as there is no protection. This hampers the diffusion of the knowledge, and capability building (Aubert, 2004).

Absorption capacity, knowledge and learning in developing countries

The *absorption capacity*, which is related to the availability of a knowledgeable and skilled labour force with flexibility to adopt, adapt and subsequently further develop new technologies and management styles, is low in many developing countries. Evidence shows that the more highly educated and skilled the workforce is, the easier it is to introduce, diffuse and use new technologies (World Bank, 2004). Literacy rates and educational levels in general are low in the majority of developing countries, and this forms a significant constraint for TD&I.

In most developing countries, construction skills are still acquired mainly through doing and using in an informal traditional apprenticeship system. Informal training has limitations such as: (i) learning opportunities are restricted (in the case of learning by doing); (ii) the range of skills is narrow and static; and (iii) it is difficult to have instruction in new technologies and techniques (ILO, 2001).

In many African countries, the informal apprenticeship system is not well developed. Skills are often kept within the family, clan or tribe. Not all aspects of particular skills are passed on when they are transferred to 'outsiders' (Debrah and Ofori, 2001; Abdul-Aziz, 2001). The master craftsmen who do the training and provide the team leadership may themselves have limited skills. The informal method of acquisition of technological knowledge and skill formation no longer responds to the actual demand for construction output, which often shows project delays and poor quality. The 'learning-by-doing' system is rather costly and time-consuming, with little payback. Bell (1984) suggests that firms cannot rely on learning by doing in order to develop their technological capacities; they must invest in training and other means of knowledge creation, as, in this context, skills, knowledge and understanding are acquired through mechanisms requiring a deliberate allocation of resources.

Whereas vocational training schools do exist in most developing countries, many contractors and workers, as well as clients, see expenditure on the formal training of the workforce as an unnecessary expense rather than an investment (ILO, 2001). In both developed and developing countries, the changed structure of project execution in the construction industry discussed above has raised the barriers to formal technological capability building and training. The sub-contractors who are now the real employers of workers are small firms with limited organisational and technical capacity, and lack the time and resources to provide effective training. Thus, they are reluctant to invest in their employees. Contractors are even more reluctant to invest in training because there is a good chance that they will lose the trained workers to other firms or other countries. The contractors' reluctance is also based on the fact that training costs money, which, at least in the short run, will raise the level of their bids for projects. The workers have to face insecurity of employment and high levels of unemployment; they can be persuaded to undergo training only if they are paid for 'lost time' (ILO, 2001). Moreover, clients are reluctant to contribute to payment for training costs that will benefit only future clients as it takes time to train a worker (Philips, 2000).The cyclical pattern of construction output adds to the problem: organisations do not want to pay for their workers to train in a recession, and workers tend not to have the time to train in a boom.

Policies and strategies for technology development and innovation in developing countries

Policy requirements

From the foregoing, it is evident that, all over the world, the TD&I performance in construction leaves much to be desired. However, the experience of a number of emerging countries has shown that it is possible for the developing countries to catch up in technological development. Following the theories and concepts, as well as practical evidence of TD&I in construction, political determination is needed to realise the formulation of appropriate and adequate technology policies that stimulate TD&I. Thus, policy makers should direct their programmes to (i) broaden and strengthen the innovation system, the relationships in the actor network, the institutional infrastructure and the technological regime; (ii) develop technological capabilities; and (iii) control technology and knowledge flows.

Political determination is needed to realise the formulation of appropriate and adequate technology policies that stimulate TD&I. Policies should be properly embedded in a broader set to realise an overall TD&I-based improvement in the construction industry (Dahlman and Kuznetzov, 2004). Policies should (i) emphasise the necessary actions to be taken; (ii) provide standards and targets for the measurement of success; (iii) help to determine the required technologies including the planning and implementation of the needed infrastructure; (iv) enable the design of networks of structural relationships such as the decision-making structure and communication network; and (v) enable the implementation of the policies with the necessary programmes, action plans and instruments which facilitate a TD&I-based development.

In managing TD&I, firms have to take the specific context of their operations into consideration in order to realise an appropriate and adequate formulation of their company policies and strategies. The firm needs to have full insight into the peculiar aspects of the context since the efficacy and efficiency of TD&I management largely depend on a firm's ability to respond to the factors in its operating environment which have an impact on it.

Finally, it is pertinent to note that, 'as industries differ in terms of sources of innovation and the technological and market opportunity', there is no one best way to formulate policy (Tidd, 2001, p. 173).

Policies to broaden and strengthen the innovation system

Given the current situation in construction, policies should focus on bringing about more cohesion through a central focus of TD&I and promotion of the common interests of the actors in the construction innovation system. This means that policies should be directed to bring about changes in the technological regime to tackle major bottlenecks in the innovation system of construction, particularly with respect to the diffusion of knowledge in the actor network. The areas which policies should focus on are discussed below.

First, in order to decrease the limited interaction, communication, knowledge exchange and combination of different knowledge sets (such as research, design, engineering and marketing) in firms and organisations, efforts should be made to stimulate the development of new forms of longer-term collaboration and contractual agreements between technology and knowledge suppliers and recipients based on transparent performance measurement and improvement targets to increase trust amongst the collaborating parties, sustainability of operations and economies of scale. By doing so, entrepreneurs will be encouraged to invest in product innovation, output-quality improvement and process innovation.

Second, the expectations, beliefs and knowledge concerning TD&I should be improved by two major initiatives. First, the various actors should voice their expectations about the new technologies and knowledge, and these should be taken into account and acted upon. Second, there should be active technology and knowledge exchange amongst the actors in the innovation system about the entire range of relevant issues including design and engineering specifications, user characteristics and their requirements, environmental issues, industrial development options, government policies, regulatory framework and governmental role concerning incentives for diffusion and implementation. These actions will also increase awareness of the potential of novel technologies in the market, thereby increasing the market demand for innovation.

In Chile, for example, the success of strategies which had been adopted contributed to a change in regime in industries. The main aspect of the strategies involved the establishment, in 1976, of a national agency (Fundación Chile) for the promotion of the development of novel technologies and innovation. The agency operates with a highly professional management team and private shareholders 'which do not expect an immediate return and tolerate risks' (Aubert, 2004, p. 20). It was allowed to act as a private enterprise, and it carried out activities focused on technology and knowledge development and dissemination by incubating new ventures through entrepreneurship by (i) creating innovative 'demonstration' enterprises; (ii) developing, adapting and selling technologies to clients in the private and public sectors; (iii) promoting institutional innovations and incorporating new transfer mechanisms; and (iv) disseminating technologies through seminars, specialised magazines and project support.

Policies to support technological capability building

Technological capabilities differ among countries, and sometimes even among groups of countries which are considered to share certain similarities (for example, the emerging countries such as Brazil, China and India) and by sector. For this reason, there is the need to formulate specific policy agendas which take into account the different levels of development, and existing technological and educational infrastructure, as well as paying attention to the technological capabilities in each country, or even each sector (Aubert, 2004).

In a number of *middle-income countries* which possess relatively strong technological capabilities, the technology and innovation policies can be directed (i) to support the local firms in the exploitation of the available capabilities and (ii) to provide opportunities to attract foreign capital. The specific actions would include organising transport and logistics for exports in selected niches, building on the activities of state-owned enterprises and making good use of government procurement. At the same time, reforms should be implemented in key areas such as education, finance and trade, in order to create a broader environment more conducive to innovation (Aubert, 2004). In middle-income countries which have good research capabilities but lack institutional capabilities, policies should be directed towards bringing about an effective technology diffusion mechanism by establishing agencies to promote and support for R&D and innovation, and foster better research–industry collaboration.

In *countries which lack research capabilities*, but do have some institutional capacities such as those for financing and regulating, vision and drive, owing to the presence of effective leadership and strong agencies, policies should focus on the establishment of high-technology innovations, possibly by means of importing some well-educated people (Aubert and Reiffers, 2003).

In *low-income countries* which have limited technology, research and institutional capabilities, policies should focus on basic investment in the development of a technology infrastructure with reforms in the educational system at all levels, the creation of an investment promotion (and control) agency and the launching of pilot projects to demonstrate relatively basic innovations and novel technologies, and promote and support their diffusion in the country.

Aubert (2004) argues that the indigenous knowledge in developing countries which is not documented, and which is based on experience, accumulated knowledge and skills, should be seen by policy makers as a specific valuable asset and a source of local innovation that is relevant and well adapted to the needs of the communities. This knowledge has been transferred over generations, being part of the culture of the local communities. The strategy, then, is to identify, integrate and implement the indigenous knowledge and know-how in TD&I by making use of copying and borrowing new advanced technologies and knowledge, including those developed elsewhere, and merging them with the existing indigenous technologies and knowledge to adapt them to fit the local cultural and physical contexts. In construction, the technology gap between local firms and their foreign counterparts could be filled, for example, through joint ventures, sub-contracting, hiring of consultants, R&D collaboration, importation or local purchase of new equipment and materials, as well as non-conventional channels such as firms obtaining knowledge from newly hired workers as a result of their experience with previous employers, reverse engineering and reverse brain-drain (Ofori, 1996; Raftery *et al.*, 1998; Pack, 1999; Radosevic, 1999; Ofori and Chan, 2001). *Joint ventures* seem to offer the best possibilities for international transfers of technology and knowledge (Chow, 1985; World Bank, 2004).

Policies to control technology and knowledge flows

Policies which focus on attracting technology flows from regions which are in geographical and cultural proximity can effectively facilitate the absorption of the new technologies and knowledge in the host country thanks to similarities in technological regimes (including expectations, beliefs and practices) in south–south flows.

Policies aimed at promoting healthy competition between local firms and transnational corporation affiliates could take the form of either offsetting the market power of transnational corporation affiliates and suppliers of foreign brands through fiscal measures or assisting national firms to build their own brands and technological capability. Protectionist policies promote market-seeking FDI inflows. Protection accorded to local production from imports creates locational advantages and can force the exporting firms to undertake market-defending local production. The emerging countries in East Asia have consciously sought technologies with a selective policy, which has allowed the technology-importing local firms an autonomous path of expansion by independent decision making.

Regional economic co-operation among developing countries might help them to exploit the potential of intra-regional FDI and gain from efficiency-seeking restructuring of their industries. An institutional framework for promoting south–south FDI flows can be fruitful. It should include information dissemination facilities and a special fund for financing feasibility studies and for venture-capital-type financial support of projects based on developing country investments and technologies. Networking of chambers of commerce should also be promoted.

In low-income countries, such those as in sub-Saharan Africa, the World Bank promoted the development of 'competitive platforms' to provide a 'package' of support services relating to communications, negotiations, technical services (such as quality control and testing), training and so on (Table 8.2). This can compensate for the lack of technological capabilities

Table 8.2 Legal instruments for controlling international technology and knowledge flows

Type	Instrument
General regulations	Specific technology development programmes and policy plans
	Foreign exchange control regulations
	Regulations on specific branches or sectors
	Fiscal policies
	Anti-trust regulations (limitation to monopolistic and oligopolistic tendencies)
Specific laws	Foreign investment laws and policies
	Industrial property relations (such as duration, freedom, obligations of patents, trademarks)
Specific technology transaction regulations	Registration of transferred technologies
	Screening of foreign technologies
	Control of direct costs
	Restrictions on technology transfer transactions (including settlement of disputes, output of sales)

Source: Egmond (2006) adapted from ESCAP (1989).

which are also needed for the proper sourcing and selection of the required technologies and knowledge.

International interventions

Particular barriers in the international market need to be addressed in multi-lateral negotiations. They include domestic regulations, technical standards, licensing and qualification requirements and procedures, restrictions on movement of persons, government procurement practices, tied aid, subsidies, and transfer of technology provisions in contracts for the supply of goods or services (UNCTAD, 2000a).

Investment incentives offered by industrialised countries distort the pattern of FDI and tend to be in conflict with the interests of the developing countries in several ways. The international community should restore the flow of official resources (UNCTAD, 2000a). Moreover, efficient international aid programmes should help innovators in the construction industries in developing countries by offering appropriate integrated and coherent packages of support which have a maximum degree of flexibility.

Conclusion

Although the share of global technology and knowledge inflows and outflows contributed by the developing countries has increased in recent years, the bulk of technology flows originate from a handful of relatively more developed and technologically more dynamic countries. Many developing countries still depend on a sustained acquisition of technologies and knowledge from abroad while making little use of the existing domestic stock of technology. Moreover, technology and knowledge flows do not automatically produce beneficial results for the recipient firms. In the long run, the social benefits of the indigenous generation of technology and knowledge might be preferable.

TD&I are often isolated actions by a single actor, whereas the construction innovation system in developing countries shows weak or even non-existent linkages among the actors. Technological capability building then takes place on a non-durable basis and is easily lost when it is not further diffused and assimilated in the innovation system. A major factor that determines the importation of technology and knowledge and the effectiveness of local usage of the imported technology and knowledge in a developing country is the local technological and absorptive capacity, especially the availability of a knowledge stock. It helps foreign firms investing in the country to exploit and increase their competitive advantage (Radosevic, 1999). However, the technological capabilities in construction showed deficiencies. Evidence shows that the actual technology flows in collaborations between local enterprises and foreign firms take place only on a limited scale. Moreover, the new technology and knowledge are not effectively diffused throughout the local construction industry.

Foreign firms collaborate only with local contractors and sub-contractors which have established track records and capabilities. Technological capability building amongst local contractors working with foreign firms predominantly takes place as a transfer of know-how (skills) by means of the learning-by-doing-and-using mechanism. The problem here is that the acquisition of this knowledge requires dedicated investment in training, which involves extra time and cost on projects that are carried out with generally low profit margins.

The technological regime and routines in construction in developing countries, the expectations, beliefs, norms, values and motives of the local communities, often do not correspond to those of the foreign suppliers of new technologies and knowledge. In some cases, this even leads to non-acceptance of the new technologies and knowledge. Thus, the national setting in many developing countries forms a major obstacle to TD&I. There are also deficiencies in the physical infrastructure (the communication, transportation, energy and water supply network) in many countries.

Therefore, the governments of the developing countries should complement their efforts to attract FDI with incentives and policy liberalisation, with a focus on improving the nations' human resources and the institutional technology infrastructure, and developing local entrepreneurship. They should also give emphasis in their national development policies to initiatives for creating a stable macro-economic framework and conditions conducive to productive investments to support the development process in the countries.

References

Abbott, P. G. (1985) *Technology Transfer in the Construction Industry*. Economist Intelligence Unit: London.

Abdul-Aziz, A.-R. (2001) Foreign workers and labour segmentation in Malaysia's construction industry. *Construction Management and Economics*, 19 (8), 789–798.

Abernathy, W. J. and Utterback, J. M. (1978) Patterns of industrial innovation. *Technology Review*, 80, 41–47.

Aubert, J. (2004) *Promoting Innovation in Developing Countries: A Conceptual Framework*. World Bank Institute: Washington, DC.

Aubert, J. E. and Reiffers, J. L. (eds) (2003) *Knowledge Economies in the Middle East and North Africa toward New Development Strategies*. Papers from the World Bank Forum on Knowledge for Development in the Middle East and North Africa, 9–12 September 2002, Marseilles, France. World Bank: Washington, DC.

Barlow, J. (1999) From craft production to mass customisation: innovation requirements for the UK housebuilding industry. *Housing Studies*, 14 (1), 23–42.

Barlow, J. (2000) Innovation and learning in complex offshore construction projects. *Research Policy*, 29 (7–8), 973–989.

Beausang, F. (2003) *Third World Multinationals: Engines of Competitiveness or New Form of Dependency?* Palgrave: London.

Bell, M. (1984) Learning and the accumulation of industrial technological capacity in developing countries. In M. Fransman and K. King (eds) *Technological Capability in the Third World*. Macmillan: London.

Bera, S. and Gupta, S. (2009) *South–South FDI vs North–South FDI: A Comparative Analysis in the Context of India*. ICRIER Working Paper No. id:2143. http://ideas.repec.org/d/icriein.html

Bertelsen, P. and Müller, J. (2003) Changing the outlook: explicating the indigenous systems of innovation in Tanzania. In M. Mammo (ed.) *Putting Africa First: The Making of African Innovation Systems*. Aalborg Universitetsforlag: Aalborg.

Bijker, W., Hughes, T. and Pinch, T. (1987) *The Social Construction of Technological Systems: New Directions in the Sociology and History of Technology*. MIT Press: Cambridge, MA.

Blayse, A. and Manley, K. (2004) Key influences on construction innovation. *Construction Innovation*, 4 (3), 1–12.

Breschi, S. and Malerba, F. (1997) Sectoral systems of innovation: technological regimes, Schumpeterian dynamics and spatial boundaries. In C. Edquist (ed.), *Systems of Innovation*. Frances Pinter: London.

Breschi, S., Malerba, F. and Orsenigo, L. (2000) Technological regimes and Schumpeterian patterns of innovation. *Economic Journal of the Royal Economic Society*, 110 (463), 388–410.

Bresnen, M., Goussevskaia, A. and Swan, J. (2004) Embedding new management knowledge in project-based organizations. *Organization Studies*, 25 (9), 1535–1555.

Briscoe, G. H., Dainty, A. R. J., Millett, S. J. and Neale, R. H. (2004) Client-led strategies for construction supply chain improvement. *Construction Management and Economics*, 22 (2), 193–201.

Callon, M. (1995) Techno-economic networks and irreversibility. In J. Law (ed.) *A Sociology of Monsters: Essays on Power, Technology and Domination*. London: Routledge.

Carillo, P. (1994) Technology transfer: a survey of international construction companies. *Construction Management and Economics*, 12, 45–51.

Carlsson, B. and Stankiewicz, R. (1991) Functions and composition of technological systems. *Journal of Evolutionary Economics*, 1, 93–118.

Chaminade, C. and Vang, J. (2008) Globalisation of knowledge production and regional innovation policy: supporting specialized hubs in developing countries. *Research Policy*, 37 (10), 1684–1696.

Chow, K. F. (1985) *Construction Joint Ventures in Singapore: A Management Guide to the Structuring of Joint Venture Agreements for Construction Projects*. Butterworths: Singapore.

Cohen, M. D., Burkhart, R. Dosi, G. Egidi, M. Marengo, L. Warglien, M. and Winter, S. (1996) Routines and other recurring action patterns of organizations: contemporary research issues. *Industrial and Corporate Change*, 5, 653–698.

Coriat, B. and Dosi, G. (1998) Learning how to govern and learning how to solve problems: on the co-evolution of competences, conflicts and organizational routines. In A. Chandler, P. Hagström and Ö. Sölvell (eds) *The Dynamic Firm: The Role of Technology, Strategy, Organization, and Regions*. Oxford University Press: Oxford.

Cyert, R. M. and March, J. G. (1992) *A Behavioral Theory of the Firm*, 2nd edition. Blackwell Business: Oxford.

Dahlman, C. and Kuznetzov, Y. (2004) *Chile: Towards a Pragmatic Innovation Agenda*. Presentation prepared at the request of the Ministry of Finance of Chile, in preparation for a Development Policy Review for Chile. November 2004. http://info.worldbank.org/etools/docs/library/201232/ChileNovember2904.pdf

Debrah, Y. and Ofori, G. (2001) The state skills formation and productivity enhancement in the construction industry: the case of Singapore. *International Journal of Human Resource Management*, 12 (2), 184–202.

Dosi, G. and Winter, S. (2002) Interpreting economic change: evolution, structures and games. In M. Augier and J. G. March (eds) *The Economics of Choice, Change and Organization*. Edward Elgar: Cheltenham.

Dosi, G., Pavitt, K. and Soete, L. (1990) *The Economics of Technical Change and International Trade*. Harvester Wheatsheaf: London.

Dosi, G., Hobday, M. and Marengo, L. (2000a) *Problem-Solving Behaviours, Organisational Forms and the Complexity of Tasks*. LEM Working Paper No. 6. Sant'Anna School of Advanced Studies: Pisa.

Dosi, G., Coriat, B. and Pavitt, K. (2000b) *Competences, Capabilities and Corporate Performance*. Final Report Dynacom Project. http://www.sssup.it/~LEM/Dynacom/files/DFR.pdf

Dosi, G., Nelson, R. R. and Winter, S. (2000c) *The Nature and Dynamics of Organizational Capabilities*. Oxford University Press: Oxford.

Dosi, G., Orsenigo, L. and Sylos-Labini, M (2005) Technology and the economy. In Neil J. Smelser and Richard Swedberg (eds) *Handbook of Economic Sociology*. Princeton University Press: Princeton, NJ.

Drewer, S. (1997) Construction and development: further reflections on the work of Duccio Turin. *Proceedings of the First International Conference on Construction Industry Development, Singapore, 9–11 December.*

Dubois, A. and Gadde, L.-E. (2002) The construction industry as a loosely coupled system: implications for productivity and innovation. *Construction Management and Economics*, 20 (7), 621–631.

Dulaimi, M. F., Ling, F. Y. Y., Ofori, G. and De Silva, N. (2002) Enhancing integration and innovation in construction. *Building Research and Information*, 30 (4), 237–247.

Dunning, J. H. (1981) *International Production and the Multinational Enterprise*. Allen & Unwin: London.

Dunning, J. H. and Narula, R. (2004 *Multinational and Industrial Competitiveness: A New Agenda*. Edward Elgar: Cheltenham.

Egmond, E. L. C. van (1999) *Technology Mapping for Technology Management*. Delft University Press: Delft.

Egmond, E. L. C. van (2004) International technology and knowledge flows for technological capabilities in building in developing countries. *Proceedings International Symposium on Globalization and Construction*, CIB W 107, Bangkok, November.

Egmond, E. L. C. (2006) International technology transfer. Lecture notes, TU Eindhoven.

Egmond, E. L. C. van (2009) Innovation and transfer by strategic niche management. In R. McCaffer (ed.) *Proceedings of the International Conference on Global Innovation in Construction*, September 2009, Loughborough, UK.

Egmond, E. L. C. van and Erkelens, P. A. (2007) Technology and knowledge transfer for capability building in the Ghanaian construction industry. In *Proceedings CIB World Building Congress*, Cape Town, South Africa, 14–18 May.

Egmond, E. L. C. and Kumaraswamy, M. M. (2003) Determining the success or failure of international technology transfer. *Industry & Higher Education*, 17 (1), 51–59.

Egmond, E. L. C. van and Oostra, M. (2008) Knowledge management to foster innovation in construction. In M. C. Naaranoja, A. F. H. J. den Otter, M. Prins, A. Karvonen and V. Raasaka (eds) (2008) *Performance and Knowledge Management: Proceedings of CIB Joint Conference of W096 Architectural Management and W102 Information and Knowledge Management in Building*, June 2008, Helsinki, Finland.

Egmond-deWilde de Ligny, E. van and Erkelens, P. (2008) Construction technology diffusion in developing countries: limitations of prevailing innovation systems. *Journal of Construction in Developing Countries*, 13 (2), 43–64.

ESCAP (1989) A framework for technology based development planning: an overview of the framework for technology-based development. In *Technology Atlas Project*, Vol. I. United Nations ESCAP: Bangkok.

ESCAP (1994) *Application and Extension of the Technology Atlas*. United Nations: New York.

Freeman, C. (1982) *The Economics of Industrial Innovation*. Francis Pinter: London.

Geels F. W. (2004) From sectoral systems of innovation to socio-technical systems Insights about dynamics and change from sociology and institutional theory. *Research Policy*, 33, 897–920.

Gelb, S. and Black, A. (2004) Globalisation in a middle-income economy: FDI, production and the labour market in South Africa. In W. Milberg (ed.) *Labor and the Globalization of Production*. Macmillan: London.

Gyadu-Asiedu, W., Egmond, E. L. C. van and Scheublin F. J. M. (2007) Project performance measurement for productivity improvement. In *Proceedings CIB World Building Congress*, Cape Town, South Africa, 14–18 May.

Hippel, E. von (1988) *Sources of Innovation*. Oxford University Press: New York.

Hodgson, G. (1993) *Economics and Evolution: Bringing Life Back into Economics*. Polity Press: Cambridge.

IEA (2009) CO_2 *Emissions from Fuel Combustion*. OECD/IEA: Paris.

ILO (2001) *The Construction Industry in the Twenty First Century: Its Image, Employment Prospects and Skill Requirements*. ILO: Geneva.

Javernick, A., Levit, R. E. and Scott, W. R (2007) *Understanding Knowledge Acquisition, Integration and Transfer by Global Development, Engineering and Construction Firms*. Collaboratory for Research on Global Projects Working Paper #0028. Stanford University: Stanford, CA.

Kadefors, A. (1995) Institutions in building projects: implications for flexibility and change. *Scandinavian Journal of Management*, 11 (4), 395–408.

Katz, J. (ed.) (1987) *Technology Generation in Latin American Manufacturing Industry*. Macmillan: London.

Keegan, A. and Turner, J. R. (2002) The management of innovation in project-based firms. *Long Range Planning*, 35 (4), 367–388.

Kline, S. J. and Rosenberg, N. (1986) An overview of innovation. In R. Landau and N. Rosenberg (eds) *The Positive Sum Strategy: Harnessing Technology for Economic Growth*. National Academy Press: Washington, DC.

Kumaraswamy, M. M. (1995) Technology exchange through joint ventures. In *Proceedings*, International Conference on 'Technology Innovation and Industrial

Development in China and Asia Pacific towards the 21st Century', Hong Kong, November.

Kumaraswamy, M. M. and Shrestha, G. B. (2002) Targeting technology exchange for faster organisational and industry development. *Building Research and Information*, 30 (3), 183–195.

Lall, S. (1985) Trade in technology by a slowly industrializing country: India. In N. Rosenberg and C. Frischtak (eds) *International Technology Transfer: Concepts, Measures and Comparisons*. Praeger: New York.

Lall, S. (1987) *Learning to Industrialize*. Macmillan: London.

Lall, S. (1992) Technological capabilities and industrialization. *World Development*, 20 (2), 165–186.

Lall, S. (1996) *Learning from the Asian Tigers*. Macmillan: London.

Lall, S. (2003) *Industrial Success and Failure in a Globalizing World*. Working Paper No. 102, QEH Working Paper Series. University of Oxford: Oxford.

Lall, S. and Pietrobelli, C. (2005) National technology systems in sub-Saharan Africa. *International Journal of Technology and Globalisation*, 1 (3/4), 311–343.

Lee, T. and Tunzelman, N. von (2004) A dynamic analytic approach to national innovation systems: the IC industry in Taiwan. *Research Policy*, 34, 425–440.

Lundvall, B. A. (1993) *National Systems of Innovation*. Frances Pinter: London.

Lundvall, B. and Gu, S. (2006) China's innovation system and the move toward harmonious growth and endogenous innovation. *Innovation: Management, Policy & Practice*, 8 (1–2), 1–26.

Macozoma, D. S. (2002) Secondary construction materials: a resource pool for future construction needs. *Proceedings of the Concrete for the 21st Century Conference*, Midrand, South Africa, March.

Malerba, F. (1999) Sectoral systems of innovation and production. Paper presented to DRUID conference on national innovation systems, industrial dynamics and innovation policy, Rebild, Denmark, 9–12 June.

Malerba, F. (2002) Sectoral systems of innovation and production. *Research Policy*, 31 (2), 247–264.

Malerba, F. (ed.) (2004) *Sectoral Systems of Innovation: Concepts, Issues and Analyses of Six Major Sectors in Europe*. Cambridge University Press: Cambridge.

Manley, K. (2001) Frameworks for understanding interactive innovation processes. *International Journal of Entrepreneurship and Innovation*, 4 (1), 25–36.

March, J. G. (1988) *Decision and Organization*. Basil Blackwell: Oxford.

March, J. G. (1991) Exploration and exploitation in organizational learning. *Organization Science*, 1 (2), 71–87.

March, J. G. and Simon, H. (1993) Organizations revisited. *Industrial and Corporate Change*, 2 (3), 299–316.

Marsili, O. (2001) *The Anatomy and Evolution of Industries: Technological Change and Industrial Dynamics*. Edward Elgar: Cheltenham.

Metcalfe, S. (1995) Technology systems and technology policy in an evolutionary framework. *Cambridge Journal of Economics*, 17, 25–46.

Miozzo, M. and Dewick, P. (2002) Building competitive advantage: innovation and corporate governance in European construction. *Research Policy*, 31 (6), 989–1008.

Miozzo, M. and Dewick, P. (2004) *Innovation in Construction*. Edward Elgar: Cheltenham.

Moavenzadeh, F. (1978) Construction in developing countries. *World Development*, 6 (1), 97–116.

Nam, C. H. and Tatum, C. B. (1997) Leaders and champions for construction innovation. *Construction Management and Economics*, 15 (3), 259–270.

Narula, R. and Marin, A. (2005) *Exploring the Relationship between Direct and Indirect Spillovers from FDI in Argentina*. Research Memoranda 024. MERIT, Maastricht Economic Research Institute on Innovation and Technology. Maastricht.

Nelson, R. R. and Pack, H. (1999) The Asian Miracle and modern growth theory. *Economic Journal*, 109 (457), 416–436.

Nelson, R. and Winter, S. (1982) *An Evolutionary Theory of Economic Change*. Harvard University Press: Boston, MA.

Nelson, R. and Winter, S. (2002) Evolutionary theorizing in economics. *Journal of Economic Perspectives*, 16 (2), 23–46.

Ofori, G. (1990) *The Construction Industry: Aspects of its Economics and Management*. Singapore University Press: Singapore.

Ofori, G. (1996) International contractors and structural changes in host–country construction industries: case of Singapore. *Engineering, Construction and Architectural Management*, 3 (4), 271–288.

Ofori, G. (2000) The knowledge-based economy and construction industries in developing countries. *CIB W107 1st International Conference Proceedings: Creating a Sustainable Construction Industry in Developing Countries*, Stellenbosch, 11–13 November.

Ofori, G. and Chan, S. L. (2001) Factors influencing development of construction enterprises in Singapore. *Construction Management and Economics*, 19 (2), 145–154.

Pack, H. (1999) *Modes of Technology Transfer at the Firm Level*. Prepared for the World Bank research project 'Microfoundations of International Technology Diffusion'. University of Pennsylvania, November.

Patel, P. and Pavitt, K. (1997) The technological competencies of the world's largest firms: complex and pathdependent, but not much variety. *Research Policy*, 26 (2), 141–156.

Pavitt, K. (1999) *Technology Management and Systems of Innovation*. Edward Elgar: Cheltenham.

Pavitt, K. (1984) Sectoral patterns of technical change: towards a taxonomy and a theory. *Research Policy*, 13 (6), 343–374.

Philips, P. (2000) A tale of two cities: the high-skill, high-wage and low-skill, low-wage growth paths in US construction. Paper presented to International Conference on Structural Change in the Building Industry's Labour Market, Working Relations and Challenges in the Coming Years, Institut Arbeit & Technik, Gelsenkirchen, Germany, 19–20 October.

Polanyi, M (1967) *The Tacit Dimension*. Anchor Books: New York.

Prahalad, C. K. (2004) *The Fortune at the Bottom of the Pyramid: Eradicating Poverty through Profits*. Pearson Prentice Hall: Upper Saddle River, NJ.

Pries, F. and Janszen, F. (1995) Innovation in the construction industry: the dominant role of environment. *Construction Management and Economics*, 13 (1), 43–51.

Radosevic, S. (1999) *International Technology Transfer and Catch-up in Economic Development*. Edward Elgar: Cheltenham.

Raftery, J., Pasadilla, B., Chiang, Y. H., Hui, E. C. M. and Tang, B. S. (1998) Globalization and industry development: implications of recent developments in the construction sector in Asia. *Construction Management and Economics*, 16, 729–737.

Rogers, E. M. (1995) *Diffusion of Innovations*, 4th edition. Free Press: New York.

Roodman, D. M. and Lenssen, N. (1995) *A Building Revolution: How Ecology and Health Concerns are Transforming Construction*. Worldwatch Paper 124. Worldwatch Institute: Washington, DC, March.

Rosegger, G. (1996) *The Economics of Production and Innovation: An Industrial Perspective*. Butterworth-Heinemann: Oxford.

Rosenberg, N. (1976) *Perspectives on Technology*. Cambridge University Press: Cambridge.

Schot, J. W. and Rip, A. (1996) The past and future of constructive technology assessment. *Technology Forecasting and Social Change*, 54, 251–268.

Slaughter, E. S. (1998) Models of construction innovation. *Journal of Construction Engineering and Management*, 124 (3), 2262–32.

Slaughter, E. S. (2000) Implementation of construction innovations. *Building Research and Information*, 28 (1), 2–17.

Ssegawa, J., Ngowi, A. B. and Kanyeto, O. (eds) (2001) Conference summary. In *Proceedings of the 2nd International Conference of the CIB Task Group 29 on 'Challenges Facing the Construction Industry in Developing Countries'*, November 2000, Gaborone, Botswana.

Stewart, F. (1979) *International Technology Transfer: Issues and Policy Options*. World Bank Staff Working Paper No. 344. World Bank: Washington, DC.

Stewart, F. and James, J. (1982) *The Economics of New Technology in Developing Countries*. Frances Pinter: London.

Stiglitz, J. (1999) Knowledge as a Global Public Good, July 1999, Oxford Scholarship Online Monographs, pp. 308–326.

Tatum, C. B. (1991) Incentives for technological innovation in construction. In L. M. Chang (ed.) *Preparing for Construction in the 21st Century: Proceedings of the Construction Conference, New York, ASCE*.

Teece, D. J., Pisano, G. and Shuen, A. (1997) Dynamic capabilities and strategic management. *Strategic Management Journal*, 18 (7), 509–533.

Thomassen, M. A. (2003) *The Economic Organization of Building Processes*. Unpublished PhD thesis, BYG-DTU, Technical University of Denmark.

Tidd, J. (2001) Innovation management in context: environment, organization and performance. *International Journal of Management Reviews*, 3 (3), 169–183.

Tidd, J., Bessant, J. and Pavitt, K. (2005) *Managing Innovation*. Wiley: Chichester.

UNCTAD (2000a) Report of the Expert Meeting on National Experiences with Regulation and Liberalisation: Examples in the Construction Services Sector and Its Contribution to the Development of Developing Countries. TD/B/COM.1/32. TD/B/COM.1/EM12.3.

UNCTAD (2000b) Regulation and Liberalization in the Construction Services Sector and Its Contribution to the Development of Developing Countries. Note by the UNCTAD Secretariat. Geneva, October.

UNCTAD (2006) *World Investment Report 2006, FDI from Developing and Transition Economies: Implications for Development*. United Nations New York.

UNCTAD (2009) *World Investment Report 2009*. United Nations: New York.

Walker, D., Hampson, K. and Ashton, S. (2003) Developing an innovative culture through relationship-based procurement systems. In D. Walker and K. Hampson (eds) *Procurement Strategies*. Blackwell: Oxford, UK.

Wells, J. (2001) Construction and capital formation in less developed economies: unravelling the informal sector in an African city. *Construction Management and Economics*, 19, 267–274.

Wenger, E. (1998) *Communities of Practice: Learning, Meaning, and Identity*. Cambridge University Press: New York.

WIPO (World International Property Organization) (2010) *World Intellectual Property Indicators 2010*. www.wipo.int/export/sites/www/ipstats/en/statistics/patents/pdf/941_2010.pdf

World Bank (1984) *The Construction Industry: Issues and Strategies in Developing Countries*. World Bank: Washington, DC.

World Bank (2004) *Patterns of Africa–Asia Trade and Investment: Potential for Ownership and Partnership*. World Bank: Washington, DC.

Yeung, H. (1994) Third world multinationals revisited: a research critique and future agenda. *Third World Quarterly*, 15 (2), 297–317.

9 Human resource development in construction

Ellis L. C. Osabutey, Richard B. Nyuur and Yaw A. Debrah

Introduction

The construction industry has demonstrated unique and complex linkages to other sectors of a nation's economy. It can generate employment and has the potential to generate income in the location of the construction project and thus alleviate poverty (Ofori, 2002). The industry goes beyond the provision of shelter and infrastructure and provides human and local material resources for the development and maintenance of buildings, housing and physical infrastructure. The attraction of foreign direct investment (FDI) is intrinsically linked to infrastructure development. As most developing countries clamour for FDI, rapid infrastructure development has become prevalent, but most of these countries lack the requisite complement of professional and technical expertise and skills required to construct modern and complex infrastructure. Given this, huge and complex infrastructural projects in developing countries are predominantly handled by foreign firms with minimum or no involvement of local professionals. This adversely affects the development of local capacity for the construction and maintenance of such infrastructure, with consequent inability of local professionals to adequately maintain or repair such infrastructure.

Debrah and Ofori (2005, 2006) observe that globalisation has increased competition for local firms and local professionals in emerging and developing countries because they lack requisite professional and managerial competences to compete with foreign firms and foreign professionals. Foreign participation in infrastructure development is not necessarily bad for developing countries, but decades of overdependence to the extent that local professionals either are denied the opportunity or are unable to maintain and repair infrastructure constructed by foreign firms is evidence of poor human resource development (HRD) and capacity building (Osabutey, 2010). This exacerbates the skill development deficiencies in some developing countries.

The absence of requisite human resources in quality, quantity and variety is expected to adversely affect the planning and implementation of construction projects in developing countries. The dilemma is not limited to the dearth of skilled personnel at all levels and across a broad range of crafts and professions; there are also major deficiencies at the managerial level (Imbert, 1990). Debrah and Ofori (2001a) observed that many developing countries face shortages of skilled construction personnel. In an economic boom, increased construction activity might require the migration of skilled and unskilled workers from other developing countries. They indicated that low-skilled workers were more prone to workplace accidents and this emphasises the importance of HRD in the construction sector in developing countries. It has been noted that HRD is an integral part of construction industry development (Ofori, 1994a). The CIB (1999, p. xiii) defined construction industry development as 'a deliberate and managed process to improve the capacity and effectiveness of the construction industry to meet the national economic demand for building and civil engineering products, and to support sustained national economic and social development objectives'. This means that HRD and capacity building are integral to construction industry development.

HRD of professionals in developing countries has not received the attention it deserves in the literature despite the role of competent professionals in economic development (Debrah and Ofori, 2005, 2006). Osabutey (2010) also observed that training facilities and skill development programmes for middle-level and lower-level manpower in the construction industries in developing countries were dwindling, collapsing, inadequate or non-existent. Capacity building at this level has also been neglected in the literature. Osabutey (2010) further noted the absence of requisite technology and knowledge within local construction firms in developing countries. In addition, overdependence on foreign firms for the design, construction and maintenance of infrastructure is detrimental to the development of local capacities and local firms. Chatterji (1990) observed that the technology and knowledge transfer process must aim at achieving higher capacity while simultaneously reducing reliance on foreign contractors and imported resources.

Current participation of construction firms from other emerging or rapidly industrialising countries, other than western construction firms, in the markets of developing countries and the effects on local capacity has also not received adequate attention in the literature. In addition, HRD in the construction industries of developing countries outside South-East Asia and perhaps South America has not received adequate attention. Finally, construction HRD in Africa has not been adequately studied. Therefore, it is important to improve the understanding by governments, practitioners and researchers, among others, of HRD trends in developing countries in the current context and the linkages of HRD to construction industry development and national development. It is necessary to review the theory and practice and also examine what has been done on the subject in research, identify the current challenges and discuss what needs to be done as well as suggest an agenda for future research.

Human resource development and firm performance

HRD, as a focus of academic enquiry, draws on a wide range of disciplines and is more or less related to a range of management ideas such as strategic management, human resource management (HRM) and leadership (Stewart, 2005). This wide perspective of HRD in both academic foundation and application of professional practice provides a rationale, if not an explanation, for the lack of a universal definition of HRD (Lee, 2003; Stewart, 2005). However, the salient features of predominant definitions reveal that HRD is an integrated use of organisational learning interventions such as training and development, career development, and organisation development with the specific aim of improving skills, knowledge and understanding, as well as improving individual and organisational performance and effectiveness (McLean and McLean, 2001; Swanson, 2001; Sydhagen and Cunningham, 2007). Recent literature suggests that HRD is connected with, and is explained by, the concepts of learning organisation, 'chaordic' enterprise, organisation learning, organisation development and knowledge management (Stewart, 2005; Raiden and Dainty, 2006).

According to Raiden and Dainty (2006), the concept of learning organisation refers to an organisation that facilitates the learning of all its members in that new and expansive patterns of thinking are nurtured and people continually learn how to learn together, thereby enabling the organisation to continually change itself in accordance with the existing operating circumstances. Nyhan and colleagues (2004) also suggest that a learning organisation exhibits four characteristic features: (i) coherence between the formal organisational structure and informal culture, and between organisational goals and individual employee needs; (ii) challenging work; (iii) support and provision of opportunities for learning; and (iv) partnership between vocational education, formal training and informal HRD. Referring to Steers's (1988) definition of organisational development, Reynolds and Ablett (1998) suggest that it is an ongoing system-wide development approach that seeks to improve both productivity and efficiency on one hand and the quality of working life on the other. Stewart (2005) further notes that organisational development programmes have characteristics that agree with the principles of learning organisations.

In explaining the concept of a 'chaordic' enterprise, Fitzgerald and van Eijnatten (1998) posit that it is an enterprise in which chaos and order are simultaneously managed and maintained intentionally so as to ensure a dynamic balance of learning within the organisation. According to other researchers, observable features of such an organisation include discontinuous growth, organisational consciousness, connectivity, flexibility, continuous transformation and self-organisation (van Eijnatten, 2004; Raiden and Dainty, 2006). Stewart (2005) further explains that the idea of learning organisation has been taken over by knowledge management. Arguably, 'learning' is central to all the concepts mentioned above and they could be viewed as constructs with the same meanings. According to Stewart (2005), the adoption of any of the terms in the wider literature by researchers and writers is based on strong political as well as intellectual attachments to particular concepts and terms.

The relationship between HRD and firms' organisational performance has been extensively studied and has generated a considerable volume of literature. Therefore, HRD has been seen as an inherent and important element in improving the performance of companies, industries and economies. In other words, it is recognised as important in national, organisational and individual growth as well as in the globalisation process (Stewart, 2007; Sydhagen and Cunningham, 2007). Others also consider HRD as a key to sustainable economic development of developing countries (Paprock, 2006; Li and Nimon, 2008). In an analysis of a construction organisation's approach to HRD, Raiden and Dainty (2006) also found that elements of learning organisations, dimensions of organisational learning and features of the 'chaordic' enterprise theory could all be observed in the organisation's HRD programme. This emphasises the importance of HRM/HRD to performance improvement during construction projects.

Contributions, linkages, complexities and challenges of the construction industry

The construction industry contributes to a nation's gross domestic product (GDP) and employment generation for both skilled and unskilled workers. It is estimated that the industry contributes 5–8 per cent of the GDP of developed countries and 3–5 per cent of that of developing countries (Turin, 1973; Ofori and Han, 2003). Debrah and Ofori (2001b) note that the construction industry in Singapore contributes around 8.5 per cent of the country's GDP, employing a total of 300,000 people (professional and technical staff as well as skilled and unskilled site workers), which is about 8–9 per cent of the population. Ng and colleagues (2009) also observe that Hong Kong commits about US$12 billion to its construction industry annually, which contributes around 6 per cent of GDP and employs 280,000 people. Thus, the construction industry is an important pillar of the economies of developing countries as the nations go through the organic process of development (Ng et al., 2009).

The performance of the construction industry is essentially linked to national development and prosperity. Therefore, developing countries should commit resources to programmes to enhance the industry's performance, particularly in the area of developing the human resources needed for the growth and development of the sector. There is the need to ensure that professionals and tradespersons and their firms possess the requisite knowledge and skills to meet the challenges within the industry and contribute to national development. The importance of the sector is also reflected in its multi-disciplinary nature, requiring both technical and managerial capabilities as well as creating employment for skilled and unskilled workers within the national economy.

Characterised as a labour-intensive, project-based and location-specific industry within which individual projects are custom-built to the specifications of clients (Wild, 2002; Loosemore *et al.*, 2003), the construction industry requires not only the setting up of temporary organisational structures at sites but also the employment of professionals, technicians and skilled and unskilled site workers to form project teams to execute projects (Debrah and Ofori, 2001a; Fellows *et al.*, 2002). Atkins and Gilbert (2003) note that a construction project environment is characterised by groups of individuals working together for short periods of time before being disbanded and redeployed elsewhere within the organisation. In addition, the industry is considered to be often filled with crises, uncertainty and suspense, and the short-term interaction of project teams requires the combined and effective use of superior technical knowledge, skills and expertise with appropriate behaviour and co-ordination. The identification, assessment and maintenance of these core competencies will ensure that projects are executed to set objectives (Dainty *et al.*, 2005; Debrah and Ofori, 2005) and this emphasises the importance of HRM/HRD. The quality, quantity and variety of the labour force will enable the practitioners in the construction industry to undertake a wide range of varied and complex construction projects to required standards.

Considered as one of the most dynamic and complex industrial environments (Wild, 2002; Loosemore *et al.*, 2003), the construction industry has witnessed significant changes within organisations over the past three decades, with professionals within the industry increasingly in short supply since the early 1990s. Moreover, the industry is considered to suffer from low graduate retention rates, particularly among persons in the mid-career age group (27–44 years of age) as they move into other sectors in order to get more training or promotion, or to enhance their career prospects (Young, 1991; Dainty *et al.*, 2000). This trend is prevalent in developing countries, where technical manpower is not given the recognition it deserves. Osabutey (2010) observed that many construction graduates are not retained in the construction industry because of lack of recognition and poor remuneration. These graduates often drift into banking and other business sectors at a very early stage of their careers because they are offered far higher remuneration; this is a serious drain on the capacity building and capability of the construction industry in some developing countries.

Dainty and colleagues (2000) observe that construction professionals need to be maintained and retained, and require careful management and development to contribute effectively to the industry. In addition, they need to acquire a set of competencies relevant to the management, co-ordination and execution of construction projects in an increasingly competitive and globalising construction environment. This emphasises the importance of HRD in the construction industries in developing countries.

Dainty and colleagues (2005) studied the predictability of managers' job performance based on their behaviours and found that superior-performing managers acquire more specific and key behaviours or competencies that underpin effective management performance than average-performing managers. These indicate that, to ensure superior performance, construction industry professionals cannot depend solely on their technical skills but also need other core competencies, which can be obtained through training and development. Others observe that, in many countries, educated construction professionals are available but the industry still faces a critical shortage of managers with up-to-date skills and competencies required to work in a globalised competitive economy. This view is espoused by Debrah and Ofori (2005), who reveal that the attainment of formal qualifications in construction-related fields is necessary but not sufficient for construction industry professionals and employees to develop the competencies required to compete in a globalised competitive environment. In order for the construction industries in developing countries to optimally enhance their performance, and contribute to national development, training and skill development of the labour force is critical (Debrah and Ofori, 2001b). Design professionals, clients' project managers and contractors' managerial personnel require training on relevant organisational structures, enhancement of their social and public relations skills, and education on the management of a large labour force (Ofori *et al.*, 1996). This highlights the priority which should be given to much-needed HRD within the sector to enable construction practitioners at relevant levels to acquire the occupational competencies, organisational competencies and managerial competencies to effectively manage projects and the organisations (Debrah and Ofori, 2005).

Despite the evidence that the adoption of HRD increases organisational performance, Brandendurg and colleagues (2006) argue that its application to the construction industry has received little attention among construction companies. The inadequate volume of literature on HRD in the construction industry also shows that HRD is underutilised within the construction industry. For instance, Dainty and colleagues (2000) referred to the work of Hancock and colleagues (1996) and Young (1988) to reveal that, although large construction companies understand the concepts of HRD, only around half of them actually practise it. They further observed that 75 per cent of construction companies have no career development policies to allow individual employees to compare their personal career needs with those of their organisation. They further supported their characterisation of the lack of HRD in construction companies by pointing to the findings of Mphake (1989), which also reveal that only 17 per cent of large construction companies have formal management development policies. The view that HRD is given less importance by the majority of construction companies is demonstrated by the findings that 28 per cent of human resource (HR) managers in the sector occupy board positions, compared with 54 per cent in other private-sector companies (Druker *et al.*, 1996; Dainty *et al.*, 2000).

However, other authors reason that inherent variability in the construction product and place, autonomy of operational managers at project levels, the need to cope with unexpected changes, and the susceptibility to economic fluctuations make the applicability of HRM practices including HRD within construction companies problematic (Hendry, 1995; Druker and White, 1996a; Huang *et al.*, 1996; Dainty *et al.*, 2000). Arguably, these features of the industry should rather call for emphasis on HRM/HRD programmes and practices within construction organisations. The rationale here is that an organisation with well-trained, well-developed and motivated employees will find novel ways of handling these challenges. In addition, regarding HR managers as strategic partners in the management and execution of construction projects will go a long way to reduce the movement of construction professionals into other industrial sectors. Accordingly, construction companies, despite the problems the industry faces in implementing effective HRM/HRD policies, must find more effective ways of rewarding and developing their workforce if they are to avoid losing their best employees to competitors or other industries (Druker and White, 1996b; Dainty *et al.*, 2000). The studies discussed so far suggest that the rewarding of employees must not be limited to the offer of increased financial remuneration, but should extend to the fulfilment of the psychological contracts with employees that involve training and further development. This issue is discussed further below.

The HRM/HRD debate

The HRM literature (Cunningham and Hyman, 1999; Mccarthy *et al.*, 2010) indicates that line and divisional managers control the majority of the issues pertaining to employees' development. This trend reduces the involvement and influence of the HR departments and increases the power of line managers to unfairly manipulate their employees' development to the detriment of the individual employee and the organisation at large. In particular, when line or operational managers are faced with tight budgets and deadlines, they perceive HRM/HRD programmes and practices as wasteful, costly and unimportant. This exacerbates the HRM/HRD problem and negatively affects the overall performance of the construction companies and the industry as a whole. Moreover, the lack of coherence of HRM systems leaves employees underinformed and unsure of the existence of opportunities that they could exploit to meet their career needs and expectations (Dainty *et al.*, 2000). The study by Raiden and colleagues (2004) on HRD in construction organisations found that companies that demonstrated significant commitment towards strategic HRD were able to retain their staff and improve overall organisational performance.

The discussion thus far reveals that the majority of construction companies find it difficult to commit to HRD programmes towards continuous skill development of their employees. The reasons cited also show that managers of construction companies feel that the industry's HRM/HRD requirements are different and that training employees is a costly function which can potentially make employees more attractive to the firms' competitors (Debrah and Ofori, 2001b; Raiden and Dainty, 2006). It is possible to contend that training and developing employees will make them sufficiently efficient to do more with less and thus reduce the cost to organisations in the long run. Having established the need for HRD in organisations, the other side of the HRD debate to consider focuses on who bears the responsibility for HRD; is it the organisations themselves, the state, or both? There has been overwhelming support in the literature for state intervention in skills development, particularly in developing countries (Keep and Mayhew, 1995; Finlay and Niven, 1996; Debrah and Ofori, 2001b).

Definition of 'developing country'

'For operational and analytical purposes', the World Bank (2010, p. xxiii) classifies economies into three main groups using gross national income (GNI) per capita (computed by the World Bank Atlas method). The classification is lower income (GNI per capita up to US$975 in 2008), middle income (GNI per capita more than US$975 but less than US$11,906 in 2008) and high income (GNI per capita US$11,906 or more in 2008). The World Bank clarifies that low- and middle-income economies are often referred to as developing countries and emphasises that the term is used for convenience and in no way implies that the economies in the group are at similar levels of development (World Bank, 2010). This means that the definition of 'developing' country can be problematic, and that countries categorised as such are dissimilar in resource endowment, levels of socio-economic development and development potential. The United Nations and other related international institutions and agencies use a similar categorisation. The term 'developing' countries in this chapter refers to low- and middle-income countries. For example, countries such as Brazil, China, Ghana, India, Kenya, Mexico, Pakistan, Sri Lanka and Tanzania, mentioned in this chapter, are at different stages of development. For example, considering countries as destinations of FDI, whereas Brazil, India, China, Mexico and South Africa can be considered strategic opportunity markets, countries such as Ghana, Kenya and Tanzania are considered long-term opportunity markets (Vital Wave Consulting, 2009). Again, China and India, for example, have comparatively improved their construction industries and their companies are involved in international construction projects around the world, whereas most countries in sub-Saharan Africa are dependent on foreign construction firms.

State intervention in skills development

Drawing on the experience of Singapore's construction industry, Debrah and Ofori (2001b) suggest that governments of developing economies should assume a key role by adopting an 'activist' HRD policy in order to develop the skills of workers in the construction industry. They emphasise that the nature and importance of the construction industry makes it imperative that national governments play an active and perhaps pivotal role in HRD. Their study established that a determined government which develops appropriate skills formation policies can increase the skills of its construction workforce. They go further to point out that the skills development of the industry's workers, particularly in developing countries, cannot be left to the companies in the industry alone, or to chance. This could be done by the state continually working with organisations and offering grants and training programmes to upgrade the skills and competencies of professionals and skilled and unskilled workers.

The plausible argument of state intervention in HRD made by Debrah and Ofori (2001b) is supported by the important contribution the construction industry makes to the national development of developing countries. Ofori and Chan (2001) isolated key factors that influence the development of the construction industry and technology transfer and it emerged that the role of the government is the most important one. Ofori (1994b) emphasised that a government as a major client can influence technology and knowledge transfer and local capacity building, and explained how the government in Singapore actively promoted joint ventures between local and foreign contractors, as shown in Box 9.1. Osabutey (2010) observed that, in most developing countries, central and state governments and government agencies are responsible for over 80 per cent of construction output

---

Box 9.1 Government actively influencing construction technology and knowledge transfer

The Singapore government actively promoted technology and knowledge transfer through joint ventures between local and foreign contractors. Under the Preferential Margins Scheme (PMS) initiated in 1980, local firms and joint ventures with at least 25 per cent net local equity participation were offered a preferential margin up to 5 per cent of the bid or S$5 million, in public-sector tenders.

Successful applicants were required to submit, to the Construction Industry Development Board (CIDB), a detailed programme and periodic reports on technology and knowledge transferred to their local employees. Firms were faced with the possibility of being made to refund the preference offered if they failed to achieve the targets set. The technology and knowledge to be transferred were clearly specified in the project's general conditions of contract.

Source: Ofori (1994b)

and this means that policies to improve capacity building will work if there is adequate commitment from the government.

However, the question that remains is: to what extent should the state intervene in skill formation and development? Building on the work of Ashton and Sung (1994) and Goodwin (1997), Brown and colleagues (2001) identified different models of state intervention to develop skills in the economy. These include the High Skills Society model used in Germany, the High Skills Manufacturing model used by the Japanese government, the Low Skills/High Skills model of the UK and the Developmental State model of Singapore. However, the applicability of all these models to other countries has been questioned (Kraak, 2002). The Developmental State model of Singapore initially put forward by Ashton and Sung (1994) has been employed in guiding further work in other developing countries (see Goodwin, 1997) and in the construction industry (see Debrah and Ofori, 2001b). These scholars found that the approach to skills development in the construction industry in Singapore is consistent with the process analysed by Ashton and Sung (1994) and reveal that, whereas training in other industries focuses only on workers already in employment, the peculiar nature of the construction industry makes government intervention in the training and upgrading of the skills both of those already in employment and those not yet employed important.

Many authors indicate that the Singaporean Developmental State model is most suitable for the construction industries in developing countries, and several of them have cited the example of Singapore to explain how the presence of foreign firms can be used as a means of maximising learning and technological acquisition and capacity building (Aryee, 1994; Ofori, 1994b; Goodwin, 1997; Lall, 2003). Aryee (1994) suggests that government can effectively lead the process of human resource creation, allocation and utilisation. Lall (2003) and Ofori (1994b) cite Singapore as an example to illustrate how foreign firms can be used to develop local capacity. Box 9.2 illustrates how government and foreign firms co-operate in Singapore's industrial policy (and Box 9.1 gives a related policy in the construction industry). The responsibility of the government and its agencies with regards to HRD in developing countries is therefore self-evident. Thus, it is important that industry stakeholders embrace this challenge. It is also important that researchers (who are also stakeholders) need to examine current issues within the industry and how these HRD challenges could be met. The question this chapter further seeks to analyse is: should learning and for that matter HRD not be the responsibility of firms operating in developing countries as well? If indeed the construction industry is described as a 'chaordic' learning organisation, then perhaps there is a role for firms themselves.

Osabutey (2010) observed that for effective technology and knowledge transfer to occur in developing countries, even with the right government policies, foreign and local firms as well as other institutions have significant

---

Box 9.2 Government and MNC cooperation on training: the case of Singapore

Government policies play a crucial role in the development of co-operation and co-ordination between government and multi-national corporations (MNCs), and among MNCs and local firms. Singapore's collaborative schemes for long-term HRD with MNCs are worthy of emulation.

The management of industrial policy and FDI targeting in Singapore was centralised in the Economic Development Board of the Ministry of Trade and Industry. The Economic Development Board was given the mandate to co-ordinate all activities relating to industrial competitiveness and FDI. It was vital that local skills and capabilities were raised to match the needs of foreign investors.

Source: UNCTAD (2003)

---

roles to play. He noted that both the HRM/HRD and knowledge management/organisational learning systems within both local and foreign firms were necessary for effective technology and knowledge transfer. The term 'chaordic' enterprise encapsulates key features such as discontinuous growth, organisational consciousness, connectivity, flexibility, continuous transformation and self-organisation (van Eijnatten, 2004). Discontinuous growth refers to cyclical organisational development from birth to growth, stability, decline and instability and then back to growth in a continuous cycle. Development and learning are seen as discontinuous in this process (van Eijnatten, 2004). The nature of learning especially within a globalised world requires firms in developing countries to manage this discontinuous process themselves in order to enhance performance and learning. Organisational consciousness emphasises the importance of the organisational 'mind' as the driving force for change. Connectivity places emphasis on the organisation as part of a wider system (van Eijnatten, 2004).

Organisational consciousness and connectivity in developing countries emphasises the need for consciousness that predisposes organisations towards learning. In this regard, Osabutey (2010) observed that perhaps in developing countries, where local firms are less developed, key roles can be played by the better organised local professional bodies within an integrated system that can facilitate the creation of the organisational mind. Continuous transformation of a 'chaordic' enterprise requires building mechanisms that enable the initiation of change during cyclical change. Self-organisation refers to the need for a collective vision that is shared by all towards direct thought and action (van Eijnatten, 2004). Here, it is important to emphasise that, although van Eijnatten (2004) views the organisation as the unit of analysis, this chapter takes the view of a 'chaordic' industry. Perhaps that description suits the construction industry in developing countries even more. For effective HRD in developing countries, an institutional framework that connects relevant government institutions to contractors and consultants, professional bodies and other stakeholders is needed.

The key characteristics of a 'chaordic' enterprise are explored in relation to developing countries in recent research by Osabutey (2010). Loosemore and colleagues (2003) confirm the discontinuous growth in the construction environment characterised by increased or reduced construction activity during economic growth and decline respectively. As indicated earlier, developing countries have no alternative but to continuously develop infrastructure to meet developmental needs. Osabutey (2010) observed that, in Ghana, most local construction firms did not embrace short-term (reactive) or long-term (strategic) training and development. One reason given was that contractors were not assured of winning projects in future. Often, the winning of construction contracts was influenced not by competencies alone but also by political influences and affiliations. This is another reason why local firms are not encouraged to invest in HRD and knowledge management systems. Debrah and Ofori (2006) argued that corruption was inimical to HRD in some developing countries and that quality standards were often not enforced because corruption is partly responsible for the marginalisation of local contractors as the procurement or tendering procedures and processes were not always transparent.

With regards to organisational consciousness, learning is based on the emphasis placed on the importance of a collective vision as the driving force for change. This requires a strategic policy that provides a clear direction in terms of employee development at the firm level. Strategic management and strategic HRM are absent among most local construction firms in most developing countries (Ayirebi, 2005; Osabutey, 2010). Most local firms do not have learning and development as an important element of their operational activities. Therefore, training budgets are almost non-existent. Connectivity places emphasis on the nature of an organisation in relation to how it also forms part of a wider system. In most local construction firms in developing countries, the organisational structures are difficult to decipher and there are no direct links with other various stakeholders, such as the government, industry associations, professional institutions and educational institutions, which can enhance learning within the industry as a whole. Osabutey (2010) posited that, for learning and HRD to occur effectively within the industry, there was the need for a synergistic stakeholder-driven regulatory infrastructure. This connectivity is needed in order to realise effective HRD in developing countries.

Flexibility is important for organisations within the construction industry because each project is, in one way or another, different from the next. Continuous learning, as well as continuous change management, becomes necessary. Therefore, learning should be adaptive and should respond to changing requirements which could be either client specific or environment specific. This requires new working practices and procedures, but this is where most local firms fall short. They have not invested adequately in HRD and, therefore, they are unable to sufficiently adapt to new challenges. Thus, local firms are not equipped for continuous transformation. They lack the capability and capacity to handle new types of projects. They have not developed enough to take up business opportunities, and have failed to gain knowledge from outside their organisations. The spirit of self-organisation is missing. Therefore, local organisations lack self-development and are unable to improve, and to embrace change. The local organisations are slowly but inexorably driving themselves into irrelevance, and the entire industry in most developing countries stands to deteriorate further in human capacity. Construction HRD in developing countries cannot be achieved without a vibrant local industry. Therefore, in developing countries, state intervention and local firm HRD incentive systems should be integrated to improve HRD in the industry.

As mentioned earlier, there is a link between HRD and the concepts of 'chaordic' enterprise, organisational learning, knowledge management, learning organisation and organisational development. Stewart (2005) asserts that change programmes that transform organisations on the principles of the learning organisation ideal have features common to organisational development, which is integral to organisational HRD. Therefore, in order to develop HRD in developing countries, formulated state intervention policies should further seek to develop local organisations. This calls for a paradigm shift in national HRD systems in developing countries towards more integrative systems. The era of globalisation implies that governments should also seek to improve the competitiveness of their local firms through incentives that encourage organisational development; the new HRD must encompass a process that seeks to develop other related concepts within local construction firms in developing countries. Local firms should also be encouraged to embrace new technology and knowledge that can help them to acquire or create, disseminate and utilise knowledge to improve their competitiveness.

Previous research on HRD in construction

Debrah and Ofori (2005) observed that Tanzanian construction professionals lack the competences required to compete for, and manage, projects in the liberalised and globalised markets which are prevalent worldwide. They highlighted the absence of occupational and organisational competences among the local professionals. Their study suggested that there was the need to establish sustainable training programmes for managers and professionals. The findings of Debrah and Ofori (2005) were consistent with similar HRD studies in the Tanzanian construction industry (see, for example, Hollway, 2001). They also encapsulate the fact that in most of the countries in sub-Saharan Africa the absence of training to equip employees with relevant contemporary competences can be attributed to the lack of competitiveness of local professionals and local firms (Kamoche, 2002). This means that training would improve the competitiveness of local firms and local professionals. Debrah and Ofori (2005) further emphasise that, after formal education and training, professionals require continuing professional development through appropriately instituted and managed training programmes to bring them into alignment with modern management and professional practices to enhance their competitiveness.

Debrah and Ofori (2006) further explore the reasons behind the absence of systematic training programmes for professionals within the industries in developing countries. They observed that, even for the limited training which then existed, only engineers and contractors benefited; there was no recognisable equivalent training for architects and quantity surveyors. Their study also revealed that existing programmes were fragmented and poorly managed and co-ordinated. The research captured the current and desired approaches to training professionals as shown in Table 9.1. A major finding was that these deficiencies can be attributed to poor funding. The study then suggests an industry-specific levy and emphasised that, in developing countries, there is the need to integrate the administration of related funds with the training programmes to ensure success. The authors stress the importance of structured training and express the view that government policy initiatives could be key drivers for success. They highlight mandatory continuing professional development (CPD), which is a condition for the renewal of practising licences for professional architects and engineers in Singapore, as an example that most developing countries could follow to enhance competencies and, subsequently, the competitiveness of their firms.

Jayawardane and Gunawardena (1998) address HRD issues in the construction industry in Sri Lanka by discussing the occupational structure and characteristics of the construction workforce. They studied middle- and lower-level manpower in the industry, and found that in developing countries the manpower deficiencies at those levels were more alarming than at the higher levels. The study revealed that in Sri Lanka the industry is dominated

Table 9.1 Current and desired approaches to training professionals

Current approach	Desired approach
No overarching policy and strategy for meeting HRD needs of professionals	Offering training based on strategic HRD needs of professionals
No co-ordinated HRD infrastructure within the industry	Instituting integrative and collaborative HRD programmes for professionals
No sustainable funding for HRD	Setting up funding mechanism for HRD
No national accreditation or recognition system	Establishing national accreditation and recognition systems

Source: Debrah and Ofori (2006).

by unskilled workers (50.4 per cent). There are six traditional skills: masons (32.5 per cent of the workforce), carpenters (10.5 per cent), electricians (1.9 per cent), bar benders (1.6 per cent), painters (1.5 per cent) and plumbers (1.2 per cent). Anecdotal evidence indicates that the situation is not different in most developing countries. Jayawardane and Gunawardena (1998) compared this narrow variety of workforce skills with the existence of over 18 main skills in the United States. They further reported that, in Sri Lanka, the National Trade Test (NTT) (equivalent to the National Vocational Qualifications [NVQs] in the UK) is used to evaluate both theoretical knowledge and practical ability. NVQ is a 'competence-based' national qualification. It involves learning through practical, work-related tasks designed to help individuals develop the skills and knowledge to do a job effectively. (NVQs are based on national standards for various occupations which indicate what a competent person in a job could be expected to do.)

Jayawardane and Gunawardena (1998) reported that, on average, 33 per cent of skilled workers were aware of the NTT, only 2 per cent had sat the test and only 1 per cent had actually obtained the qualifications. This is the trend in most developing countries and it depicts poor HRD. Indeed, in some countries, there are no such qualifications or certifications for local craftsmen and artisans (Osabutey, 2010). This means that systematic manpower development at the artisan level has long been neglected, resulting in a lack of competence even at the basic levels of the construction industries in many developing countries. Jayawardane and Gunawardena (1998) contrasted their findings with those of Rowings and colleagues (1996), who found that, in a comparable workforce in the United States, 74 per cent had completed craft training programmes, 73 per cent had finished apprenticeship programmes and 76 per cent had received on-the-job training.

In many developing countries, there are no reputable institutions for training lower-level personnel; on-the-job training is largely inadequate and the personnel at that level remain largely unskilled. Therefore, they produce poor construction outputs. Most developing countries have failed to establish institutions that train and certify skilled construction workers. For example, an estimated 95 per cent of construction workers in the Philippines acquire their skills through informal apprenticeships (Yuson, 2001). Some 85 per cent of craftsmen in Egypt are trained through traditional apprenticeships (Assaad, 1993). The International Labour Organisation (ILO) (2001) observed that in most developing countries, such as Brazil, India, Kenya and Mexico, construction skills are acquired mainly through informal apprenticeship systems. Vocational training schools do exist in most of the countries but many workers and contractors see formal training as a cost rather than an investment. The ILO (2001) further notes that informal training has limitations, and observes that in many African countries the informal apprenticeships are not well developed, with 'master craftsmen' often lacking requisite skills themselves (Fluitman, 1989). In Asia, it has been noted that informal apprenticeships are often kept within families or clans as it is believed that they may be diluted when passed on to 'outsiders' (Debrah and Ofori, 1997; Abdul-Aziz, 2001). The ILO (2001) observed that informal methods of skill acquisition can come under particular strain when there is a sudden and/or sustained increase in construction activity or when there is pressure from clients for better quality and faster completion times. These requirements become prevalent when economic growth is accompanied by social changes as observed in newly industrialised countries.

Most of the construction HR literature indicates the prevalence of casual or temporary employees and associated high turnover, even in developed countries such as the UK and United States. As a result, most construction workers are reluctant to invest in their own training and contractors are not keen to invest because of the likelihood of losing trained workers to other firms. The cyclical pattern of construction demand and output adds to the problem (Philips, 2000).

Role of training institutions

Training institutions in developing countries are either under-resourced or not empowered to play an active role in manpower development. Obudho (2008) revealed that, in Kenya, the training of construction technicians did not adequately fulfil the performance requirements of the construction industry and there was a need to align training with the current needs of the industry. This requires collaboration between the industry and the training institutions to match the syllabi of the courses to changing technology and knowledge needs of the industry. In India, the existing curriculum in most universities leaves most graduates with inadequate technology and knowledge skills to use modern construction technology and management techniques (Ahmad *et al.*, 2008). In Pakistan, the curriculum of engineering and construction universities, especially at the postgraduate level, does not meet the needs of the industry because of the absence of interaction between the industry and the universities (Khalid Huda *et al.*, 2008).

In a needs assessment study to determine capacity building in the construction industry in Ghana, Egmond and Erkelens (2007) examined the industry, the educational system in general and the education and training of construction practitioners in particular. They examined the capabilities of teaching staff, teaching materials and facilities in two polytechnics and found that the present level of education and the expertise of the teaching staff at the polytechnics were below the level required. They observed that the polytechnics currently show an inability to combine theoretical education with practical exposure in order to produce qualified graduates for direct absorption into the industry. There were specific needs for investment in laboratories and workshops, information technology facilities, educational materials and other resources. Their work identifies the root cause of the human capital deficiencies in the construction industries in most developing countries and the need for a more integrated policy towards manpower planning and development in all sectors, and the construction industry in particular.

There is the need for policies that ensure that educational and training institutions are adequately resourced and equipped to fully train practitioners for the industry. Policies should also engender avenues that can enhance collaboration between these institutions and the industry. Although the cyclical nature of construction activity does not encourage the hiring of permanent employees in high proportions and has therefore resulted in the lack of attention to training in most construction firms in developing countries, the HRD needs of construction practitioners and workers are evident. First, the technology and knowledge gap between professionals in developed and developing countries is widening. Second, the middle- and lower-level workforce lack formal training and proper certification and capacity building. Finally, national educational policies appear to have ignored HRD for the construction industry.

Individuals at the lower levels of the construction workforce cannot afford to acquire the needed training themselves and firms are not encouraged to train them. This lack of training adversely affects the quality of work within the industry in developing countries. This makes it necessary for some foreign firms to seek labour beyond some of the countries where they operate. For example, Chinese contractors often justify such employment practices with costs related to productivity (Alden and Davies, 2006). This increases construction costs, and it is clear that developing countries can reduce such costs if adequate investments are made towards improving HRD at all levels of the construction industry. This trend emphasises the need for government to actively engage with practitioners and other stakeholders to improve HRD within the sector.

Current issues and challenges in construction

Globalisation and emerging technology and knowledge have changed the dynamics of the world economy. Inherent in this dynamism is the presence of foreign construction firms in developing countries. On globalisation, Ofori (2007) suggested that research should aim to equip firms in developing countries with the ability to adopt a strategic response to the presence of foreign firms in their home markets in order to benefit from doing business with them, learning from them, competing with them, collaborating with them and adopting them as role models. They could also adopt their procedures and practices as benchmarks. Egmond and Erkelens (2007) looked beyond the construction industry and sought to examine tertiary institutions that educate personnel for the middle and higher levels of the workforce. They also note that the literature points to a lack of capabilities and technology and knowledge base for the construction industries in many developing countries. They observe that, for many local construction firms in developing countries, globalisation has become synonymous with a sustained acquisition of new technology and knowledge from abroad. For this reason, there has been minimal development and exploitation of the existing domestic technology and knowledge stock.

Osabutey (2010) notes that governments in developing countries, in seeking to attract FDI, have focused primarily on the development of the needed physical infrastructure to the detriment of long-term capacity building, for example through technology and knowledge transfer. In addition, education and training in construction-related programmes are important influencing factors in the development of construction industries, but have largely been ignored in most developing countries. This constitutes a major element of the capacity-building problem within the industry. This trend limits the ability of the locally trained workforce to adequately learn from the presence of foreign firms because of evident technology and knowledge gaps. This links the skill development problem in developing countries to the poor educational infrastructure, out-of-date teaching materials and out-of-date and ineffective teaching and training methods. For example, the students and trainees often receive little or no practical exposure during the entire period of education and training. Osabutey (2010) adds that the overdependence on foreign professionals and construction firms, coupled with competency deficiencies of local professionals and artisans, leads to very high costs to the developing countries.

In addition, Osabutey (2010) and UNCTAD (2003) observed that the productivity of the local workforce in developing countries is often low and this increases project costs. Osabutey (2010) further noted that skill deficiencies among local artisans and operators often gave foreign firms valid grounds not to use local manpower. The nature of construction projects means that most local firms have no time to develop a local construction workforce. He noted that the manpower gaps within the construction industries in developing countries were continuously increasing because there is a policy vacuum. There is a need to understand why most developing countries have failed to develop the quality, quantity and variety of human capital for their construction industries. The consequences of the neglect of the industry are clear but research has failed to address the urgency of an inadequately equipped workforce in the construction industries of developing countries. There is the need to study HRD within a framework of local firm, technology and knowledge development of national construction industries. It is necessary to search for the missing link between governments, educational and training institutions, and industry.

Evidence in some developing countries such as Ghana indicates that the owners and directors of most of the local construction firms are non-professionals; and a considerable number have little basic education. These individuals lack both technical and managerial skills and do not appreciate the need for investing in HRM/HRD (Osabutey, 2010). These groups of construction practitioners in developing countries may be the greatest challenge towards enhancing HRD and capacity building within the industry because their backgrounds are likely to influence their perceptions of, and attitudes towards, HRD. It would be appropriate for the government to introduce a policy to empower and enable educational institutions to either train construction practitioners or offer programmes to ensure that the practitioners are trainable.

Conclusion and implications for future research

Construction industry development has a strong link with active HRD. Globalisation has intensified competition and the need for developing countries to improve their technology and knowledge base as well as managerial skills. Most construction firms across both developed and developing countries often neglect HRM/HRD. However, there is evidence to suggest that HRM/HRD can improve the performance of construction firms. The size and scale of operations of most local construction firms in developing countries indicate that they are unable to take the responsibility for HRD of the construction workforce and that the governments of those countries have an active role to play in order to foster capacity building. In each country, the government can use its position as a major client to formulate and implement policies that can enhance national HRD. These policies should integrate the activities of key stakeholders and must seek to re-establish the missing link in most developing countries between policy makers, educational and training institutions, and industry.

Policies should also be introduced to promote and effect the involvement of the foreign firms operating in the countries in the development of local capacity for future infrastructure development, maintenance and repair. There is also the need to equip and challenge local professionals in this regard to reduce overdependence on foreign firms. Professional institutions and practitioners should be involved in curriculum development for tertiary, technical and vocational institutions offering programmes related to the construction industry. There should be incentives that can encourage both foreign and local firms to support internships for students in tertiary, technical and vocational institutions. These institutions should also organise systematic training programmes for local firms. The contract-awarding institutions can use managerial competence, technology and knowledge and evidence of training as a prerequisite for the award of contracts; but this would depend on the transparency of the procedures and processes for the award of contracts.

The chapter also highlights 'chaordic' learning organisation as akin to HRD in construction organisations. It further recognises the link between HRD and the concepts of organisational learning, knowledge management, learning organisations and organisational development. The effective implementation of these concepts is linked with up-to-date information technology systems. Anecdotal evidence suggests that most developing countries lack requisite information technology infrastructure. It appears clear from the discussion that twenty-first-century HRD means that state intervention in skill development should encompass policies seeking to build local capacity through incentives that can encourage local firms to embrace modern and proactive HRD. This is the challenge of HRD in the construction firms in developing countries in a globalised competitive construction environment.

Researchers should seek to explain why most developing countries have been unable to follow examples from countries such as Singapore in the development of local capacity and technology and knowledge transfer. The knowledge of policy makers, practitioners and researchers needs to improve: to recognise the importance of HRD and capacity building as the additional challenges of globalisation, particularly in developing countries. This is because the policies that worked in the 1980s for countries such as Singapore may need modification before they are applied in other developing countries, because of a marked change in the international construction scene. There is the need to understand why in certain parts of the developing world, such as sub-Saharan Africa, local capacity building has been ignored and the technology and knowledge gaps between foreign and local firms are widening. This trend is detrimental to national development. In particular, it is necessary to examine how the presence of construction firms from new emerging and industrialised economies in developing countries can enhance HRD. There is the need to improve how well policy makers in developing countries understand the importance of the construction industry and the need to develop the industry. They should also be made aware of the need for policies which consider feedback from the industry in the development of educational programmes, and integrate the syllabi of academic programmes and short courses, as well as other activities of the educational and training institutions, with those of practitioners towards improving local manpower at all levels.

References

Abdul-Aziz, A.-R. (2001) *Site Operations in Malaysia: Examining the Foreign–Local Asymmetry*. Unpublished report for ILO: Geneva.

Ahmad, T., Masood, A., Muhammad, A. and Mital, V. P. (2008) Issues in curriculum development for MTech in construction management in developing countries. In *First International Conference on Construction in Developing Countries: Advancing and Integrating Construction Education, Research and Practice*, NED University of Engineering and Technology, 4–5 August.

Alden, C. and Davies, M. (2006) A profile of the operations of Chinese multinationals in Africa. *South African Journal of International Affairs*, 13 (1), 83–96.

Aryee, S. (1994) The social organisation of careers as a source of sustained competitive advantage: the case of Singapore. *International Journal of Human Resource Management*, 5 (1), 67–88.

Ashton, D. and Sung, J. (1994) *The State, Economic Development and Skill Formation: A New Asian Model?* Working Paper 3. Centre for Labour Market Studies, Leicester University: Leicester.

Assaad, R. (1993) Formal and informal institutions in the labour market, with application to the construction sector in Egypt. *World Development*, 21 (6), 925–939.

Atkins, S. and Gilbert, G. (2003) The role of induction and training in team effectiveness. *Project Management Journal*, 34 (2), 44–52.

Ayirebi, D. (2005) Strategic planning practice of construction firms in Ghana. *Construction Management and Economics*, 23, 163–168.

Brandendurg, S. G., Haas, C. T. and Byrom, K. (2006) Strategic management in human resources in construction. *Journal of Management in Engineering*, 22 (2), 89–96.

Brown, P., Green, A. and Lauder, H. (2001) *High Skills: Globalisation, Competitiveness and Skill Formation*. Oxford University Press: Oxford.

Chatterji, M. (ed.) (1990) *Technology Transfer in the Developing Countries*. Macmillan: London.

CIB (1999) Managing construction industry development in developing countries. In *First Meeting of the CIB Task Group 29*, Arusha, Tanzania. CIB: Rotterdam.

Cunningham, I. and Hyman, J. (1999) Devolving human resource management responsibilities to the line. *Personnel Review*, 28 (1&2), 9–27.

Dainty, A. R. J., Bagilhole, B. M. and Neale, R. H. (2000) The compatibility of construction companies human resource development policies with employee career expectations. *Engineering, Construction and Architectural Management*, 7 (2), 169–178.

Dainty, A. R. J., Cheng, M. and Moore, D. R. (2005) Competency-based model for predicting construction project managers' performance. *Journal of Management in Engineering*, 21 (1), 1–9.

Debrah, A. Y. and Ofori, G. (1997) Flexibility, labour subcontracting and HRM in the construction industry in Singapore: can the system be refined? *International Journal of Human Resource Management*, 8 (5), 690–709.

Debrah, A. Y. and Ofori, G. (2001a) Subcontracting, foreign workers and job safety in Singapore construction industry. *Asia Pacific Business Review*, 8 (1), 145–166.

Debrah, A. Y. and Ofori, G. (2001b) The State, skill formation and productivity enhancement in the construction industry: the case of Singapore. *International Journal of Human Resource Management*, 12 (1), 184–202.

Debrah, Y. A. and Ofori, G. (2005) Emerging managerial competence of professionals in the Tanzanian construction industry. *International Journal of Human Resource Management*, 16 (8), 1399–1414.

Debrah, Y. A. and Ofori, G. (2006) Human resource development of professionals in an emerging economy: the case of the Tanzanian construction industry. *International Journal of Human Resource Management*, 17 (3), 440–463.

Druker, J. and White, G. (1996a) Constructing a new reward strategy: reward management in the British construction industry. *Employee Relations*, 19, 128–146.

Druker, J. and White, G. (1996b) *Managing People in Construction*. IPD: London.

Druker, J., White, G., Hegewisch, A. and Mayne, L. (1996) Between hard and soft HRM: human resource management in the construction industry. *Construction Management and Economics*, 14, 405–416.

Egmond, E. van and Erkelens, P. (2007) Technology transfer for capacity building in the Ghanaian construction industry. In *CIB World Building Congress*, 14–18 May, 2007, Cape Town, South Africa.

Eijnatten, F. M. van (2004) Chaordic system thinking: some suggestions for a complexity framework to inform a learning organization. *Learning Organisation*, 11 (6), 430–449.

Fellows, R., Langford, D., Newcomber, R. and Urry, S. (2002) *Construction Management in Practice*, 2nd edition. Blackwell: London.

Finlay, I. and Niven, S. (1996) Characteristics of effective vocational education and training policies: an international comparative perspective. *International Journal of Vocational Education and Training*, 4 (1), 5–22.

Fitzgerald, L. A. and Eijnatten, F. M. van (1998) Letting go for control: the art of managing in chaordic enterprise. *International Journal of Business Transformation*, 1 (4), 261–270.

Fluitman, F. (1989) *Training for Work in the Informal Sector*. ILO: Geneva.

Goodwin, J. (1997) *The Republic of Ireland and the Singaporean Model of Skill Formation and Economic Development*. Working Paper 14. Centre for Labour Market Studies, Leicester University: Leicester.

Hancock, M. R., Yap, C. K. and Root, D. S. (1996) Human resource development in large construction companies. In D. A. Langford and A. Retik (eds) *The Organization and Management of Construction, vol. 1.* Proceedings of the CIB W65 conference, University of Strathclyde.

Hendry, C. (1995) *Human Resource Management: A Strategic Approach to Employment*. Butterworth-Heinemann: Oxford.

Hollway, A. M. (2001) Experiences and challenges of the NCC's consultants training programme. In *Proceedings of Construction Industry Forum 2001: Construction Industry Development for Tanzania Social and Economic Transformation – Forum Report*, Dar es Salaam, 18–20 April.

Huang, Z., Olomolaiye, P. O. and Ambrose, B. (1996) Construction company manpower planning. In A. Thorpe (ed.) *Proceedings of the 12th Annual ARCOM Conference*, September, Sheffield Hallam University, Vol. 1.

ILO (2001) *The Construction Industry in the Twenty-First Century: Its Image, Employment Prospects and Skill Requirements*. ILO: Geneva.

Imbert, I. D. C. (1990) Human issues affecting construction in developing countries. *Construction Management and Economics*, 8, 219–228.

Jayawardane, A. K. W. and Gunawardena, N. D. (1998) Construction workers in developing countries: a case study of Sri Lanka. *Construction Management and Economics*, 16, 521–530.

Kamoche, K. (2002) Introduction: human resource management in Africa. *International Journal of Human Resource Management*, 13 (7), 993–997.

Keep, E. and Mayhew, K. (eds) (1995) *Evaluating the Assumptions that Underlie Training Policy: Acquiring Skills*. Cambridge University Press: Cambridge.

Khalid Huda, S. M., Farooqui, R. U. and Saqib, M. (2008) Finding ways for enhancing postgraduate level education in construction management in Pakistan. In *First International Conference on Construction in Developing Countries: Advancing and Integrating Construction Education, Research and Practice*, Karachi, Pakistan, 4–5 August.

Kraak, A. (2002) High skills, globalisation, competitiveness and skill formation. *Journal of Education and Work*, 15 (4), 485–488.

Lall, S. (2003) *Reinventing Industrial Strategy: The Role of Government Policy in Building Industrial Competitiveness*. QEH Working Paper Series, No. 111. University of Oxford: Oxford.

Lee, M. (ed.) (2003) *HRD in a Complex World*. Routledge: London.

Li, J. and Nimon, K. (2008) The importance of recognising generational differences in HRD policy and practices: a study of workers in Qinhuangdao, China. *Human Resource Development International*, 11 (2), 167–182.

Loosemore, M., Dainty, A. R. J. and Lingard, H. (2003) *Managing People in Construction Projects: Strategic and Operational Approaches*. E&FN Spon: London.

McCarthy, A., Darcy, C. and Grady, G. (2010) Work–life balance policy and practice: understanding line manager attitudes and behaviors. *Human Resource Management Review*, 20, 158–167.

McLean, G. N. and McLean, L. (2001) If we can't define HRD in one country, how can we define it in an international context? *Human Resource Development International*, 8 (4), 449–465.

Mphake, J. (1989) Management development in construction. *Management Education and Development*, 18, 223–243.

Ng, S. T., Fan, R. Y. C., Wong, J. M. W., Chan, A. C., Chiang, Y. H., Lam, P. T. I. and Kumaraswamy, M. (2009) Coping with structural change in construction: experiences gained from advanced economies. *Construction Management and Economics*, 27 (2), 165–180.

Nyhan, B., Cressey, P., Tomassini, M. and Poell, R. (2004) European perspectives on the learning organization. *Journal of European Industrial Training*, 28 (1), 67–92.

Obudho, S. O. (2008) Building construction technicians training: its relevance to the modern construction industry in Kenya. In *First International Conference on Construction in Developing Countries: Advancing and Integrating Construction Education, Research and Practice*, Karachi, Pakistan.

Ofori, G. (1994a) Formulating a long-term strategy for the construction industry of Singapore. *Construction Management and Economics*, 12, 213–217.

Ofori, G. (1994b) Construction industry development: role of technology transfer. *Construction Management and Economics*, 12, 379–392.

Ofori, G. (2002) *Developing the Construction Industry in Ghana: The Case for a Central Agency*. Discussion Paper. Ministers of Roads and Transport and Works and Housing of Ghana: Accra.

Ofori, G. (2007) Guest editorial: construction in developing countries. *Construction Management and Economics*, 25, 1–6.

Ofori, G. and Chan, S. L. (2001) Factors influencing development of construction enterprise in Singapore. *Construction Management and Economics*, 19, 145–154.

Ofori, G. and Han, S. S. (2003) Testing hypotheses on construction and development using data on China's provinces. *Habitat International*, 27 (1), 37–62.

Ofori, G., Hindle, R. and Hugo, F. (1996) Improving the construction industry in South Africa: a strategy. *Habitat International*, 20 (2), 203–220.

Osabutey, E. L. C. (2010) *Foreign Direct Investment, Technology and Knowledge Management in the Construction Industry in Africa: A Study of Ghana*. Unpublished PhD thesis, University of Wales, Swansea: Swansea.

Paprock, K. E. (2006) National human resource development in the developing world: introductory. *Advances in Developing Human Resources*, 8 (1), 12–27.

Philips, P. (2000) A tale of two cities: the high-skill, high wage and low-skill, low-wage growth paths in US construction. In *International Conference on Structural Change in the Building Industry's Labour Market, Working Relations and Challenges in the Coming Years*, Institut Arbeit und Technik: Gelsenkirchen, Germany.

Raiden, A. B. and Dainty, A. R. J. (2006) Human resource development in construction organisations: an example of 'chaordic' learning organisation? *Learning Organisation*, 13 (1), 63–79.

Raiden, A. B., Dainty, A. R. J. and Neale, R. H. (2004) Exemplary human resource development (HRD) within large construction contractors. In F. Khosrowshahi (ed.) *20th Annual ARCOM Conference*, Heriot Watt University, Edinburgh, 1–3 September.

Reynolds, R. and Ablett, A. (1998) Transforming the rhetoric of organisational learning to the reality of the learning organisation. *Learning Organisation*, 5 (1), 24–35.

Rowings, J. E., Federie, M. O. and Birkland, S. A. (1996) Characteristics of the craft work force. *Journal of Construction Engineering and Management ASCE*, 122 (1), 83–90.

Steers, R. M. (1988) *Introduction to Organisational Behaviour*. Scott Foreman: Glenview, IL.

Stewart, J. (2005) The current state and status of HRD research. *Learning Organisation*, 12 (1), 90–95.

Stewart, J. (2007) The future of HRD research: strength, weaknesses, opportunities, threats and actions. *Human Resource Development International*, 10 (1), 93–97.

Swanson, R. A. (2001) Human resource development and its underlying theory. *Human Resource Development International*, 4 (3), 299–312.

Sydhagen, K. and Cunningham, P. (2007) Human resource development in sub-Saharan Africa. *Resource Development International*, 10 (2), 121–135.

Turin, D. A. (1973) *The Construction Industry: Its Economic Significance and Its Role in Development*, 2nd edition. University College Environmental Research Group: London.

UNCTAD (2003) *Investment Policy Review: Ghana. United Nations Conference on Trade and Development*, Geneva: United Nations.

Vital Wave Consulting (2009) *Emerging Markets Definition and World Market Segments*. Vital Wave Consulting: Palo Alto, CA.

Wild, A. (2002) The unmanageability of construction and the theoretical psycho-social dynamics of projects. *Engineering Construction and Architectural Management*, 9 (4), 345–351.

World Bank (2010) *World Development Indicators*. World Bank: Washington, DC.

Young, B.A. (1988) *Career Development in Construction Management*. Unpublished PhD thesis, University of Manchester Institute of Science and Technology, Manchester.

Young, B. A. (1991) Reasons for changing jobs within a career structure. *Leadership and Organisational Development Journal*, 12, 12–16.

Yuson, A. S. (2001) *The Philippine Construction Industry in the 21st Century: Is There a Globalisation of the Local Construction Industry?* Report for the Sectoral Activities Department, ILO and the International Federation of Building and Woodworkers: Geneva.

10 Contractor development

Winston M. Shakantu

Introduction

The construction industry plays an indispensable role both directly and indirectly in the economic growth and long-term development of any country. The industry makes a huge contribution to efforts to improve the quality of life of peoples. Its products, buildings and other structures change the nature, function and appearance of towns and the countryside (DETR, 2000; van Wyk, 2004). Construction also provides governments in many parts of the world with economic, regulatory and public-sector policy and capacity delivery mechanisms which can be used to obtain various broad objectives (Hillebrandt, 2000; Fellows *et al.*, 2002; van Wyk, 2004). However, as the UK reviews of the construction industry (see, for example, Latham, 1994; Egan, 1998) have consistently shown, the industry is highly fragmented. In any country, the construction industry is composed of thousands of firms, most of which are small, medium and micro-enterprises (SMMEs). Ashworth (2006) posits that, in the UK, 95 per cent of construction firms employ fewer than a dozen people. This shows that the construction industry everywhere relies on the large number of small firms to undertake its geographically distributed productive activities.

SMMEs: a definition

The term 'SMME' conjures up a wide range of imagery. From the definition adopted in some countries, both the sole proprietor and a firm with 499 employees, and everything in between, are part of the SMME family (Little, 2005). There are several difficulties relating to attempts to define a 'small' business enterprise. The yardstick for delineating enterprises by size is usually one or more of the following: total number of employees, value of fixed assets, paid-up capital, annual turnover, annual volume of physical production. The nature of activity determines the viable and normal economic operating size (UNCHS, 1996; Seeley, 1997, p. 5). Therefore, there is no single definition (Burke, 2006). From whatever definition that is adopted, SMMEs encompass a wide range of firms, from established traditional family businesses down to self-employed informal enterprises.

The firms in the SMME sector can be classified into four broad groups: survivalist, micro, small and medium enterprises.

- The survivalist enterprise level is the lowest level of entry into the SMME sector and is where the largest number of enterprises are found. Survivalist enterprises mainly undertake 'survival-type' construction activities, which consist of very small projects. Survivalist enterprises mostly utilise family labour and seldom employ more than five people (Seeley, 1997, p. 6; Rebello, 2005). Survivalists do possess the minimum skills required to do the work at hand.
- The micro-enterprise level is where very small businesses, which usually involve the owner, some family members and not more than five paid employees, operate (DTI, 1995; Department of Public Works, 2006). They often work in the informal sector as they usually lack 'formality' in terms of business licences, value-added tax (VAT) registration, formal business premises, operating permits and accounting procedures (DTI, 1995). Most micro-enterprises are run by owners who have some skills in the particular business area in which they operate. The average annual turnover of micro-enterprises varies widely, and is up to R1,000,000 (approximately €125,000) in South Africa (Department of Public Works, 2006).
- Small enterprises are businesses that employ 6–60 people and generate between R1.1 million and R12 million per annum (Department of Public Works, 2006). They are usually owner managed and are likely to operate from business or industrial premises, be tax registered and meet other formal registration requirements (Berry *et al.*, 2002; Jewel *et al.*, 2005; Rebello, 2005). They also employ skilled personnel to carry out the work required (Shakantu *et al.*, 2006).
- Medium enterprises are a wide range of firms. These companies can employ from 61 to about 300 employees and generate between R12.1 million and R60 million per annum (Seeley, 1997, p. 5; Department of Public Works, 2006). Medium enterprises are also still usually owner/manager controlled (DTI, 1995; Jewell *et al.*, 2005; Rebello, 2005). Like small enterprises, medium enterprises employ skilled people (Shakantu *et al.*, 2006).

Table 10.1 presents a summary of the four categories of SMMEs.

The majority of SMMEs, comprising the survivalist and micro-enterprises, are concentrated at the lowest end of the scale in terms of size and capacity. The South African Construction Industry 2004 Status Report gives support to this statement by revealing that the emerging contractors enter the market at the lower end, and in the general building contracting category (cidb, 2004, p. 27).

Table 10.1 Categories of SMMEs

Category	Annual average turnover (million rand)	Number of employees
Survivalist enterprises	Variable but < 1	< 5
Micro-enterprises	~ 0–1	1–5
Small enterprises	~ 1.1–12	6–60
Medium enterprises	~ 12.1–60	61–300

Source: adapted from Department of Public Works (2006) and Burke (2006).

Role of SMMEs in the economy

Bearing in mind the difficulties with (and differences in) definition across countries, SMMEs are not peculiar to developing countries, but play an important role in the economies of all countries. For instance, the OECD (2004) reported that SMMEs account for over 95 per cent of enterprises and 60–70 per cent of total employment, and generate a large share of new jobs. UNCHS (1996) reported that SMMEs make up 91–93 per cent of industrial enterprises in the South-East and East Asian countries of Malaysia, Singapore, Taiwan and Thailand. They contribute 35–61 per cent of employment, and account for 22–40 per cent of total value added. Therefore, in summary, SMMEs are socially and economically important to all countries in that they typically represent 95 per cent of all enterprises and provide up to 50 per cent of employment. They also produce about 50 per cent of the gross domestic product (GDP) of most countries, and up to 55 per cent of all technical innovations (Burke, 2006).

SMMEs are important to almost all economies in the world, but especially to the developing countries and, within that broad category, particularly to those with major employment and income distribution challenges (Palma, 2005). UNCHS (1996) and Palma (2005) observe that both cross-sectional and time-series data show that the process of industrialisation usually begins with the rapid growth of small-scale enterprises. However, the role of such companies continues to be important throughout and beyond industrialisation because, at any level of a country's development, some activities that are needed involve few or no economies of scale (Palma, 2005).

Context of development of construction SMMEs

Many writers stress the importance of adequate construction capacity for national development (see, for example, UNCHS, 1996; UN-HABITAT, 2009). The cyclical relationship between an efficient and effective construction industry and national socio-economic development is also important (see, for instance, Ministry of Infrastructure, 2009, p. 8). Thus, construction SMMEs are a vital component of every country's economy. Moreover, construction SMMEs are important in that they undertake the numerous small, relatively simple, scattered and often isolated projects which are necessary for economic development and the social uplift of people in the rural areas and within urban communities (UNCHS, 1996; Fellows *et al.*, 2002; Ibrahim *et al.*, 2010).

Construction is a complex activity which is influenced by many factors. Several of these factors constitute problems for, and constraints on, enterprises involved in construction. Therefore, it is vital that contractors be capable of completing the necessary volume of the construction items required for socio-economic progress, and do this within the specified time, at reasonable and predictable cost, and with good quality, paying attention to the health and safety of the public and the workers on site, as well as using materials, equipment and methods that are least harmful to the environment (UNCHS, 1996; Barrie and Paulson, 2001; Harris *et al.*, 2006; Ashworth, 2006). However, according to UNCHS (1996), in developing countries, local contractors with such capabilities are in short supply. Moreover, when available, local contractors are ineffective and inefficient and also face a wide array of problems and constraints relating to their own management, organisational and technical skills, as well as the resources they require, the regulations they should comply with, and procedures adopted by clients and consultants to administer their work (UNCHS, 1996; Ministry of Infrastructure, 2009; Ibrahim *et al.*, 2010). Most of these local contractors are SMMEs. Therefore, it is vital that ways be found to stimulate the emergence of small contractors in developing countries, and support the continuous development and upgrading of such SMMEs (Ibrahim *et al.*, 2010).

Justification for construction SMME development

Owing to the importance of SMMEs in the economy, the promotion of such enterprises is a crucial component of governments' strategies to create employment opportunities, and foster economic growth and national development (DoF, 1996; Lewis, 2001). SMMEs provide a vehicle through which most of the underprivileged, who lack financial resources and skills, can typically gain access to economic opportunities (Gounden, 2000; Chandra *et al.*, 2001). Siddiqi (2005) posits that SMMEs are the potential engines of wealth creation, value reorientation, job creation and poverty eradication. This is particularly crucial for developing countries which are characterised by the legacy of big business dominance and huge unequal distribution of wealth (Gounden, 2000; Hirsch, 2005; Palma, 2005). The economic and social rationales for promoting construction SMMEs are also highlighted by Kesper (2000), Palma (2005) and Shakantu and colleagues (2006), who suggest that such enterprises use more labour-intensive techniques and tend to have a high capacity for the absorption of labour.

It is a policy of governments in many developing countries to promote job creation by providing support to small businesses and by promoting entrepreneurs (Rebello, 2005). The typical structure of the construction sector characterised by a preponderance of small firms bodes well for the policy of promoting SMME entrepreneurs (Shakantu *et al.*, 2006). Similarly, its labour-intensive nature and low barriers to entry signify the sector's potential for job creation. From the economic point of view, it is generally believed that optimising the contribution of SMMEs to employment and economic development can be translated into the following broad objectives:

- raising the rate of formation of new micro-enterprises with growth potential and increased contribution to investment, employment and income generation;
- increasing the rate of economic ownership;
- increasing the rate of graduation of micro-enterprises into SME businesses;
- raising the performance of existing micro-enterprises with a view to increasing their competitiveness; and
- decreasing the undesirably high mortality rate of micro-enterprises.

UNHCS (1996) also posits that there are some technical reasons for supporting the developmental effort with regard to construction SMMEs. Some of the more pertinent reasons it highlighted are as follows:

- Small-contractor development effort tailored to improvement of performance in quantity, time, cost and quality would assist the construction delivery system to reduce the high volumes of unmet needs.
- As small contractors mainly fulfil the construction needs of poorer sections of communities, any development efforts, especially those that would result in the reduction of the high cost of construction, would bring direct benefits to a large section of the population in developing countries.
- It is often necessary to build small projects in isolated parts of each country, which only SMME contractors are able and willing to undertake.
- The quality of work of construction SMMEs needs to be upgraded to enable them to give greater value for clients' money in the long term.
- The reliance on imported inputs should be reduced; with improved expertise, construction SMMEs can help to achieve this.

For small businesses to contribute effectively to the economy, they should evolve into efficient, well-organised, technically competent and well-managed operations which are able to respond to opportunities and challenges in their environment (UNCHS, 1996; Shakantu and Kajimo-Shakantu, 2007). They should be able to offer reliable products with dependable delivery and conformance to quality. They should be price-competitive and continuously improve upon their performance. They should focus on cost effectiveness, integrated quality actions, customer responsiveness, information technology management and human resource management.

Given the significant socio-economic role of SMMEs in developing countries, Dlungwana and Rwelamila (2004) suggest that there should be increased effort in the programmes that promote contractor development. Greater amounts of resources should also be dedicated to this task. Well-structured contractor development models and supportive procurement programmes should be implemented in order to improve technical and managerial skills, knowledge and, hence, competitiveness of contractors. Dlungwana and Rwelamila (2004) posit that there are some benefits that could accrue from effective contractor development. These include increased competitiveness, sustainable business growth, good environmental management and socio-economic development of the countries.

Theoretical framework

Many authors suggest that governments are involved in four main economic activities: providing a legal and institutional framework for all economic activities; redistributing income through taxation and spending; providing public goods and services; and purchasing goods, services and public works from the private sector (see, for example, Thai, 2001).

Although governments usually conduct these four economic activities passively, it sometimes becomes essential for them to intervene in economic affairs. There are two main reasons that increasingly justify state intervention in the economy. One is to correct instances of market failure, that is efficiency, and the other is to achieve a more equitable distribution of income and wealth, that is equity (Przeworski, 1998; Cook, 1999; Muradzikwa *et al.*, 2004; cidb, 2009, 2010). In the second instance, government acts as an active participant in the market and, broadly, in the economy. It is an important customer for public works, goods and services, and also a supplier of goods and services, many of which are vital for the functioning of the economy (Morand, 2003; Bayat *et al.*, 2004; Mohamed, 2005; cidb, 2009, 2010). Throughout the world, governments have generally accepted the responsibility of ensuring the delivery of services such as education, health, security, transport, and basic utilities such as water, sanitation and electricity. In order to provide these services, the governments invest in building and maintaining infrastructure such as schools, housing, hospitals, roads, bridges, power plants, airports and ports (Rogerson, 2004; Govender and Watermeyer, 2001).

Therefore, public procurement is an important activity for any government, and this investment contributes significantly to the gross domestic product (GDP) (Morand, 2003; Rogerson, 2004; McCrudden and Gross, 2006). Morand (2003) found that, on average, public procurement contributes about 10–15 per cent to GDP in most European countries, and 20 per cent in Latin America, whereas other studies found that the figures are about 12–25 per cent in western Europe (International Council for Local Environmental Initiatives, 2002) and about 10 per cent in the United States (Bajari and Tadelis, 2001; Marion, 2005).

Public procurement has two goals: primary and secondary (Arrowsmith, 1995; McCrudden, 1995; Thai, 2001). The primary goal relates to the sourcing of services, goods and works to achieve good governance. The secondary goal, argues McCrudden (1995, p. 5), relates to the use of public procurement to achieve a wide range of economic and social objectives including (i) stimulating economic activity; (ii) protecting the national industry against foreign competition; (iii) improving the competitiveness of certain industrial sectors; (iv) remedying regional disparities; and (v) achieving more directly social objectives such as (a) fostering job creation to promote the use of local labour, (b) encouraging equality of opportunity between men and women, (c) improving environmental quality, (d) promoting the increased utilisation of the disabled in employment and (e) prohibiting discrimination against minority groups.

Watermeyer (2002, p. 210) observes that 'procurement provides business and employment opportunities and, depending on how it is structured and conducted, can be used as an instrument of government policy to facilitate social and economic development'. As a major client to most of the sectors of the economy, the government can leverage its large purchasing capability to influence the behaviour of economic participants and outcomes (Dayal *et al.*, 2004; McCrudden, 2004; Thompson, 2004), particularly in sectors such as defence and construction. Indeed, many studies suggest that several developed and developing countries use procurement as a vehicle to achieve a variety of socio-economic objectives (Arrowsmith, 1995; Govender and Watermeyer, 2001; Rogerson, 2004). Examples of such countries include Botswana, Brazil, Canada, India, Malawi, Malaysia, Namibia, Nigeria, the Philippines, Singapore, South Africa, Spain, Sri Lanka, Tanzania, the UK, the USA and Zambia (Govender and Watermeyer, 2001; Sowell, 2003; Rogerson, 2004). Public procurement is used to promote objectives such as labour standards and employment creation (Boston, 1999; Watermeyer, 2002; cidb, 2009, 2010). Ohlin (1992) and cidb (2009) note that governments play an important role in socio-economic development because, in principle, they have a coercive power that can further the socio-economic transformation process. Hence, it is common to find public procurement being used in some countries, for instance South Africa, to support contractor development (cidb, 2009, 2010).

Impetus for contractor development

UNCHS (1996) posits that the lack of appreciation of the importance of construction in the development process has led to only a small number of systematic efforts being made in developing countries to create sizeable, efficient and effective national construction industries.

However, recently some developing countries have implemented initiatives for improving upon the performance of local construction contracting firms, either as part of, or in isolation from, comprehensive programmes for improving the capacity and capability of the construction industries (see for instance Ministry of Infrastructure, 2009).

The momentum for action to improve the performance of local construction SMMEs has been a recognition of the difficulties associated with infrastructure delivery (Ministry of Infrastructure, 2009). A beneficial consequence of improving infrastructure has been the potential it creates to improve the business environment for existing enterprises and to promote investment in new business activities (Mohamed, 2005; Siddiqi, 2005). As increased economic activity associated with infrastructure development cascades through the economy and across the sectors, an even larger volume of spending on infrastructure ensues (Venter, 2006). At this rate of infrastructure and macro-economic expansion, the construction industry becomes extended beyond its limits and starts to face capacity constraints (cidb, 2006). The significant expansion of large-scale projects consequently creates a supply gap at the lower end of the market: the small municipal infrastructure projects (Campbell, 2005; Creamer, 2005; Mowson, 2005). It is at this level that SMMEs, especially survivalists and micro-enterprises, enter the construction market, and that is where most of them operate throughout their corporate lives (cidb, 2004; Hauptfleisch *et al.*, 2005).

Prerequisites for success in small contractor development

Need for entrepreneurship

Entrepreneurship is an important success factor in any business endeavour (Pearce and Robinson, 2007). A number of researchers (Ofori, 2000; Dlungwana and Rwelamila, 2004; Flanagan, 2007) agree that the high failure rate of local contractors in developing countries is largely due to the lack of entrepreneurship. This lack of entrepreneurship results in the inability of SMMEs to seek and exploit business opportunities, and to innovate. In addition, lack of entrepreneurship limits the ability of SMMEs to successfully innovate, create opportunities for themselves, and grow their enterprises (Gerber, 2008). Burke (2006) suggests that the key entrepreneurial traits are ability to spot and respond to available business opportunities; self-confidence to pursue and acquire work; being passionate about the firm's product or service offered to clients; networking; determination; and appetite for risk. Considering that there are few barriers to entry into the construction industry, and that, as a consequence, competition within the industry is keen, companies require entrepreneurship in order to have a sustainable workflow and, hence, good growth prospects (Rebello, 2005; Cheetham and Mabuntana, 2006; Longenecker *et al.*, 2006).

Need for expertise

Construction, even at the level of small projects, is technically challenging and requires a wide range of skills to ensure success. Therefore, construction SMMEs should have above-average decision-making, technical, and cost and time management skills. Among other areas, they should have expertise in financial analysis, loss control, preparation of claims, administration of safety, and environmental management. However, developing countries generally suffer from a skills shortage as a result of many years of underinvestment in human capital development (Kapur and McHale, 2005). For this reason, a large percentage of the populations of these countries are either unskilled or have very limited skills. Thus, education and training policies need to be tailored towards improving labour employability, productivity and flexibility.

Access to finance

SMMEs generally, but particularly survivalist and micro-enterprises, suffer from lack of financial resources, which has a negative impact on their ability to grow and develop. Shakantu and colleagues (2006) report that the targeted programmes launched by governments for increasing access to finance, such as those in South Africa, have recorded only limited success because of low awareness and usage by the intended beneficiaries. The problems of insufficient access to finance are further compounded by high interest rates. In addition, reports show that SMMEs have difficulties in gaining access to information, lack market exposure and are discriminated against by financial institutions because they lack sufficient collateral to back their applications for loans (Gounden, 2000).

Govender and Watermeyer (2001) point out that the requirement for a performance bond presents significant financial obstacles for SMMEs. As a result of the greater risk factor which they are considered to present to the insurers, SMMEs are forced to obtain their performance bonds at significantly higher rates than larger enterprises. In South Africa, even the changes to bond requirements such as reduction of the bond amount to between 2.5 and 10 per cent, depending on the risk classification of a contract, have barely reduced this problem (Govender and Watermeyer, 2001). The inadequacy of the external finance which they can obtain at the critical growth and transformation stages of SMMEs hinders the effort of the enterprises with growth potential to expand (Nissanke, 2001). Therefore, although governments have made some effort to increase the access of SMMEs to finance, by and large the majority of those companies have very limited access to external finance.

Business angel finance and venture capital are two suggested alternatives to funding from the banks for emerging micro-enterprises which do not have collateral or equity resources that are sufficient, or in the conventional form, to enable them to qualify for loans from the banks. These initiatives have, to a limited extent, proved useful to bridge the gap in funding for micro-enterprises. For example, angel financing has been used successfully in South Africa, the UK and the United States (Wong, 2002; Siddiqi, 2005; Botes, 2009).

Access to work opportunities

In addition to the issues highlighted above, many SMMEs face problems of lack of market exposure or access to work opportunities. As there are large numbers of SMMEs in the construction industry, competition within this part of the market is intense, leading to difficulty, particularly for new entrants, in having a sustainable workflow. This inability to sustain workflow has a negative impact on the ability of the enterprises to create sustainable employment and economic empowerment (Rebello, 2005; Cheetham and Mabuntana, 2006; cidb, 2006).

In response to this problem, the governments of many countries have attempted to provide work opportunities through preferential procurement, but these are limited on account of the large number of SMMEs in each country. Given that not many SMMEs can access public-sector contracts owing to the keenness of the competition for the limited work opportunities, sub-contracting tends to provide a viable option for SMMEs to deal with the problems they face in obtaining the volume and regularity of workflow they desire.

Sub-contracting arrangements occur between micro-enterprises and medium and small enterprises or between micro-enterprises and large firms. This can provide a route through which SMMEs can acquire experience, build up finances and also develop business linkages, and eventually enter the formal economic sector. Large firms undertaking major infrastructural contracts can help micro-enterprises to grow and develop by sub-contracting some of their work to them (Siddiqi, 2005; cidb, 2009, 2010).

Supportive regulatory environment

Legislation is one of the factors of the external business environment which affect the performance of enterprises. Kesper (2000) and Bhorat and colleagues (2002) allude to the costs and inflexibility associated with compliance with labour and employment laws. Thus, a commonly perceived constraint for SMMEs is the set of labour laws which are said to raise the cost of employment, artificially prolong retrenchments or corrective action with respect to human resources, and do not allow companies to have adequate flexibility, especially in the determination of wage levels and arrangement of the working time (Bhorat *et al.*, 2002). It follows that any change in the legislative framework disrupts the operations of businesses. As a result, enterprises feel a profit squeeze, and this has an impact on their willingness to create additional jobs, as well as their ability to grow.

Other critical success factors for construction SMME development

The project and financial management ability of the contractor is a critical success factor. Other critical success factors include (i) the ability of contractors to market their services among the industry role players (Kotler, 2002; cidb, 2010); (ii) the experience and management expertise of the owner (UNCHS, 1996; cidb, 2010); (iii) the ability to maintain a good relationship with clients, suppliers and other relevant role players (Day, 1997; Kale, 1999; Winter and Preece, 2000; cidb, 2010); (iv) project management capability, marketing and supply chain relationship (Jaafar and Abdul-Aziz, 2005); (v) the understanding of contractual requirements, progressive estimating, scheduling, purchasing, knowing what has to be done and how, and being flexible enough to adjust to changing situations (UNCHS, 1996; cidb 2010); and (vi) competence and skills, adequate resources, proper timing of activity planning and performance, teamwork, effective communication, fair dealing with people, honesty and integrity (Holroyd, 2003).

Causes of failure in construction SMME development

The problems which confront construction SMMEs are not different from those faced by SMMEs in general. The key constraints to micro-enterprise development and growth include labour regulations; restrictive environmental, business, trade and tax regulations; scarcity of skilled labour; inadequate infrastructure; high tax rates and cost of capital; changing government policies; corruption; volatility of the exchange rates; crime and theft; and keen competition for limited opportunities (Bhorat *et al.*, 2002). The impact of these constraints on construction SMME development varies.

Implications of construction SMME successes and failures for industry development

In the discussion so far, the potential that lies in SMMEs and the constraints that must be addressed in order to realise that potential (growth and development) have been highlighted. This section explores some of the implications of construction SMME successes and growth constraints for construction industry development.

To achieve the nation's socio-economic objectives, the government should create an enabling environment for business and opportunities for more labour-intensive activities (cidb, 2005; DTI, 2006). The first step is to build up small businesses to bridge the gap between the formal and informal sectors of the economy. Here, the government's investment in infrastructure could be used to improve socio-economic conditions (le Roux, 2005a,b). Therefore, it is important to find ways for infrastructure investment to benefit SMMEs through the development of technical solutions to assist them (le Roux, 2005a,b). This can be achieved through increasing the levels of public-sector investment to promote small businesses and addressing issues such as the impact of regulations on labour-intensive sectors of the economy.

The second is the ramping up of central government and other public-sector enterprise investment to support the first step; as the large contractors' order books and capacity become overextended, opportunities open up for small businesses to enter the construction market (Venter, 2006). The third is the revamping of specific sectoral regulatory frameworks which hamper the development of businesses (Creamer, 2005; DTI, 2006). In addition, because the construction industry's products are of a capital investment nature, it is economically logical that the expansionary fiscal framework is aimed at increasing the demand for construction and, in turn, increasing employment creation capacity (Rebello, 2005).

At present, with double-digit unemployment rates in many countries, especially the developing nations despite good levels of economic growth, job creation seems to be an elusive concept for economic planners. According to the Global Entrepreneurship Monitor (GEM), the main reason why job creation eludes economic planners is that micro- and survivalist enterprises, by definition, are unlikely to create significant numbers of jobs (Jewell *et al.*, 2005; Von Broembsen, 2005). Moreover, there seems to be a mismatch between the policy objectives of promoting SMMEs as vehicles for poverty alleviation and that of considering them as agents of employment generation (Von Broembsen, 2005). The root cause of this mismatch is that, currently, SMME policies tend to focus on the redistribution of wealth or poverty alleviation and not so much on training and education in entrepreneurship for small business owners. This apparent mismatch compromises the potential of SMMEs to create more jobs and to nurture the majority of survivalist and micro-enterprises into profitable and sustainable SMMEs (cidb, 2006). Von Broembsen (2005) proposes that, in the short run, governments should shift their focus away from SMME programmes that support the redistribution of wealth or poverty alleviation to those that develop entrepreneurial skills and create jobs. Similarly, Kajimo-Shakantu and Root (2006) suggest that governments' priority should be to grow the economies so that they can expand and increase opportunities in which the majority of the populations can participate. Kajimo-Shakantu and Root (2006) conclude that the identification and addressing of factors in the construction sector which are crucial to enhancing economic growth is a critical success factor in these regards.

Wider developmental objectives are also cited for schemes aimed specifically at developing contractors. Watermeyer and colleagues (1995) suggest that the key objectives of a contractor development programme are to (i) optimise job creation opportunities in the construction of infrastructure and housing; (ii) encourage the creation and sustainability of small-scale enterprises; (iii) strive towards fulfilling the country's projected construction needs; and (iv) enhance the benefits accruing to the community through their involvement in construction.

From the above discussion, a virtuous cyclical link between contractor development and national development can be envisaged.

Lessons

Lessons from past contractor development schemes include (i) implementation of a comprehensive programme attending to all the needs of the small contractor; (ii) careful and systematic screening of firms to be supported; and (iii) avoiding dependency among the beneficiaries (UNCHS, 1996; cidb, 2009, 2010).

According to UNCHS (1996) a number of inferences can be made from past attempts to improve upon the performance of small contractors in developing countries. First, large companies can be used to develop small ones through technical and managerial training, demonstration of technology and stronger and integrated intra-industry linkages for the mutual benefits of both small and large firms. Second, measures for protecting the local construction industry do not, by themselves, develop the indigenous contractors; such measures should be specifically targeted at firms in a position to benefit from them. Third, contractor development programmes should be comprehensive and integrated, and seek to address the needs of SMMEs on a broad front, rather than tackling the needs of such firms on a piecemeal basis. Fourth, contractors' associations can play a key role in small-contractor development; national forums embracing all interests in construction and regional federations of associations can seek the welfare of contractors, and enable the exchange of experiences and ideas. Finally, the support and commitment of small contractors is essential to the success of contractor development programmes; interactive learning by contractors should be encouraged.

Proposals for action

The development of small contractors in each country should be pursued with a two-pronged strategy. First, efforts should be made to improve the capability and capacity of contractors. This will make them better able to deal with the problems and constraints resulting from the nature of their operating environments (cidb, 2009, 2010). The second part of the strategy is to take measures to eliminate or at least relieve the effects of these problems and constraints (Belt and Richardson, 2005). However, it should be noted that the relevant environmental factors of each country – such as its construction market; the nature of its industry; economic, administrative and political circumstances; developmental objectives; and future prospects – can affect SMME development efforts. Targets should be well defined and, wherever possible, measurable (cidb, 2009, 2010).

More specifically, the contractor development programmes should seek to (i) influence the government to formulate or appropriately alter its construction-related policies; (ii) set and publicise standards and targets to be achieved by SMME contractors; (iii) provide SMME contractors with training, and help them to gain access to funding; and (iv) provide contractors with financial incentives and advisory services, as well as information support.

Models of contractor development: the case of South Africa

The South African government's support for the development of SMMEs is well documented and is a strategic objective which the country is striving to attain. This is evinced by the government's creation of an enabling environment for SMMEs since 1995. The government has achieved this through legislation, amongst which is the National Small Business Development Act of 1996 and a host of other legislation which has been promulgated subsequently.

In 1995, the South African government adopted the White Paper on the National Strategy for Development and Promotion of Small Businesses in South Africa. The main objective of this white paper was to create an enabling environment for the accelerated growth of small enterprises following a history characterised by the dominance of large, capital-intensive companies and the continued neglect of small enterprises. A target of 10 years after which the benefits of the above-mentioned strategy would be assessed was set. A number of industry support measures have been introduced since 1994 to enhance SMME development. These include placing more emphasis on supply-side rather than demand-side measures, such as improved access to finance. In 2005, the government (through its Department of Trade and Industry) consolidated its support for SMME development initiatives through the formulation of an Integrated Small Enterprise Development Strategy.

Around 1997, a mechanism to ensure the sustainability and development of the construction industry was put in place. The aim was to create an enabling regulatory and development framework for the construction industry through the enactment of the Construction Industry Development Board Act and the publication of the White Paper on Creating an Enabling Environment for Reconstruction, Growth and Development in the Construction Industry in 1999. The documents provided for the establishment of the Construction Industry Development Board (cidb) with the principal objective of promoting and stimulating the development, improvement and expansion of the construction industry. In enabling the creation of the required new industry capacity, government also wished to promote the participation and growth of SMMEs through preferential procurement in support of historically disadvantaged sectors of South African society.

However, the many strategies targeted at the development of SMMEs, particularly those based on instruments such as preferential procurement, generally achieved little with regard to empowering SMMEs because they were implemented without well-defined skills transfer frameworks. The South African construction industry and its constituent firms continued to be regarded by clients and suppliers as a high commercial risk, and this presented further barriers to meaningful development of SMME contractors (cidb, 2003). Related to this was the lack of clear policy targets against which to measure the effectiveness of contractor support programmes. Thus, the gaps between the established contractors and the SMMEs were not being reduced.

The South African government is currently trying to reduce the gaps, through the implementation of the Integrated Strategy on the Promotion of Entrepreneurship and Small Enterprises, launched in 2007. In construction, the government, through the Department of Public Works (DPW) in collaboration with other stakeholders such as the cidb and Construction Education and Training Authority (CETA), has embarked on a variety of interventions to support SMME contractor development (cidb, DPW and CETA, 2005). The interventions are located on both the demand and supply sides. Some of the more relevant interventions are now discussed.

Access to finance and venture capital

Earlier in the chapter, business angel finance and venture capital were highlighted as two initiatives created for SMMEs in South Africa which do not qualify for loans from the banks (Siddiqi, 2005). The Small Business Act 1996 created three institutions – Ntsika Enterprise Promotion Agency, Khula Investment Finance and the National Small Business Council – to provide management support, funding and lobbying, respectively, for the enterprises in the SMME sector. In addition, the government has provided substantial amounts of start-up capital to support small businesses. This has mainly been achieved through the DTI and its various agencies such as Ntsika and the Industrial Development Corporation (DTI, 2003a,b).

Education and skills development

South Africa has a history of systematic underinvestment in human capital. This has resulted in a labour force with a skewed distribution of managerial and craft skills, workplace experience and, hence, career opportunities. To address the skills imbalance, the government enacted the Skills Development Act (SDA) of 1998, among other things. The aim of the SDA is to create the structures and framework for the National Employment and Skills Development Strategy. The implementation of the Act benefits from the work of the Human Resource Development Branch of the Department of Labour. It provides for the establishment of the National Skills Fund (NSF) and the CETA through the accumulation of monthly skills development levies of 1 per cent of the budgeted payroll, paid by employers (Kesper, 2000).

Preferential procurement and sub-contracting

Preferential procurement policies are one of the strategies which the government of South Africa is implementing to bring historically disadvantaged enterprises (HDEs) into the mainstream economy. The government sees preferential procurement in the construction industry as a vehicle through which national socio-economic objectives can be realised whilst, at the same time, contributing to the development of the industry. Preferential procurement comprises participative programmes aimed at the engagement of SMMEs owned by previously disadvantaged persons (PDPs) and increasing the volume of work available to the poor whilst simultaneously generating income for the previously marginalised sectors of society (Govender and Watermeyer, 2001).

As a practice, preferential procurement entails giving preference to HDEs in the awarding of public-sector contracts. Through preferential procurement, large construction contracts are broken down into smaller packages in order to give SMMEs the opportunity to compete for them. Using preferential procurement, a number of programmes aimed at contractor development, such as the Emerging Contractor Development Programme (ECDP), Extended Public Works Programme (EPWP), learnership programme and the DPW's Contractor Incubator Programme, have been launched, and are being implemented. These aim to stimulate the access to markets, and develop entrepreneurial skills, of targeted beneficiaries through public-sector construction contracts. Sub-contracting is considered to be one of the ways by which HDEs can participate in public-sector procurement (Siddiqi, 2005).

Broad-based black economic empowerment

Some of the preferential procurement policies are currently being implemented through the policy provisions enshrined in the Broad-Based Black Economic Empowerment (B-BBEE) Act of 2003, which provides the primary statutory framework for the promotion of black economic empowerment (BEE) in South Africa. The objectives of the Act, where the construction industry is concerned, are implemented through the Construction Sector Transformation Charter. The objectives of the charter include:

- the transformation and growth of the sector;
- improvement in the competitiveness and efficiency of the sector;
- the achievement of a substantial change in the racial and gender composition of the ownership, control and management within the sector;
- addressing the critical skills shortage and skills development with a specific focus on women; and
- enhancing entrepreneurial development and sustainable development of black economic empowered SMME construction companies through strategic partnerships.

Therefore, the charter provides a framework for the South African construction industry to address B-BBEE, enhance capacity and increase productivity. It also provides for a system of quantitative measurement, through monitoring and evaluation, of progress made towards the attainment of B-BBEE by enterprises.

National Contractor Development Programme

The National Contractor Development Programme (NCDP) was launched by the cidb (2010). The aim of the NCDP is to unlock growth constraints and to develop sustainable contracting capacity in South Africa, as well as to elevate the development of previously disadvantaged individuals and enterprises. This is achieved through the enhancement of capacity and promotion of equity ownership across the different contracting categories as well as improving skills and performance in the delivery of capital works and maintenance across the public sector. Through the NCDP, the government intends to (i) create sustainable contracting enterprises by enabling continuous work through a competitive process, thus creating the platform for sustained employment and skills development; (ii) improve the performance of contractors in terms of quality, employment practices, skills development, safety, health and the environment; and (iii) improve the business management and technical skills of these contractors.

The key principles underpinning the NCDP are that (a) government would use its entities and partnerships, and its procurement of infrastructure in order to achieve contractor development; (b) within the NCDP, good performance of contractors would be rewarded by ongoing opportunities; and (c) contractors would enter the programme based on predefined criteria and would receive support to enable contractors to exit the programme on the basis of achieving predefined criteria relating to skills, qualifications, certification, sustainability, quality and so on. In other words, the programme is designed to avoid the development of a 'dependency syndrome' among the contractors benefiting from it.

Firms participating in the NCDP would have to go through the various components of the development programme in order to become not just competent in their fields of operations, but experts, and to grow and improve their performance. The typical components are now discussed.

Construction workforce development

The development of the construction workforce is one of the key components of the NCDP. It is pursued through artisan and supervisor training and development.

Contractor development

The main component of the NCDP focuses directly on the development of contractors and comprises several sub-components starting at the emerging contractor stage and progressing to the stage which seeks to develop the contracting enterprises. The sub-components of this unit are described below.

- *Emerging contractor*: This component of the NCDP uses learnerships predominantly incorporating mentorship in which the emerging contractor learns the business side of contracting including tendering for work, pricing work, human resource management, marketing, financial management and contract administration. A budget is allocated under the programme to ensure that there is a sustainable workload for the learner contractors.
- *Enterprise development*: When the enterprises start growing, the NCDP incorporates a unit which helps them in developing markets for their services, expanding their workforce and areas of operation, accumulating capital and plant and equipment for future growth, and for the acquisition of business and technical systems. Key instruments used within this stage include the provision of work for the beneficiaries through a combination of direct contracts and joint ventures. In this stage, contractors are awarded contracts through competitive bidding, utilising appropriate procurement strategies to ensure a sustainable flow of work to the contractors within the competitive bidding environment.
- *Performance improvement*: The established enterprise introduces best practice systems for health and safety, quality management and environmental management in order to improve its performance. Key instruments used include a combination of joint ventures and direct contracts as well as the provision of support under the cidb's support schemes.

Case studies

To illustrate how contractor development is implemented in South Africa, detailed studies of three PDE contractors were undertaken. The empirical studies, using the philosophically phenomenologist paradigm and epistemologically subjectivist case study method, were conducted on a purposefully chosen number of contractors. The contractors were selected using the snowballing technique. The primary selection criteria were size and availability of information on the contractor's development profile.

A questionnaire with 11 questions was used as the data collection instrument for the research. The questions were as follows:

1   What grade of the cidb registers is your construction firm registered in?
2   What was the first grade you were registered in?
3   How long did it take to move from the first grade to where you are now?
4   What steps did you take to ensure upward movement within the cidb registration ladder?

5   Have you had any assistance from the national contractor development programme?
6   What benefits have you had from being B-BBEE supportive/compliant?
7   Have you had any direct government assistance such as contracts within the national contractor development programme?
8   What assistance have you had from the cidb?
9   What is your ratio of private versus government (state-owned enterprises [SOE])/sponsored/funded projects?
10 In a nutshell, can you please describe the development of your construction firm from a small contractor to where it is now as a giant in the construction industry?
11 What could you attribute this phenomenal growth to (within the lines of contractor development)?

The three case studies are presented in the following section.

Boshard Construction

Boshard Construction was founded in 1973 by Keith Boshard. As head of the family business, Mr Boshard fostered the ethos of a family spirit, attracting relatives and friends into the employment of the firm and investing in the training and development of all its workers. As a result of this policy, by the end of 1987, all but two of Boshard's senior executives, responsible for the daily operations, were previously disadvantaged individuals although they did not have any equity stake in the firm.

In November 1997, the directors persuaded Grinaker-LTA, one of South Africa's largest construction groups, to participate in buying out the previous owner by taking a 30 per cent stake in the firm and to finance the existing management in a buyout of the remaining shares. This resulted in a change of the shareholding of Boshard Construction to 70 per cent ownership by previously disadvantaged individuals and 30 per cent by Grinaker-LTA. The two enterprises, Grinaker-LTA and Boshard, co-operated with respect to procurement and the use of heavy plant whilst operating as two independent and autonomous entities. In March 2006, the management of Boshard bought back Grinaker-LTA's 30 per cent shareholding. Thus, currently, Boshard Construction is 100 per cent owned by previously disadvantaged individuals.

In terms of its development as a contractor, Boshard started off at the Grade 7 General Building,[1] Potentially Emerging[2] level of cidb registration (because it had existed since 1973 and had a track record). Currently, it is registered in the Grade 8 General Building, Potentially Emerging and Grade 7 Civil Engineering, Potentially Emerging category. It took the firm about five years to develop the profile required to progress from the lower to the higher grading. As Boshard did not directly benefit from the NCDP, it had to work hard to secure a substantial volume of projects, and increase its annual turnover and net value. In the pursuit of this outcome, the company was, to some extent, assisted by being compliant with the B-BBEE programme, and thus it gained access to government contracts awarded through competitive bidding. The joint managing directors, Gavin Hunter and Shaun Pearce, have been instrumental in this drive, which saw Boshard's annual turnover of R27 million in 1997 rise to R100 million in 2007. The firm's annual turnover increased to R200 million in the year ending June 2009. Gavin Hunter attributes Boshard's phenomenal growth to maintaining a good and satisfied client base, teamwork, well-motivated personnel and good levels of productivity, all of which are essential for getting repeat business from clients.

Realising how essential it is to foster other SMMEs, particularly those from previously disadvantaged backgrounds, Boshard has the practice of finding sub-contractors from that background and assisting them to purchase materials or hire equipment. This policy has resulted in their bonding such companies to them and creating many standing relationships with teams capable of producing high-quality work. Some of Boshard's sub-contractors now work almost exclusively for the firm.

The majority of Boshard's current personnel have benefited from the company's active human resource development programme involving, in particular, the transfer of skills and self-advancement. The diversity of Boshard Construction's experience puts the company in a position to handle almost any building contract and a certain amount of civil engineering work.

Boshard Construction's safety policy is to place the well-being of its employees ahead of any other consideration. To this end regular safety training is carried out and the company has a safety record far better than that of the average construction contractor. Boshard believes that health and safety is a basic human right. Therefore, Boshard requires that its personnel be competent to perform tasks that may have an impact on safety, health and environment in the workplace. Boshard has a methodology to identify hazards, the assessment of risk and the implementation of the necessary control measures.

GVK-Siya Zama Construction

GVK-Siya Zama was founded in 1994 when the Gordon Verhoef and Krause Group entered into an agreement with employees to establish an empowerment company. Therefore, GVK-Siya Zama has its roots firmly in the country's transition to a democratic South Africa. GVK-Siya Zama's vision was to develop a fledgling enterprise into a full-scale contracting operation under the hands-on management and control of its black shareholders. GVK-Siya Zama has since seen exponential growth, through prudent management and strong controls, and is now a significant contractor in its own right, with a national footprint and offices in all the major centres in South Africa.

GVK-Siya Zama operates as a group of four independently registered firms each with its own geographical base. The group's bases are in the Western

Cape, Eastern Cape, KwaZulu-Natal and Gauteng provinces. Currently, the group's components are individually registered with the cidb as follows:

- GVK-Siya Zama Western Cape in Grade 8 General Building, Potentially Emerging;
- GVK-Siya Zama Eastern Cape in Grade 8 General Building, Potentially Emerging;
- GVK-Siya Zama Gauteng in Grade 8 General Building; and
- GVK-Siya Zama KwaZulu-Natal in Grade 7 General Building.

However, in combination, the group can also tender as a Grade 9 General Building firm.

From humble beginnings, GVK-Siya Zama reached the various levels in which it has been registered in the Western Cape and three other provinces in 2005. It took approximately three years to develop from Grade 7 General Building to the current ones. To ensure an upward movement on the cidb registration ladder, GVK-Siya Zama had to increase its financial base and the contract value of projects on its order books. Moreover, it also had to seek and secure professional registration of its core engineers and project managers. Today, GVK-Siya Zama is a medium-sized building firm with an annual turnover of R740 million (in the year ending in February 2010). It employs a total of approximately 1,400 persons countrywide, which, at any given time, may be supplemented by up to 2,000 people employed by its sub-contractors.

Although the company had no direct assistance from the NCDP, it benefited from B-BBEE compliance through opportunities to competitively tender for government and SOE projects as well as those for multi-national corporations. GVK-Siya Zama's current order books comprise 60 per cent government and SOE and 40 per cent private-sector projects. The company attributes its phenomenal growth and success to the application of good management systems; investment in staff, training and upliftment; conservative, managed growth; careful selection of staff; and cultivation of a healthy corporate culture. Says Amelia Keefer, the firm's director of corporate communications:

> The company structure and systems afford uniformity of delivery countrywide, combined with the diversified strength and reach of a national organisation. Strong internal management systems and controls are continuously reviewed and upgraded to match and promote growth and ensure effective delivery.

'Siya Zama' is a Zulu expression which means 'We are trying'. Few words can explain more accurately the company culture and philosophy of this black-empowered, specialist construction company. Over time, these words have evolved into the company's slogan and motto – 'Together we try' – a tribute to transformation and investment in its greatest asset, its people. The company's growth and success signify the values of teamwork, nurturing of employees, support and development.

GVK-Siya Zama considers training to be important in countering the skills shortage, improving what it can offer to clients, and ensuring long-term sustainability, growth and retention of personnel as well as positioning itself as a preferred employer of skilled personnel. The broad-based training which the company offers includes learnerships and apprenticeships for general and semi-skilled workers, development of site management from leading hands to foremen and contracts managers, as well as training of plant operators, health and safety representatives and first aid providers. The company's management actively focuses on areas where skills shortages exist and creates development opportunities for both permanent staff and contract workers. A substantial investment in IT-based process management systems on a company-wide basis is also under way to ensure increased effectiveness in its operations.

Rainbow Construction

Rainbow Construction was established in 1995 as a B-BBEE contractor. As a result of that background, Rainbow is currently a dynamic, high-skilled black-empowered firm with a reputation for being one of the first major construction companies in South Africa to have a black majority shareholding. Rainbow Construction operates in both building and civil engineering construction. Its organisation consists of four divisions: (i) Roads and Earthworks; (ii) Infrastructure (comprising civil works, mining and industrial mining); (iii) Building; and (iv) Services. The four divisions share common services with respect to administration, human resources, finance and procurement. Each of the four divisions is led by a director who also serves on the board of directors of Rainbow Construction, along with the chairman, the chief executive officer, the chief finance officer and the procurement director.

Since its establishment, Rainbow has grown steadily over the years. The company has stayed true to its vision of developing a strong skills, plant and financial resource base to equip itself to undertake a broad range of construction projects. This development strategy, which includes a strong emphasis on BEE, has enabled Rainbow to attain a Grade 9 (with no limit on tendering capacity) registration in both the Building and Civil Engineering fields with the cidb. It also has a level 2 B-BBEE certification.

Rainbow's annual turnover has grown from about R130 billion in 2001 to R590 billion in 2009. Its net asset value has also grown from about R75 billion to R490 billion over the same period. The company attributes this growth to (i) success in winning high-profile World Cup 2010 contracts; (ii) involvement in public–private partnership projects; (iii) participation in government and SOE contracts emanating from attaining one of the highest

B-BBEE ratings within the industry; (iv) a very high rating by the cidb; and (v) development and retention of excellent relationships with the company's industry partners.

Conclusion

Owing to their size and nature, small businesses tend to be adaptive and innovative. These attributes make SMMEs the logical vehicles for the redistribution of wealth, business opportunities and the implementation of poverty alleviation programmes, whilst simultaneously generating employment (Thwala, 2005). Focusing on the development of entrepreneurship in developing contractors would complement the efforts of the Department of Trade and Industry, because both initiatives seek to promote technological know-how within the SMME sector through the deployment of skills, knowledge and innovation critical to the improvement of productivity. Developing entrepreneurship would drive competitiveness at the level of the enterprise. Therefore, it appears to be prudent to harness the potential of SMMEs to deliver the government's programmes in order to attain its policy objectives (Bruce, 2003).

However, the SMMEs face a number of constraints in their effort to enhance their competitiveness in their industry. The South African government recognises that these constraints cannot be dealt with through normal market forces and private-sector action (DTI, 1995). The government contends that SMMEs need to be assisted to overcome the obstacles that prevent them from attaining their potential. It recognises that assistance programmes for SMMEs should be in the form of capacity building, training, provision of business advice (such as financial and legal advice, and on the development of business strategies) and counselling. In addition, it is necessary to provide SMMEs with adequate access to technological and financial resources. These resources are required for buying in technical and marketing expertise and for developing production technology that is crucial to the continued survival and maturing of SMMEs (Shakantu *et al.*, 2002).

Integrated efforts at the local and national levels are essential to the success of the empowerment and development of the SMMEs. It is also important that the efforts of public and private institutions in contributing to the sustainable socio-economic development of the SMME sector in South Africa and elsewhere be effectively integrated if they are to bear fruit.

In order to meet the demand and growth targets arising from the investment of the government and parastatals in South Africa in infrastructure, the construction industry will have to double its output. To be able to do so, the industry will need more skilled workers. The promotion of micro-enterprises through targeted programmes can assist in the development of the industry's human resources. However, programmes such as the preferential/affirmative procurement policies have had limited success in achieving their objectives because they have not included entrepreneurial training components. Such training and provision of finance should be the focus of the programmes for the development of SMMEs. This will make it possible to harness the potential of the SMMEs to improve infrastructure delivery in addition to the creation of jobs and the consequent alleviation of poverty.

The example from South Africa shows the importance of an integrated set of initiatives which is, at the same time, comprehensive, and is planned and implemented from a long-term perspective, drawing on the strengths of both the public and private sectors. However, the case studies of Boshard, GVK-Siya Zama and Rainbow Construction demonstrate the difficulty of co-ordinating the varied but closely inter-related aspects of SMME development and their influencing factors.

Government policies which are assisting in the establishment of microfinance facilities, such as angel finance, should be supported by the Department of Trade and Industry, and considered as an important component of the government's strategy to reduce poverty and inequality (Rebello, 2005). Moreover, the existing initiatives merit further financial support, extension and fine-tuning in order to boost micro-enterprise development and to expand the ranks of emerging entrepreneurs and integrate them into the mainstream economy (May, 1998). It would also be appropriate for the developmental responsibilities of local government to include reviewing the potential for local interventions which support SMME development, especially those in the survivalist enterprise sub-sector.

Notes

1 On the registers of the cidb in South Africa, construction firms are registered in Grades 1 to 9, indicating the maximum limit of the value of projects they can tender for. They are also placed under various heads of type of work: General Building, Civil Engineering, Mechanical Engineering, Electrical Engineering and Specialist Works (which is also broken down further into particular works, such as Demolition and Blasting).

2 In addition to the grade and type of work, the contractor can also apply for recognition as a 'potentially emerging contractor'. This status indicates that the contractor has significant development potential, but has impediments that must be overcome. It also reflects the black majority ownership of the enterprise. Thus, a cidb-registered company can be described as 'Grade 8 General Building, Potentially Emerging'.

References

Arrowsmith, S. (1995) Public procurement as an instrument of policy and the impact of market liberalisation. *Law Quarterly Review*, 111, 235–284.

Ashworth, A. (2006) *Contractual Procedures in the Construction Industry*, 5th edition. Pearson/Prentice Hall: Harlow.

Bajari, P. and Tadelis, S. (2001) Incentives versus transaction costs: a theory of procurement contracts. *Rand Journal of Economics*, 32 (3), 387–407.

Barrie, D. S. and Paulson, B. C. (2001) *Professional Construction Management: Including Design-Construct and General Contracting*. McGraw-Hill: Singapore.

Bayat, A., Bayat, M. and Molla, Z. (2004) *Municipal Procurement and the Poor: A Case Study of Three Municipalities in the Western Cape Province.* http://www.competitionregulation.org.uk/conferences/Southafrica04/BayaytMoola.doc

Belt, V. and Richardson, R. (2005) Social labour, employability and social exclusion: pre-employment training for call centre work. *Urban Studies,* 42 (2), 257–270.

Berry, A., von Blottnitz, M., Cassim, R., Kesper, A., Rajaratnam, B. and van Seventer, D. E. (2002) *The Economics of SMMES in South Africa.* Paper presented at the 2002 TIPS Forum, December, Johannesburg.

Bhorat, H., Lundall, P. and Rospabe, S. (2002) *The South African Labour Market in a Globalising World: Economic and Legislative Considerations.* ILO Employment Paper, 2002/32. ILO: Geneva.

Boston, T. D. (1999) *Affirmative Action and Black Entrepreneurship.* Routledge: London.

Botes, B. (2009) *Business: Why Are South African Entrepreneurs Turning to Business Angels?* http://www.ecademy.com/node.php?id=135621

Bruce, B. C. (2003) *Industry Reform in South Africa: The Basis for Socio-economic Development.* Paper presented to the CIB International Conference 'Revaluing Construction', Manchester, UK, February.

Burke, R. (2006) *Small Business Entrepreneur: Guide to Running a Business.* Cosmic MBA Series. Burke Publishing: London.

Campbell, K. (2005) Skills, safety and charter priorities for civil engineering. *Engineering News,* 25 (6), 21.

Chandra, V., Moorty, L., Rajaratnam, B. and Schaefer, K. (2001) *Constraints to Growth and Employment in South Africa, Report No. 1: Statistics from the Large Manufacturing Firm Survey.* Informal Discussion Papers on Aspects of the Economy of South Africa. World Bank, Southern Africa Department: Washington, DC.

Cheetam, T. and Mabuntana, L. (2006) *Developmental Initiative towards Accelerated and Shared Growth Initiative of South Africa (ASGISA).* Department of Public Works: Pretoria.

cidb (Construction Industry Development Board) (2003) *Construction Industry Stakeholder Forum Report 2003.* cidb: Pretoria.

cidb (2004) *SA Construction Industry Status Report – 2004: Synthesis Review on the South African Construction Industry and its Development.* Discussion document. cidb: Pretoria.

cidb (2005) *Labour-Based Methods and Technologies for Employment Intensive Construction Works: A cidb Guide to Best Practice.* cidb: Pretoria.

cidb (2006) *In Focus.* Special issue, March.

cidb (2009) *Status Quo Report: SA Contractor Development Programmes, March 2009.* cidb: Pretoria.

cidb (2010) *Framework: National Contractor Development Programme, January 2010.* cidb: Pretoria.

cidb, DPW and CETA (2005) *Towards Sustainable Contractor Development 2005.* Working paper presented at National Workshop: Towards a Common Framework for Enterprise Growth and Sustainability, Johannesburg.

Cook, D. (1999) *The Role of Government, Market Failure.* Seminar Notes for Fellows, Centre for Policy Excellence Budget Policy Workshops 8 and 9. http://www.icps.com.ua/doc/w8&9_eng.doc

Creamer, T. (2005) Presidential economics. *Creamer Media's Engineering News*, 25 (8), 4–10.

Day, A. (1997) *Digital Building*. Butterworth-Heinemann: Oxford.

Department of Finance (DoF) (1996) *Growth, Employment and Redistribution: A Macro-economic Strategy*. Department of Finance: Pretoria.

Department of Public Works (2006) *Construction Sector Broad-Based Black Economic Empowerment Charter, Version 6 (Final)*. Department of Public Works: Pretoria.

DETR (Department of Environment, Transport and the Regions) (2000) *Building a Better Quality of Life: A Strategy for More Sustainable Construction*. HMSO: London.

Dlungwana, W. S and Rwelamila, P. D. (2004) *Contractor Development Models for Promoting Suitable Building: A Case for Developing Management Capabilities of Contractors*. Building and Construction Technology, CSIR: Pretoria.

DTI (Department of Trade and Industry) (1995) *White Paper on National Strategy for the Development and Promotion of Small Business in South Africa*. Department of Trade and Industry: Pretoria.

DTI (2003a) Black economic empowerment strategy. *Government Gazette*, 29616. Department of Trade and Industry: Pretoria.

DTI (2003b) *Black Economic Empowerment Strategy document*. http://www.dti.gov. gov.za/bee/complete.pdf

DTI (2006) *ASGISA*. http://www.info.gov.za/asgisa/

Egan, J. (1998) *Rethinking Construction: The Report of the Construction Task Force*. HMSO: London.

Fellows, R., Langford, D., Newcombe, R. and Urry, S. (2002) *Construction Management in Practice*. Construction Press: London.

Flanagan, R. (2007) The drivers and issues shaping the construction sector. In Y. M. Xie and I. Patnaikuni (eds) *Innovations in Structural Engineering and Construction*, Proceedings of the 4th International Conference on Structural and Construction Engineering, Melbourne, Australia, 26–28 September, Vol. 1.

Gerber, M. (2008) *Awakening the Entrepreneur: How Ordinary People Can Create Extraordinary Companies*. Collins: New York.

Gounden, S. (2000) *The Impact of the National Department of Public Works' Affirmative Procurement Policy on the Participation and Growth of Affirmable Business Enterprises in the Construction Sector*. Unpublished PhD thesis, University of Natal, Durban.

Govender, J. N. and Watermeyer, R. B. (2001) *Potential Procurement Strategies for Construction Industry Development in the SADC Region*. Unpublished paper, Department of Public Works, Pretoria.

Harris, F. and McCaffer, R. with Edum-Fotwe, F. (2006) *Modern Construction Management*. Blackwell: Oxford.

Hauptfleisch, D., Dlungwana, S. and Lazarus, S. (2005) An integrated emerging contractor development model for the construction industry. *SA Builder*, 974, 15–16.

Hillebrandt P. (2000) *Economic Theory and the Construction Industry*, 3rd edition. Macmillan: Basingstoke.

Hirsch, A. (2005) *Season of Hope: Economic Reform under Mandela and Mbeki*. University of KZN Press: Durban.

Holroyd, T. M. (2003) *Buildability: Successful Construction from Concept to Completion*. Thomas Telford: London.

Ibrahim, A. R. B., Roy, M. H., Ahmed, Z. and Imtiaz, G. (2010) An investigation of the status of the Malaysian construction industry. *Benchmarking: An International Journal*, 17 (2), 294–308.

International Council for Local Environmental Initiatives (ICLEI) (2002) Local procurement for development and sustainability. Unpublished report, International Council for Local Environment Initiatives: Bonn.

Jaafar, M. and Abdul-Aziz, A. R. (2005) Resource-based view and critical success factors: a study on small and medium sized contracting enterprises (SMCES) in Malaysia. *International Journal of Construction Management*, 5 (2), 61–77.

Jewell, C., Flanagan, R. and Cattell, K. (2005) The effects of the informal sector on construction. In *Proceedings of the ASCE Construction Research Congress, San Diego, California, 5–7 April*.

Kajimo-Shakantu, K. and Root. D. (2006) Winners and losers in preferential procurement: a conflict theory view of preferential procurement policies in the constriction industry. In *Proceedings of the 4th cidb Postgraduate Conference, Stellenbosch, 8–10 October*.

Kale, P. (1999) *Alliance Capability and Success: A Knowledge-Based Approach*. Unpublished PhD thesis, Wharton School, University of Pennsylvania, Philadelphia.

Kapur, D. and McHale, J. (2005) *The Global Migration of Talent: What Does It Mean for Developing Countries?* Center for Global Development: Washington, DC.

Kesper, A. (2000) *Failing or Not Aiming to Grow? Manufacturing SMMEs and Their Contribution to Employment Growth in South Africa*. TIPS Working Paper 15 – 2000. Trade and Industrial Policy Secretariat (TIPS), Department of Trade and Industry: Pretoria.

Kotler, P. (2002) *Principles of Marketing*, 4th edition. Pearson/Prentice Hall: New Delhi.

Latham, M. (1994) *Constructing the Team: Final Report of the Government/Industry Review of Procurement and Contractual Arrangements in the UK Construction Industry*. HMSO: London.

le Roux, H. (2005a) Construction to outpace national growth *Engineering News*, 25 (6), 21.

le Roux, H. (2005b) The struggle for entrepreneurs. *Engineering News*, 25 (22), 16.

Lewis, J. D. (2001) Policies to promote growth and employment in South Africa. *World Bank, Southern Africa Informal Discussion papers on Aspects of Economy of South Africa, 2001 World Bank Annual Forum 10–12 September 2001, Johannesburg*. Trade and Industrial Policy Secretariat, Department of Trade and Industry: Pretoria.

Little, S. S. (2005) *The 7 Irrefutable Rules of Small Business Growth*. John Wiley: Hoboken, NJ.

Longenecker, J. G., Moore, C. W., Petty, J. M. and Palich, L. E. (2006) *Small Business Management: An Entrepreneurial Emphasis*. Thomson/South-Western: Mason, OH.

McCrudden, C. (1995) *Public Procurement and Equal Opportunities in the European Community: A Study of Contract Compliance in Member States of the European Community and under European Law*. University of Oxford: Oxford.

McCrudden, C. (2004) Using public procurement to achieve social outcomes. *Natural Urban Forum*, 28 (4), 257–267.

McCrudden, C. and Gross, S. G. (2006) WTO government procurement rules and the local dynamics of procurement policies: a Malaysian case study. *European Journal of International Law*, 17 (1), 151–185.

Marion, J. (2005) *How Costly Is Affirmative Action? Government Contracting and California's Proposition 209*. Working paper, University of California, Santa Cruz.

May, J. (1998) *Poverty and Inequality in South Africa*. Presentation at the Centre for Social and Development Studies, University of Natal. http://www.gov.za/reports/1998/poverty/presentation.pdf

Ministry of Infrastructure (2009) *Rwanda National Construction Industry Policy*. Ministry of Infrastructure: Kigali.

Mohamed, S. (2005) Smarter infrastructure planning can support industrial development. *Engineering News*, 25 (16), 16.

Morand, P. (2003) SMEs and public procurement policy. *Review of Economic Design*, 8 (3), 301–318.

Mowson, N. (2005) Accelerated roll out: public infrastructure investment to rise to R111 billion in 2006. *Engineering News*, 25 (6), 14.

Muradzikwa, S., Smith, L. and de Villiers, P. (2004) *Economics*. Oxford University Press: Cape Town.

Nissanke, M. K. (2001) Financing enterprise development in sub-Saharan Africa. *Cambridge Journal of Economics*, 25, 343–367.

OECD (2004) *Promoting Entrepreneurship and Innovative SMEs in a Global Economy: Towards a More Responsible and Inclusive Globalization*. OECD: Paris.

Ofori, G. (2000) Challenges for construction industries in developing countries. In *Proceedings of the Second International Conference of the CIB TG 29*, Gaborone, Botswana, November.

Ohlin, G. (1992) Varieties of policy in the third world. In L. Putterman and D. Rueschemeyer (eds) *State and Market in Development: Synergy or Rivalry?* Lynne Rienner Publishers: Boulder, CO.

Palma, G. (2005) Four sources of de-industrialization and a new concept of the 'Dutch disease'. In J. A. Ocampo (ed.) *Beyond Reform: Structural Dynamics and Macroeconomic Vulnerability*. Stanford University Press: Stanford, CA.

Pearce, J. A. and Robinson, R. B. (2007) *Strategic Management*. McGraw-Hill/Irwin: Chicago.

Przeworski, A. (1998) The State in a market economy. In J. Nelson, C. Tilly and L. Walker (eds) *Transforming Post-Communist Political Economies*. National Academy Press: Washington, DC.

Rebello, E. (2005) Small business, SA's biggest test. *Engineering News*, 25 (1), 16–17.

Rogerson, C. M. (2004) Pro-poor local economic development in South Africa: the application of public procurement. *Urban Forum*, 15 (2), 180–210.

Seeley, I. H. (1997) *Quantity Surveying Practice*, 2nd edition. Palgrave Macmillan: Basingstoke.

Shakantu, W. M. W. and Kajimo-Shakantu, K. (2007) Harnessing the informal and formal SMME construction sectors to resolve the South African construction skills shortage. *Proceedings of the CIB 2007 World Building Congress, Construction for Development*, Cape Town, South Africa, 14–17 May.

Shakantu, W. M., Kaatz, E. and Bowen, P. A. (2002) Partnering with small, micro and medium enterprises (SMMEs) to meet socio-economic sustainability objectives. *Proceedings of the 1st International Conference of CIB W107: Creating a*

Sustainable Construction Industry in Developing Countries, 11–13 November 2002, Stellenbosch, South Africa. Buildnet/CSIR: Pretoria.

Shakantu, W. M., Kajmo-Shakantu, K., Finzi, E. and Mainga, W. (2006) Bridging the informal, formal and indigenous construction knowledge systems to resolve the construction skills shortage: an exploratory study. In *Proceedings of the 4th cidb Postgraduate Conference, Stellenbosch, 8–10 October.*

Siddiqi, M. (2005) Africa attracts growing global capital. Investment. *Africa Review of Business and Technology,* December/January, 14–17.

Sowell, T. (2003) 'Affirmative action' quotas on trial. *Capitalism Magazine.* http://www.capitalismmagazine.com/culture/racism/2340-affirmative-action-quotas-on-trial.html

Thai, K. V. (2001) Public procurement re-examined. *Journal of Public Procurement,* 1 (1), 9–50.

Thompson, F. (2004) *Black Economic Empowerment in South Africa: The Role of Parastatals and Choices for the Restructuring of the Electricity Industry.* Paper presented at the ICS/CAS International Conference 'Looking at South Africa Ten Years On', 10–12 September, London.

Thwala, W. D. (2005) *Employment Creation through the Construction and Maintenance of Public Infrastructure in South Africa through the Use of Labour-Intensive Methods: Experiences and Problems.* Paper presented at the 24th Annual Southern African Transport Conference, 11–13 July, Pretoria. Document Transformation Technologies: Pretoria.

UNCHS (1996) *Policies and Measures for Small-Contractor Development in the Construction Industry.* HABITAT: Nairobi.

UN-HABITAT (2009) *Country Activities Report 2009.* UNON/Publishing Services Section: Nairobi.

Venter, I. (2006) Construction economy set for 9.42% growth. *Engineering News,* 26 (2), 16–17, 62.

Von Broembsen, M. (2005) Small business can create a lot of work. *Sunday Times Comment,* 26 March, p. 6.

Watermeyer, R. B. (2002) The use of procurement to attain labour-based and poverty alleviation objectives. In D. J. Mason (ed.) *Proceedings of the 9th Regional Seminar for Labour-Based Practitioners: Towards Appropriate Engineering Practices and an Enabling Environment,* Maputo, Mozambique.

Watermeyer, R. B., Nevin, G., Amod, S. and Hallett, R. (1995) An evaluation of projects within Soweto's Contractor Development Programme. *Journal of the South African Institution of Civil Engineers,* 37 (2), 17–26.

Winter, C. and Preece, C. N. (2000) Relationship marketing between specialist subcontractors and main contractors: comparing UK and German practice. *International Journal for Construction Marketing,* 2 (1), 1–11.

Wong, A. Y. (2002) *Angel Finance: The Other Venture Capital.* http://papers.ssrn.com/sol3/papers.cfm?abstract_id=941228

van Wyk, L. (2004) *A Review of the South African Construction Industry.* CSIR (Boutek): Pretoria.

11 The critical role of consulting firms in the acceleration of infrastructure delivery and the improvement of the quality of life

Ron Watermeyer

Introduction

People are surrounded by economic infrastructure (fixed capital investment) including the homes in which they live, the offices and factories in which they work, schools, which are essential for the education of their children, and hospitals and clinics, which are fundamental for their health and well-being. They are also surrounded by economic infrastructure which supports the economy in its totality. Road and rail infrastructure not only enables travel between homes and places of work, schools and hospitals but also distributes goods and services to people. Border posts, harbours and airports are the physical links with neighbouring countries and the world. Dams provide water not only for human consumption, but also for agricultural and industrial purposes. Power stations generate electricity. Networks deliver water and electricity to homes, places of work, schools and hospitals and convey industrial effluent, soil water and waste water to treatment works. Economic infrastructure is foundational to a better life for all.

The failure of economic infrastructure – for example blackouts due to insufficient power supply to meet demand, or the interruption of water supply to consumers due to burst pipes or subsidence in a roadway – can not only disrupt the lives of affected people but also have economic consequences. The lack of economic infrastructure to meet the demands of an ever-growing population, for example that relating to water and sanitation or clinics and hospitals, has a negative impact on the health and well-being of communities and leads to frustration. Accordingly, the failure or lack of sufficient infrastructure puts the spotlight on government, which has a goal of delivering a better life for all.

Investment in economic infrastructure occurs in expectation of demand or in reaction to demand for capacity. When it happens, it has the following three impacts (Langenhoven, 2010):

1 an initial growth in demand for people, equipment and materials on the project, which lasts as long as it takes to create the asset;

2   a demand on resources over the lifespan of the project to maintain the asset; and

3   a productivity impact in the overall economy, either producing more or producing it better owing to more efficient infrastructure (or simply the availability of capacity such as harbour capacity and electricity).

While an asset is being created, the employment of more people and demand for equipment, plant and materials create their own impact on growth due to secondary demand coming from more disposable income from the employed, and demand for equipment and materials to produce equipment and materials for the project. This is the so-called multiplier, accelerator principle; a project has employment/demand multipliers which in themselves cause growth to increase.

The number of people directly employed per annum per US$1 million spent in the construction sector is high relative to other sectors. For example, in South Africa, around 50 persons per annum are employed per US$1 million. This is the third highest of all sectors in South Africa and is surpassed only by the agriculture and mining sectors.

Economic and technical progress, which has dominated the past two or three centuries, did not foresee the wider physical and non-physical consequences at a systems level. It was never expected that human activity would lead to impacts on a global scale that could threaten the environment and humanity's place in it. It is now becoming clear that the earth is no longer able to withstand and rebound from human activity. It has limits. A more systems view of the world is needed and solutions at a systems level need to be developed to address the following two issues of truly global proportions (Jowitt, 2009):

1   engineering the world away from an environmental crisis caused in part by previous generations in terms of greenhouse gas emissions and profligate resource use;

2   providing the infrastructure platform for an increasingly urbanised world and lifting a large proportion of the world's growing population out of poverty.

Sustainability is about maintaining a dynamic balance between the demands of people and what is ecologically possible. Accordingly, sustainable development has two key concepts: the need for sustainable habitats in which present and future generations can live healthy lives and the idea that the state of technology and social organisation, both now and in the future, imposes limits on the environment's ability to meet present and future needs. Sustainable development can also help to eradicate poverty as it is rooted in the simple concept of providing a better quality of life. It is a way of looking at all resources that will lead to a higher quality of life for the current generation, without compromising that of future generations.

Sustainable development needs to be driven from two distinctly different starting points. Affluence and over-consumption give rise to the so called 'green' agenda (earth agenda), which focuses on the reduction of the environmental impact of urban-based production, consumption and water generation on natural resources and ecosystems, and ultimately on the world's life support system (see Figure 11.1) (Watermeyer, 2006). Poverty and underdevelopment give rise to the so-called 'brown' agenda (people agenda), which focuses on the need to reduce the environmental threats to health that arise from the poor sanitary conditions, crowding, inadequate water provision, hazardous air and water pollution, and accumulations of solid waste. Developing countries face both 'green' and 'brown' issues and have many examples of extremes in both affluence and overconsumption and poverty and underdevelopment.

Forecasts for the demand for new infrastructure expressed at the convention of the American Society of Civil Engineers in Baltimore, 2004, suggested that approximately 80 per cent of the world's new infrastructure in 15–20 years' time will be constructed in developing countries. Yet the tackling of poverty and underdevelopment is being hampered by shortcomings in the delivery and maintenance of infrastructure as evidenced in a recent World Bank report entitled *Overhauling the Engine of Growth: Infrastructure in Africa*, which examined infrastructure in 24 countries that together account for 85 per cent of the gross domestic product (GDP), population and infrastructure aid flows of sub-Saharan Africa (Foster, 2008). This report found that:

- in some countries infrastructure provision is not focused where it is most needed;
- countries typically only manage to spend about two-thirds of the budget allocated to investment in infrastructure; and
- about 30 per cent of infrastructure assets are in need of rehabilitation.

Consulting firms can contribute to both the 'green' and 'brown' agendas in developing countries throughout the life cycle of a project: planning, design, procurement, construction, maintenance and deconstruction. These processes must be not only responsive to sustainable development

	GREEN AGENDA	Key concern	BROWN AGENDA	
N	Ecosystemic well-being	Key concern	Human well-being	**S**
O	Forever	Time frame	Immediate	**O**
R	Local to global	Scale	Local	**U**
T	Future generations	Concerned about	Low-income groups	**T**
H	Protect and work with	Nature	Manipulate and use	**H**
	Use less	Services	Provide more	
	Affluence and over-consumption		Poverty and underdevelopment	

Figure 11.1 The fundamental differences between the 'green' and 'brown' agendas (Watermeyer, 2006).

imperatives but also capable of delivering and maintaining infrastructure more efficiently.

Public infrastructure planning processes for the development and maintenance of infrastructure need to be strategic and focused on mandates for service delivery with a medium- and long-term mindset. Such processes need to balance needs for physical infrastructure with those for social infrastructure. Projects need to be prioritised in an objective and equitable manner as infrastructure and budget planning processes are interconnected and dependent on each other.

The design of infrastructure needs to be not only fit for purpose but also contributing to sustainable development. The design needs to take account of the anticipated results of climate change including increased incidence and severity of storms, floods, droughts, sea level rise and storm surge, as well as the impacts over its life cycle of solutions that cause change (adverse or beneficial) to economic conditions, the environment, society or quality of life (Watermeyer and Pham, 2011).

The procurement of infrastructure needs to be conducted in a fair, equitable, transparent, competitive and cost-effective manner that promotes sustainable development objectives. These include the minimising of the harmful effects of development on the local environment, the use of environmentally sound goods, materials and construction technologies, the participation of targeted enterprises and labour, and construction technologies that increase employment and alleviate or reduce poverty.

In order to understand the role that consulting firms can play in the acceleration of infrastructure delivery and the improvement of the quality of life, it is necessary to first understand:

- what constitutes a consulting firm;
- the fundamentals of construction works, that is what construction works are and the workflow associated with the planning, designing and maintenance of such works;
- the nature of a profession and the types of professions involved in planning, designing and maintaining construction works;
- the manner in which a profession is regulated; and
- the options for procuring construction works and professional services.

The issues relating to consulting firms in developing countries can then be considered against the major challenge of the day, namely accelerating infrastructure delivery and improving the quality of life.

Fundamentals of a consulting firm

A 'consultant', within the construction industry, is commonly understood to be an expert who charges a fee for providing professional advice or services in a particular field with the reasonable skill and care that is normally used by professionals providing such services. The quality of the outputs or deliverables of the service need to satisfy client requirements and expectations while the advice that is provided needs to be independent from any affiliation, economic or otherwise, which may cause conflicts between the consultant's interests and those of the client.

The defining features of a consulting firm are captured in the constitutions of various associations or codes of conduct issued by statutory bodies governing a profession. For example, the preamble to the Constitution of Consulting Engineers South Africa (2010) states that:

> Consulting Engineers South Africa is a voluntary association of firms of consulting engineers and allied professionals which:
>
> - are primarily in the business of offering independent technology-based intellectual services in the built, human and natural environment to clients for a fee
> - are managed and have their operating policies determined by people whose professional qualifications and conduct are in keeping with the requirements of this Constitution and its By-laws
> - have high professional repute and ethical standard.

The Engineers Registration Board (undated), a statutory body which is mandated to regulate engineering activities and the conduct of engineers and engineering consulting firms in Tanzania, states in its Requirements for Registration of Engineering Consulting Firm that:

> consulting firms or business names are required to have at least one of the key personnel or partner registered with the Board as consulting engineer in one of the fields of specialisation applied for registration. They are also required to have an accessible office and basic equipment and tools for carrying out engineering consulting works.

The Code of Professional Conduct of the South African Council for the Architectural Profession (2009), a statutory body that regulates the architectural profession in South Africa, stipulates that:

> - A practice may not be described as solely practising architecture unless effective control in terms of shareholding, members' interest or voting powers is in the hands of persons registered in terms of the Act.
> - Effective control of any multi-disciplinary professional firm, which also practises architecture, shall be in the hands of registered persons and of registered members of closely allied professions.
> - Every office established for the purpose of conducting an architectural practice shall be under the continuous, direct and personal supervision of a registered person.

The International Federation of Consulting Engineers (FIDIC) (2007) admits as members only associations 'representing suitably qualified and experienced individuals and firms who derive a substantial portion of their income from the provision of impartial consulting services to a client for a fee'.

Consulting firms range in size from those with a single professional to firms which have several thousand employees and provide a diverse range of services within the construction industry relating to a number of distinct professions.

Fundamentals of construction works

What are construction works?

ISO 6707-1 defines construction works as 'everything that is constructed or results from construction operations' (International Organization for Standardization, 2004, pp. 1–2). Construction works have traditionally been categorised as falling into one of two types (International Organization for Standardization, 2004):

1 *buildings i.e.* construction works that has the provision of shelter for its occupants or contents as one of its main purposes; usually partially or totally enclosed and designed to stand permanently in one place
2 *civil engineering works i.e.* construction works comprising a structure, such as a dam, bridge, road, railway, runway, utilities, pipeline, or sewerage system, or the result of operations such as dredging, earthwork, geotechnical processes, but excluding a building and its associated site works.

Construction works also include plant: machinery and heavy equipment installed for the operation of a service or processes.

Construction works invariably need to be maintained and repaired during their working lives. They may also need to be altered, refurbished or renovated.

Construction works are owned by both the public and private sectors. However, the bulk of buildings within any country are owned by the private sector whereas the bulk of civil engineering works is owned by the public sector.

Both buildings and civil engineering works require a team to plan, design, construct and maintain the works and the necessary resources, including finance, to do so.

Workflow associated with the delivery and maintenance of construction works

The construction industry is the broad conglomeration of industries and sectors which add value in the creation and maintenance of much-needed economic infrastructure. There is seldom the direct acquisition of construction works as client needs vary considerably and consequently each project meeting those needs has different characteristics. Professional services are required, as necessary, to plan, prepare a budget, conduct condition assessments of existing construction works, scope requirements in response to the owner or operator's brief, propose solutions, evaluate alternative solutions, develop the design for the selected solution, produce production information enabling construction and confirm that design intent is met during construction. Contractors are required to deliver infrastructure in accordance with the contract, specifications and other production information or to perform maintenance services frequently on infrastructure that is in use, and hand the infrastructure over to (or back to) the user upon completion of the works or services. The delivery and maintenance of construction works also need to be financed.

Accordingly the delivery and maintenance of construction works need to be planned and implemented in a controlled, logical and methodical manner. Figure 11.2 illustrates the workflow for the delivery of construction works by a national, provincial or local government (cidb, 2010a). Minor modifications in this workflow are required to the initial stages to align the workflow with the funding of a business case associated with private sector and public entity practices.

The Construction Industry Development Board (cidb) Infrastructure Gateway System (IGS) (see Figure 11.2) provides a number of control points (gates) in the infrastructure life cycle where a decision is required before proceeding from one stage to another. Such decisions need to be based on information that is provided and, if correctly done, provide assurance that a project involving the delivery or maintenance of infrastructure remains within agreed mandates, aligns with the purpose for which it was conceived and can progress successfully from one stage to the next. The cidb IGS is based on the information flow as set out in Table 11.1.

The cidb IGS permits the undertaking of groups of activities in parallel or series and results at the end of each stage in a predetermined deliverable (a tangible, verifiable work product) and a structured decision point which enables decisions to be made to determine if the project should continue to its next stage with or without any adjustments between what was planned and what is to be delivered.

It should be noted that the deliverables at the end of each stage in stages 3–6A form the basis for the scope of work of a contracting strategy and accordingly enable the services of a contractor to be procured to take the work forward as indicated in Table 11.2. (According to ISO 10845-1, a contracting strategy is a strategy that 'governs the nature of the relationship which the employer wishes to foster with the contractor, which in turn determines the risks and responsibilities between the parties to the contract and the method by which the contractor is to be paid'; International Organization for Standardization, 2010, p. 3.)

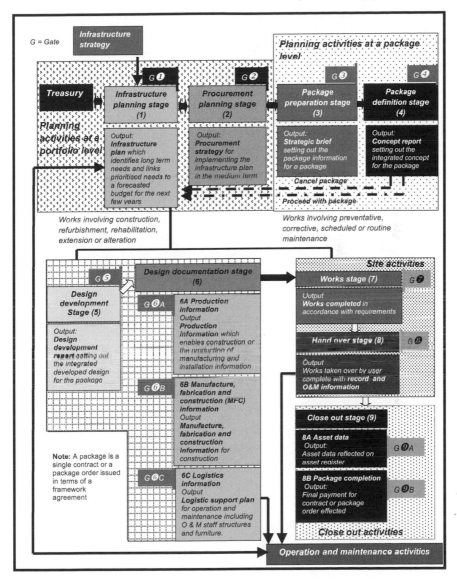

Figure 11.2 The stages and gates of the cidb Infrastructure Gateway System (IGS).

Professions commonly associated with construction works

There are six basic professions which are commonly associated with the delivery and maintenance of construction works. These are the architectural, construction management, engineering, landscape achitectural, project management and quantity surveying professions (see Table 11.3). The engineering discipline is broken down into three main disciplines – civil, electrical and

Table 11.1 cidb Infrastructure Gateway System gates and stages

Gate no.	Information (deliverable) provided for a decision to be made at a gate	Stage No.	Description
1	Infrastructure plan	1	Infrastructure planning
2	Construction procurement strategy	2	Procurement planning
3	Strategic brief	3	Package planning
4	Concept report	4	Package definition
5	Design development report	5	Design development
6A	Production information	6	Design documentation
6B	Manufacture, fabrication and construction information		
6C	Logistics information		
7	Completed works	7	Works
8	Works handed over to user	8	Handover
9A	Updated asset register	9	Close out
9B	Completed contract or package order		

mechanical engineering (see Table 11.4) – with the disciplines of electrical and mechanical engineering serving not only the constuction industry but also other industries, such as food processing, aviation and manufacturing.

A wide range of specialist engineering services are required to design modern buildings (see Table 11.5).

Regulating of a profession

The UK Inter-Professional Group (2002) defines a profession as:

> an occupation in which an individual uses an intellectual skill based on an established body of knowledge and practice to provide a specialised service in a defined area, exercising independent judgment in accordance with a code of ethics and in the public interest.

The purpose of regulating a profession is to assure the quality of professional services in the public interest. Regulating a profession involves the setting of standards of professional qualifications and practice, the keeping of a register of qualified persons and the award of titles, determining the conduct

Table 11.2 Scopes of work for different contracting strategies

Contracting strategy		Basis for scope of work	
Type	*Description*	*Document title*	*Outline of typical contents*
Management contractor	Contract under which a contractor provides consultation during the design stage and is responsible for planning and managing all post-contract activities and for the performance of the whole of the contract	Strategic brief[a]	Project objectives, business need, acceptance criteria and client priorities and aspirations
Design and construct	Contract in which a contractor designs a project based on a brief provided by the client and constructs it	Concept report	Detailed brief, scope, scale, form, and integrated concept for the project
Develop and construct	Contract based on a scheme design prepared by the client under which a contractor produces drawings and constructs it	Design development report	Detailed form, character and function. Definition of all components in terms of overall size, typical detail, performance and outline specification, as relevant
Design by employer	Contract under which a contractor undertakes only construction on the basis of full designs issued by the employer	Production information	Final detailing, performance definition, specification, sizing and positioning of all systems and components enabling either construction (where the contractor is able to build directly from the information prepared) or the production of manufacturing and installation information for construction

Note
a The scope of the work could also be based on the concept report, the design development report or the production information, depending on the allocated design responsibilities.

of registrants, the investigation of complaints and disciplinary sanctions for professional misconduct.

There are a number of international approaches for the regulation of a

Table 11.3 The broad scope of services offered by the construction professions

Profession	Typical scope of services offered
Architectural	Detailed planning or designing and reviewing of buildings, spaces and structures and associated site works by the creative organisation of materials and components with consideration to mass, space, form, volume, texture, structure, light, shadow, materials and the project brief General inspection of such works to confirm that the design assumptions are valid, the design is being correctly interpreted and the work is being executed generally in accordance with the designs, appropriate construction techniques and sound practice
Construction management	The management of the physical construction processes associated with construction projects relating to: • the management, co-ordination and integration of the detail design development process within the time, cost and quality parameters established for the project; • the establishment and implementation of procurement, health and safety and environmental strategies during construction; • the management, administration and integration of construction works contracts; and • the management and administration of the project close-out in a manner that enables the effective use of the project for its intended purpose
Engineering	The designing, detailed planning, optimisation or condition assessment of: • engineering services including water supply, sanitation, electricity supply, and transportation, telecommunication and stormwater systems; • fire protection, structural, electrical or mechanical systems for buildings and facilities; • site works necessary for access to, earthworks for, servicing of buildings and facilities and the drainage thereof; • building services including heating, cooling and ventilation systems, steam generation and distribution, drainage systems, fire installations and water supply systems including: • the general inspection of such works to confirm that the design assumptions are valid, the design is being correctly interpreted and the work is being executed generally in accordance with the designs, appropriate construction techniques and sound practice; • addressing of the potential adverse sustainable development impacts of such works including the application of engineering principles and methods in the identification, analysis, evaluation, treatment and monitoring of risk
Landscape architectural	The detailed planning or the designing and reviewing of the construction of outdoor and public spaces for human use and enjoyment or for environmental conservation and rehabilitation with due care, skill and diligence to achieve environmental, socio-behavioural, or aesthetic outcomes or any combination thereof

Table 11.3 continued

Profession	Typical scope of services offered
Project management	The overall planning, co-ordination and control of an infrastructure project from inception to completion aimed at meeting a client's requirements in order to produce a functionally and financially viable project that will be completed on time within authorised cost and to the required quality standards with due care, skill and diligence such that: • decisions converge on the achievement of the client's objectives; • the various elements of the project are properly co-ordinated; • the project includes all the work required, and only the work required, to complete the project successfully; • the timely completion of the project is facilitated; • the project is completed as far as is reasonably possible, within the budget that is agreed from time to time with the client; • the project satisfies the needs for which it was undertaken; • effective use of the people involved with the project is made; • timely and appropriate generation, collection, dissemination, storage, and ultimate disposition of project information occurs; • the systematic identification, analysis, and response to project risk occurs; and • all contracts associated therewith including those relating to professional services, are administered in accordance with the provisions of such contracts
Quantity surveying	The independent and impartial estimation and control of the cost of constructing, rehabilitating and refurbishing infrastructure with due care, skill and diligence by means of one of more of the following: • accurate measurement of the works; • comprehensive knowledge of various financing methods, construction systems, forms of contract and the costs of alternative design proposals, construction methods and materials; • the application of expert knowledge of costs and prices of work, labour, materials, plant and equipment required

profession, examples of which may be found in developing countries, including (ICE, 2005):

1 *Licensing*: Licensing can be statutory or non-statutory. An area of work restricted by statutory licensing cannot be undertaken by an unlicensed person. Non-statutory licensing provides the public with lists of approved persons competent to work in a particular area, which can also be undertaken by non-licensed persons.

2 *Registration*: Regulation of a profession involves the setting of standards, the keeping of a register of qualified persons and the award of titles. Regulation may be statutory (for example, regulations set through

Table 11.4 Basic engineering disciplines within the construction industry

Discipline	Principal activities
Civil	The planning, design, construction, maintenance and operation of works comprising: • structures such as a buildings, dams, bridges, roads, railways, runways or pipelines; • transportation, water supply and treatment, drainage and sewerage systems; • the result of operations such as dredging, earthworks and geotechnical processes; • waste disposal; and • sea defences and coastal protection
Electrical	The planning, design, construction, operation and maintenance of works comprising: • power generation; • electrical distributions systems and reticulations; • electrical installations within buildings and structure used for the transmission of electricity from a point of control to a point of consumption; and • communication lines, telephones and IT networks
Mechanical	The planning, design, construction, operation and maintenance of works comprising: • heating, cooling, ventilation and refrigeration systems; • escalators, lifts, cranes and hoists and materials handling and conveyor systems; • steam generation and distribution and centralised hot water systems; • compressed air, gas and vacuum systems; • fire protection systems; and • plant required for the operation of engineering systems through the application of engineering sciences including mechanics, solid mechanics, thermodynamics and fluid mechanics

an act of parliament) or non-statutory (for example, regulations set by the governing body of the profession or trade). If it is non-statutory, governing bodies can only use civil action to prevent non-registrants from using the title and cannot restrict any area of work to registrants. Statutory regulation normally involves a statutory register and the protection of title by law and sometimes, but not always, the statutory reservation of an area of work to registrants; in other words, to work in the area without being on the register would be an offence in law. If statutory regulation reserves an area of work, it has the same effect as statutory licensing which seeks to restrict an area of work to those who are approved persons.

3 *Specialist lists*: The non-statutory voluntary listing of professionals who have met a defined standard of competence in a specialist area, typically administered by a professional or trade body.

Table 11.5 Engineering specialisations (service areas) in the design of buildings

Service area	Principal activities
Acoustic design	Plan, design and review the construction of buildings and building components to achieve acoustical outcomes
Civil engineering	Plan, design and review the construction of site works comprising a structure such as a road, pipeline or sewerage system or the results of operations such as earthworks or geotechnical processes
Electrical engineering	Plan, design and review the installation of the electrical and electronic systems for and in a building or structure
Fire safety	Plan, design and review the fire protection system to protect people and their environments from the destructive effects of fire and smoke
Geotechnical engineering	Specialist advice on the behaviour of earth materials Geotechnical design including bearing capacity and settlement analysis; slope stability assessments; piling design and dewatering Geotechnical site investigations in the context of existing or proposed works or land usage
Mechanical engineering	Plan, design and review the construction, as relevant, of the gas installation, compressed air installations, thermal and environmental control systems, materials handling systems or mechanical equipment for and in a building
Structural engineering	Plan, design and review the construction of buildings and structures or any component thereof to ensure structural safety and structural serviceability performance during their working life in the environment in which they are located when subject to their intended use in terms of one or more of the following: • external and internal environmental agents; • maintenance schedule and specified component design life; • changes in form or properties
Wet services	Plan, design and review the construction, within and around buildings, drainage installations, which are intended for the reception, conveyance, storage or treatment of sewage, and water installations, which convey water for the purpose of firefighting or consumption within a building

Broadly speaking, licensing authorises eligible persons to practise in a specific area, registration recognises demonstrated achievement of a defined standard of competency, and specialist lists indicate peer-recognised competence in a particular area. All these forms of regulation are linked to codes of conduct. Serious breaches of a code of conduct can lead to the withdrawal of a licence, the loss of a title or the removal of the transgressor's name from a specialist list, on either a temporary or a permanent basis.

Qualifications and professional registration with regulatory bodies may, in many countries, be categorised as falling into generic tracks. For example, the engineering profession is divided into three tracks, namely engineer, engineering technologist and engineering technician, each aimed at people with different skills sets (see Table 11.6) (Watermeyer, 2010a). The precise names of the titles awarded to registered persons may differ from country to country; for example, the Engineering Council, which has its headquarters in London, registers the three tracks as Chartered Engineer, Incorporated Engineer and Technician Engineer, whereas Engineers Ireland registers Chartered Engineer, Associate Engineer and Engineering Technician. In some countries, only the engineer or the engineer and engineering technologist tracks are registered. In others, the registration of engineering technicians has only recently been embarked upon.

Researchers at Duke University, USA (Gereffi *et al.*, 2005), have taken a different view regarding engineering tracks. They see two main tracks for engineering graduates, namely dynamic engineers and transactional

Table 11.6 Capabilities of those registered in the different engineering tracks

Track	Profile of capabilities of person registered in the track
Engineer	Aimed at those who will: • use a combination of general and specialist engineering knowledge and understanding to optimise the application of existing and emerging technology; • apply appropriate theoretical and practical methods to the analysis and solution of engineering problems; • provide technical, commercial and managerial leadership; • undertake the management of high levels of risk associated with engineering processes, systems, equipment, and infrastructure; and • perform activities that are essentially intellectual in nature, requiring discretion and judgement
Engineering technologist	Aimed at those who will: • exercise independent technical judgement at an appropriate level; • assume responsibility, as an individual or as a member of a team, for the management of resources and/or guidance of technical staff; • design, develop, manufacture, commission, operate and maintain products, equipment, processes and services; • actively participate in financial, statutory and commercial considerations and in the creation of cost effective systems and procedures; and • undertake the management of moderate levels of risks associated with engineering processes, systems, equipment and infrastructure
Engineering technician	Aimed at those who: • carry supervisory or technical responsibility; • are competent to exercise creative aptitudes and skills within defined fields of technology; • contribute to the design, development, manufacture, commissioning, operation or maintenance of products, equipment, processes or services; and • create and apply safe systems of work

engineers. Dynamic engineers are those who are capable of abstract thinking, solving high-level problems using scientific knowledge, thrive in teams, work well across international borders, have strong interpersonal skills and are capable of leading innovation. On the other hand, transactional engineers, although they may possess engineering fundamentals, are not seen to have the experience or expertise to apply this knowledge to complex problems. They are viewed as being largely responsible for 'rote and repetitive tasks in the workforce'. The Duke University researchers observed that one of the key differentiators of the two types of engineers is their education. Most dynamic engineers have a minimum of a four-year engineering degree from nationally accredited or highly regarded institutions whereas transactional engineers often obtain a sub-baccalaureate degree (associate, technician or diploma awards) rather than a bachelor's degree in less than four years but in more than one. They do, however, point out that it is not a hard and fast rule that sub-baccalaureate programmes do not produce transactional engineers as in the last 50 years a number of science and technology leaders have emerged with little or no traditional education.

Procuring the services of contractors and consulting firms

Construction procurement

As mentioned above, in the construction industry there is seldom the direct acquisition of construction works as client needs vary considerably. Consequently each project meeting those needs has different characteristics. Professional services are required to fulfil a wide range of specialist tasks on each construction project.

Construction procurement accordingly includes:

1 *professional service contracts* for project management, construction monitoring, planning, design, optimisation and condition assessments and specialist investigations;
2 *service contracts* to repair and maintain infrastructure or components thereof and equipment used to provide and maintain infrastructure;
3 *supply contracts* for equipment, materials, products, components, assemblies, fuel and consumables; and
4 *engineering and construction works contracts* to design, as required, erect, construct, maintain, install, rehabilitate, renovate or demolish infrastructure.

Non-construction procurement, on the other hand, typically includes supply and service contracts for direct acquisitions which involve standard, well-defined and scoped services, off-the-shelf items and readily available commodities, where an immediate choice can generally be made in terms of the cost of goods or services satisfying specified requirements. Accordingly, there are several fundamental differences between construction and non-construction procurement. As a result, construction procurement practices and procedures have evolved within the construction industry.

Evolution of the approach to the delivery of construction works in the UK

The evolution of the approach to the delivery of construction works in the UK may be used to illustrate the dynamism of construction procurement practices and procedures mentioned above. According to Barnes (1999), virtually no civil engineering works was carried out in the UK after the Romans left until the seventeenth century, the two notable major works being the Exeter Ship Canal (1567) and the drainage of the Fens. This all changed between the 1760s and the 1850s. John Smeaton, who is often regarded as the founder of civil engineering and whose largest project was the Forth and Clyde Canal linking the east side of Scotland to the west, developed his approach to managing works. In 1768, he set down his management scheme for the construction phase with detailed tables of responsibility. His team comprised the engineer in chief, the resident engineer and the 'surveyors' for the various geographical sections working under him, and contractors (as opposed to direct labour) who executed the instructions of the engineer. This 'master–servant' model has remained in use for the majority of construction works projects for more than 200 years and is still used on projects managed in the traditional way.

Sir Joseph Bazalgette, who was responsible for constructing the major sewer projects and the embankments on the Thames in London, developed a standard form of contract in the 1860s which was adopted by the Metropolitan Board of Works. This form of contract remained as the principal model for contracts for more than 100 years and was the model for the first edition of the Institution of Civil Engineers (ICE) contract published in 1945. This ICE form of contract has in turn served as a model for forms of contract in many parts of the developing world, entrenching the master–servant relationship developed by Smeaton.

What has happened over the last 150 years is that the basic interaction between engineer and contractor has mutated from 'master and servant' to a simple collaboration between two specialist contributors. The boundaries of the traditional interactions have also moved in response to the same pressures to improve. In earlier times, for example, contractors made none of the decisions about what was to be built and all the decisions about how it was to be built. Operational and commercial relationships now permit the boundary between design and construction to be placed anywhere that preferences might dictate. The UK Office of Government Commerce's (2006) Common Minimum Standards require procurement strategies and contract types to support the development of collaborative relationships between the government client and its suppliers and facilitate the early appointment of integrated supply teams. This Standard also states that 'traditional, non-integrated procurement approaches should not be used unless it can be clearly shown that they offer best value for money – this means, in practice they will seldom be used' (p. 2).

Modern construction procurement systems

ISO 10845-1 defines procurement as the 'process which creates, manages and fulfils contracts' (International Organization for Standardization, 2010, p. 4). Procurement commences once a need for goods, services, engineering and construction works or disposals has been identified and ends when the goods are received, the services or engineering and construction works are completed or the asset is disposed of. Accordingly, a procurement system comprises processes which are underpinned by methods and procedures, which are informed and shaped by societal goals and the policy of an employer. At the same time, these processes, procedures and methods should provide a means by which risk relating to corrupt and fraudulent practices, fruitless and wasteful, irregular and unauthorised expenditure and overspending is minimised.

A procurement system needs to be developed around a set of outcomes or objectives which reflect societal or employer (or both) expectations. These objectives, in effect, establish overarching performance requirements for the system. However, these high-level performance requirements need to be interpreted both qualitatively and quantitatively. At the same time, a means of verifying that the established requirements have been complied with needs to be put in place.

ISO 10845 enables a construction procurement system which is fair, equitable, transparent, competitive and cost-effective to be established. It put in place the fundamental building blocks of a system, namely processes, procedures and methods (see Table 11.7). Construction procurement policy which aligns with national legislation is required to implement these standards. Such policy needs to designate persons to manage and control procurement activities, describe the manner in which procurement transactions are to be managed and controlled, and address a range of topics to make standard generic methods, procedures and processes organisation specific.

ISO 10845 contains a number of techniques and mechanisms associated with targeted procurement procedures, all of which are designed to promote or attain the participation of targeted enterprises and targeted labour in contracts (Watermeyer, 2000, 2004). These procedures, which are of particular interest to developing countries, relate to the:

- measurement and quantification of the participation of targeted groups through monetary transactions with such groups;
- definition and identification of targeted groups in a contractually enforceable manner;

Table 11.7 An outline of the content of the different parts of ISO 10845: Construction procurement

Part	Outline of content
1: Processes, methods and procedures	Describes processes, methods and procedures for the establishment within an organisation of a procurement system that is fair, equitable, transparent, competitive and cost-effective. It: • describes generic procurement processes around which an employer can develop its procurement system; • establishes basic requirements for the conduct of an employer's employees, agents, board members and office bearers when engaging in procurement; • establishes the framework for the development of an employer's procurement policy, including any secondary procurement policy; and • establishes generic methods and procedures for procurements, including those pertaining to disposals Informative guidance is provided on the establishment and management of procurement processes, an approach for obtaining best-value procurement outcomes and various types of procurement procedures including targeted procurement procedures
2: Formatting and compilation of procurement documentation	Establishes, in respect of supply, services and engineering and construction works contracts, at both main and subcontract level: • a format for the compilation of calls for expressions of interest and tender and contract documents; and • the general principles for compiling procurement documents
3: Standard conditions of tender	Sets out standard conditions of tender which: • bind the employer and tenderer to behave in a particular manner; • establish what a tenderer is required to do in order to submit a compliant tender; • make known the evaluation criteria to tenderers; and • establish the manner in which the employer conducts the process of offer and acceptance and provide the necessary feedback to tenderers on the outcomes of the process
4: Standard conditions for the calling for expressions of interest	Sets out standard conditions for the calling for expressions of interest which: • bind the employer and respondent to behave in a particular manner; • establish what is required for a respondent to submit a compliant submission; • make known to respondents the evaluation criteria; and • establish the manner in which the employer conducts the process of calling for expressions of interest

Table 11.7 continued

Part	Outline of content
5: Participation of targeted enterprises in contracts	Establishes key performance indicators, in the form of contract participation goals, relating to the engagement of specific target groups (labour and enterprises) on a contract for the provision of services or engineering and construction works. A contract participation goal may be used to measure the outcomes of a contract in relation to the engagement of target groups to establish a target level of performance for the contractor to achieve or exceed in the performance of a contract
6: Participation of targeted partners in joint ventures in contracts	
7: Participation of local resources in contracts	Sets out the methods by which the key performance indicator is measured, quantified and verified in the performance of the contract in respect of two different targeting strategies
8: Participation of targeted labour in contracts	

- unbundling of contracts either directly so that targeted enterprises can perform the contracts as main contractors or indirectly through resource specifications which require main contractors to engage target groups as sub-contractors, service providers or suppliers within the supply chain or as joint venture partners;
- granting of evaluation points in the assessment of expressions of interest or tenders (preferences) should respondents or tenderers satisfy specific criteria or undertake to achieve certain goals or key performance indicators in the performance of the contract;
- provision of financial incentives for the attainment of key performance indicators in the performance of the contract;
- creation of contractual obligations to engage target groups in the performance of the contract – for example sub-contract a percentage of the work to or contract goods or services from targeted enterprises, enter into joint venture with targeted enterprises, sub-contract specific portions of a contract to targeted enterprises in terms of a specified procedure or perform the works in a manner such that targeted labour is employed; and
- evaluation of procurement outcomes, that is the monitoring of the attainment of socio-economic deliverables at a contract level.

A number of contracting systems for engineering and construction works are embodied in the standard international forms of contract published by the ICE (the New Engineering Contract [NEC], such as NEC3) and FIDIC. These forms of contract contain terms that collectively describe the rights and obligations of contracting parties and the agreed procedures for the administration of their contract. Such contracts form an integral part of the construction procurement system as they establish processes, procedures and methods for dealing with matters that may arise during the performance of the contract.

Pricing strategies (i.e. strategies which are adopted to secure financial offers and to remunerate contractors in terms of the contract) have also evolved from a lump sum to provide the works to target cost contracts (see Table 11.8). Target cost contracts not only enable framework contracts (i.e. agreement between an employer and one or more contractors, the purpose of which is to establish the terms governing contracts to be awarded during a given period, in particular with regard to price and, where appropriate, the quantity envisaged) to be entered into but also enable the client to know where the money is being spent, reward strong contractor performance, share financial risk between the client and the contractor and promote collaboration or a culture whereby both parties have a direct interest in decisions that are made regarding the cost and timing of the contract. Target cost contracts also enable early contractor involvement in a project while the design is still being developed.

Table 11.9 outlines the contracting and pricing strategies that are embedded in the FIDIC and NEC3 engineering and construction works contracts. FIDIC also has a professional service contract. The NEC3 system, on the other hand, provides a range of different contract types which are suitable for use across the full spectrum of works, services and supply contracts ranging from major framework contracts or major projects to minor works contracts or the purchasing of readily available goods.

Table 11.8 Range of commonly encountered pricing strategies

Priced contract		Cost-based contract	
Activity schedule (lump sum)	*Bill of quantities*	*Cost reimbursable*	*Target cost*
The contractor undertakes to break the scope of work down into activities related to a programme and price each activity as a lump sum, which he is paid on completion of the activity. The total of the activity prices is the lump sum price for the contract work	The bill of quantities lists the items of work and the estimated or measured quantities and rates associated with each item to allow contractors to be paid, at regular intervals, an amount equal to the agreed rate for the work multiplied by the quantity of work actually completed	Contract in which the contractor is paid for his actual expenditure plus a percentage or fee	Cost-reimbursable contract in which a target cost is estimated and on completion of the works the difference between the target cost and the actual cost is apportioned between the employer and contractor on an agreed basis

Table 11.9 Contracting and pricing strategies embedded in the FIDIC and NEC3 contracts for construction works

	Consideration	NEC3 Engineering and Construction Contract (ECC)	NEC3 Engineering and Construction Short Contract (ECSC)	FIDIC Conditions of Contract for Construction and Building and Engineering Works Designed by the Employer (Red Book)	FIDIC Conditions of Contract for Plant and Design (Yellow Book)	FIDIC Conditions of contract for EPC Turnkey Projects (Silver Book)	FIDIC Short Form of Contract General Conditions (Short Form)
Contracting strategy	Design by employer	√	√	√	×	×	√
	Management contract	√	×	×	×	√	×
	Design and construct	√	√	×	√	√	√
	Develop and construct	√	√	×	√	√	√
Pricing strategy	Activity schedule	√	√	×	√	√	No standard provisions
	Bill of quantities	√	√	√	Not fully developed	×	
	Cost reimbursable	√	×	×	×	×	
	Target cost	√	×	×	×	×	

Construction procurement strategy

ISO 10845-1 defines procurement strategy as 'the selected packaging, contracting, pricing and targeting strategy, and procurement procedure for a particular procurement'. Construction procurement strategy up until fairly recently has not been well developed within the construction industry, particularly in developing countries. This has been largely because of the widespread usage of the master and servant approach to delivery and the traditional contracting arrangements, which require that the design and specifications be adequately developed and approved by clients before tenders are invited. Tenders are invariably priced as lump sums or in terms of a bill of quantities, prior to the award of a construction works contract, and the consulting firms that design and manage the project are paid a percentage of the cost of construction. As a result, the only choice has been which standard form of contract to use: local or international.

Strategy in the delivery and maintenance of construction works may be considered to be the skilful planning and managing of the delivery process. If correctly applied, it identifies the best way of achieving objectives and value for money, whilst taking into account risks and constraints. It involves a carefully devised plan of action which needs to be implemented. It is all about taking appropriate decisions in relation to available options and prevailing circumstances in order to achieve optimal outcomes. Construction procurement strategy is the combination of the delivery management strategy (the manner in which needs are met), contracting arrangements (risk allocations and requirements for outsourced professional services) and procurement arrangements (procurement and targeted procurement procedures) for a particular procurement. An outcome of this process is the identification of packages (grouping together or breaking down of projects into packages for delivery under a single construction works contract or a package order issued in terms of a framework agreement) and contracts for professional services.

Figure 11.3 outlines the activities associated with the development of a construction procurement strategy for a single project, a programme of projects or a portfolio of projects (cidb, 2010b; Watermeyer, 2010b).

Improving efficiencies in the delivery and maintenance of construction works

A critical issue to consider is: what culture should be fostered in the contractual relationships that are entered into? Alternatively, what strategy should be adopted?

Design has become increasingly fragmented over the last 100 years and each specialised participant now tends to work in an isolated silo, with no real integration of the expertise and collective wisdom of the participants. Project success requires that this fragmentation be addressed directly in

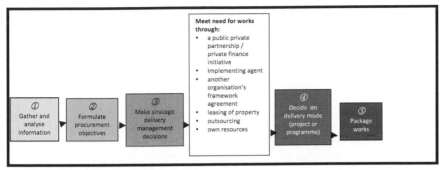

Part 1: Develop a delivery management strategy

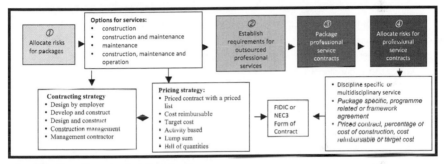

Part 2: Decide on the contracting arrangements

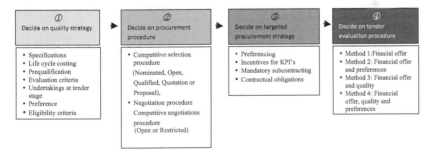

Figure 11.3 Developing a construction procurement strategy.

order to provide higher value and less waste. Research has shown that there are several critical keys to success, namely a knowledgeable, trustworthy and decisive facility owner/developer, a team with relevant experience and chemistry assembled as early as possible but certainly before 25 per cent of the project design is complete, and a contract that encourages and rewards organisations for behaving as a team (Lichtig, 2006).

A cultural change along the lines of that outlined in Table 11.10 is required to improve and optimise the delivery of infrastructure (Watermeyer, 2010b).

Table 11.10 Culture shift that is necessary to improve and optimise delivery of infrastructure

From	To
Adversarial master–servant relationship	Collaboration between two experts
Fragmentation of design and construction	Integration of design and construction
Allowing risks to take their course, or extreme and inappropriate risk avoidance, or risk transfer	Active, collaborative risk management and mitigation
Meetings focused on the past: what has been done, who is responsible, claims etc.	Meetings focused on 'How can we finish project within time and available budget?'
Develop project in response to a stakeholder wish list	Deliver the optimal project within the available budget
'Pay as you go' delivery culture	Discipline of continuous budget control
Constructability and cost model determined by design team and cost consultant only	Constructability and cost model developed with contractor's insights
Short-term 'hit-and-run' relationships focused on one-sided gain	Long-term relationships focused on maximising efficiency and shared value
Procurement strategy focused on selection of form of contract	Selected packaging, contracting, pricing and targeting strategy and procurement procedure aligned with project objectives
Project management focused on contract administration	Decisions converge on the achievement of the client's objectives
Training is in classrooms unconnected with work experience	Capability building is integrated within infrastructure delivery
Lowest initial cost	Best value over life cycle

The choice of the contracting system for a development programme can facilitate or frustrate performance in terms of the required project outcomes. Ideally, the selected form of contract should enable equitable long-term collaborative relationships, be sufficiently flexible to enable risks to be effectively managed, reward performance and focus the parties on attaining the project outcomes throughout the supply chain. The system should also cover the full range of contract types, namely supply, services and works, and provide back-to-back sub-contracts.

Location, distribution and usage of professional resources within the construction industry

Current UNESCO data show that developed, industrialised countries have between 20 and 50 scientists and engineers per 10,000 population, compared with around five scientists and engineers on average for developing countries, down to no more than one scientist or engineer for some poorer African countries (Marjoram, 2010).

Recent studies have pointed towards a number of issues relating to demand (needs) and supply (numbers) for engineering professionals. For example, the South African Institution of Civil Engineering (SAICE) has attempted to quantify the technological challenge facing developing countries and countries with dual economies by providing an indication of the ratio of engineers to population of a number of developed and developing countries, based on an extensive desktop survey and the contacting of various institutions and registering bodies. Although the data were compiled from a number of sources of varying detail and reliability, the statistics, some of which are reproduced in Table 11.11, when linked to per capita gross national income, illustrate a rough linkage between the economy of a country and the number of qualified engineers per capita.

Table 11.11 indicates that, despite the perception that South Africa is technologically stronger than other African countries, this is not necessarily the case. South Africa has a significantly higher per capita income than countries such as Tanzania and Namibia yet, according to SAICE's figures, South Africa's ratio of population to engineer is not significantly better than that of Zimbabwe, Namibia, Tanzania and other less developed countries (Watermeyer, 2006).

The age profile of the members of SAICE and the Institution of Structural Engineers (IStructE) in 2005 is shown in Figure 11.4 (Watermeyer, 2006). These profiles suggest that the numbers of civil and structural engineers have not kept pace with the population growth and have for the last few decades declined. Although the skewed age profile may in part be attributed to the introduction of computer technology, the low numbers of young engineers in the face of a growing world population is of concern.

Table 11.11 International registered engineer to population statistics

Country	Population per registered engineer (after SAICE, 2005)	Per capita gross national income, US$ (World Bank, 2005)
Norway, Finland, Denmark, Canada	≤ 200	28,390–52,0303
Sweden, Germany, France, Ireland	201–300	30,090–35,770
Japan, UK, USA, Australia, Hong Kong	301–500	37,120–41,400
Malaysia, Chile	501–1,000	4,650–4,910
Singapore, Korea, Hungary, Romania	1,001–3,000	2,920–24,220
South Africa	3,001–5,000	3,630
Sri Lanka, Tanzania, Namibia	5,001–7,500	330–2,370
Swaziland, Zambia, Ghana	≥ 12,000	380–1,660

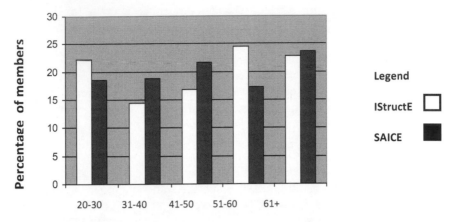

Figure 11.4 Age profile of SAICE and IStructE members (2005).

Lawless (2005) found that there are no civil engineers, technologists or technicians whatsoever employed in 34 per cent of South Africa's local municipalities and 9 per cent of district municipalities. Only one civil technician was employed in 18 per cent of the local municipalities and 9 per cent of district municipalities whereas 16 per cent of local municipalities and 13 per cent of district municipalities employed only technologists and technicians under the age of 35. Only 19 per cent of local municipalities and 53 per cent of district municipalities have at least one civil engineer in their employ. Despite these low levels of qualified persons within municipalities, and high levels of vacancies within all spheres of government, no concomitant rise in salaries occurred in response to supply and demand.

SAICE's figures are alarming if the traditional preplanned approach to delivery is the order of the day, as is the case currently in South Africa, whereby:

1 The design and specifications are adequately developed and approved by clients before tenders are invited so that the design meets the client's requirements closely and the contract when awarded can proceed without major change, delay or disruption.
2 For every project, consulting firms are appointed, briefed, directed and overseen by a gradually disappearing cadre of skilled client staff.
3 Unbundling strategies aimed at reducing the size of contracts in order to target small or local enterprises in response to social and economic imperatives, place increased demands on the client's resources to manage and oversee these small contracts.

Research indicates that this traditional approach to infrastructure delivery works best when clients have sufficient in-house capabilities and capacity to either undertake the design themselves or to brief professional service providers and to oversee the design process. In South Africa, the current reality is that departments and municipalities are experiencing difficulties in attracting and retaining suitably qualified staff while consulting firms involved in the delivery of infrastructure rarely have completed designs and associated documentation at the time that the contractor is appointed. This has resulted in:

- severely stressed departmental and municipal oversight resources;
- a crisis management culture which cuts corners in the planning processes;
- the fragmentation of design and construction, with aspects such as constructability and cost modelling determined by the design team and cost consultant only;
- tasks being allowed to take their course in a relatively uncontrolled manner, sometimes resulting in extreme and inappropriate risk avoidance or risk transfer;
- a 'pay as you go' culture in which significant cost overruns are the order of the day;
- private-sector-driven projects with perverse incentives (for example fee rates as a percentage of the value of the works); and
- a history of underexpenditure and poor service delivery.

A rough comparison of the distribution of engineers and technologists in South Africa in 1967 and 2005 can be made by comparing the figures published by the South African Human Science Research Council and SAICE (see Table 11.12). What is evident from Table 11.12 is that there has been a major flow of technologists and engineers from the public sector to the consulting sector over time. What is alarming is that, despite this shift in resources from the public sector to the private sector, the approach to managing and delivering projects remained largely unchanged (Watermeyer and Thumbiran, 2009).

Table 11.12 Change in distribution of technologists and engineers in South Africa over time

	Percentage	
Employer	*1967*	*2005*
State-owned enterprises	12	6
Government including provincial	12	4
Local government	15	10
Consultants	31	51
Industry or business	28	23
Academia	2	6

Accordingly, there are two distinctly different strategies to address the current lack of service delivery and poor project outcomes in South Africa. The first seeks to significantly increase the numbers of built environment professionals within the public sector to effectively and efficiently manage and oversee the current approach to delivery. The second harnesses the capability and capacity of consulting firms to deliver infrastructure using a radically different delivery process which matches the capabilities and capacities of government within the confines of their appetite for risk and looks to efficiencies to overcome capacity constraints.

In South Africa, the application of construction procurement strategy along the lines of that outlined in Figure 11.3 has led to some very interesting outcomes. In one large-scale pilot project, the eThekwini Metropolitan Council replaced 2,800 km of ageing asbestos cement pipes over a three-year period using four main contractors, four consulting firms providing design services and one consulting firm providing programme and project management services. These contractors and consulting firms were contracted in terms of cost-based pricing strategies following competitive tendering processes. Target prices associated with each water district were negotiated only after the scope of work and socio-economic deliverables had been finalised (Watermeyer *et al.*, 2009). Up to 4,000 temporary workers (unemployed persons) were employed on the programme to excavate trenches and were, in accordance with the requirements of the South African government's Expanded Public Works Programme, rotated every four months to allow others to benefit financially from the construction activities. Sixteen small contractors were targeted for development through this project and were engaged as sub-contractors (or 'co-contractors') to the main contractors. Independent mentors were appointed to assist these small contractors in setting up internal systems to enable them to grow their businesses and to make them more sustainable. All these contractors improved their cidb contractor grading designations during this three-year period and significantly increased their turnovers. Selected workers received training in pipe laying. All workers received HIV/AIDS training. Workplace experience was provided by consulting firms providing design services to enable eThekwini staff members to gain suitable experience to facilitate their professional registration. The demands placed on the eThekwini Metropolitan Council for the implementation of this project were minimal: only one senior project manager/engineer was involved in it.

The eThekwini water mains project demonstrated that it is possible to harness the capacity of the private sector to deliver projects without compromising any organisational objectives. This raises the question of what is needed to increase the quantum of construction works delivered and the quality of the maintenance of construction works in developing countries: numbers or systems? The eThekwini project showed that it is possible to deliver large projects successfully through procurement approaches which are designed to harness capacity that exists in the private sector and which places low demands on client resources to manage the project.

Both skills and systems are required to efficiently and effectively deliver and maintain infrastructure. A skill may be regarded as an ability to do something well. Skills are developed through training and experience. On the other hand, a system may be regarded as a method or set of procedures for achieving something or a process for obtaining an objective. As such, systems comprise an organised assembly of detailed methods, procedures and routines that are established or formulated to carry out a specific activity, perform a duty or solve a problem. Such organised assemblies are united and regulated by interaction or interdependence within a logical plan linking the various parts. A system is simply an established way of doing things and provides order and a platform for the methodical planning in one's way of proceeding.

Systems are underpinned by:

1 *processes* – a succession of logically related actions occurring or performed in a definite manner which culminates in the completion of a major deliverable or the attainment of a milestone;
2 *procedures* – the formal steps to be taken in the performance of a specific task, which may be evoked in the course of a process; and
3 *methods* – a documented, systematically ordered collection of rules or approaches.

Systems, processes, procedures and methods can be standardised and documented for common and repeated use for the achievement of the optimum degree of order in a given context. This in turn provides a solid platform for effective skills development as it permits staff to work in a uniform and generic manner and training interventions to be developed to capacitate those engaged in the performance of various activities.

Improvements in outcomes can be realised should procurement and delivery management systems be developed around the systems that are currently available; for example the cidb Infrastructure Gateway System, the ISO 10845 system for construction procurement and the NEC3 contracting system. This will not only reduce the demands on those overseeing delivery but also provide the framework within which consulting firms will be engaged in the delivery process.

Opportunities for consulting firms

The traditional approach to the delivery of public-sector construction works that evolved during Victorian times in the British Empire required the government (typically through a department of works) or local authorities (typically through engineering departments) to plan, design and maintain construction works. Usually, only the construction of new works was outsourced to the private sector, initially on a labour-only basis. Over time, design and management services were also outsourced to consulting firms, but only when internal capacity was lacking. Today, in many parts of the world, most non-core activities are commonly outsourced.

A government, in all its spheres, should not outsource its planning functions, which need to link to its developmental objectives and mandates from the people it serves, that is the infrastructure planning, procurement planning, package planning and package definition stages shown in Figure 11.2. Where it lacks capacity, skills can be procured. Accordingly, consulting firms can provide planning services at both a portfolio and a package level. In order to remove conflicts of interest and perceptions of perverse incentives, such firms should not be involved in the implementation of packages. Accordingly, there is an opportunity for consulting firms to specialise in the planning at both a portfolio level and package level should the public sector put in place a well-documented and thought-through system which ensures that projects which are proceeded with have a fitness of purpose.

Consulting firms can also be called upon during the planning stage to provide specialist services which are necessary to mitigate project risks and to determine the feasibility of projects in specific locations

If a robust planning system at both a portfolio and package level is put in place which establishes the deliverables at the end of each stage as set out in Table 11.13, it is relatively easy to procure the services of consulting firms and contractors relating to the management, design and construction to implement packages in terms of the stages set out in Table 11.14 (cidb, 2010a).

A brief is a working document which specifies at any point in time the relevant needs and aims, resources of the client and user, the context of the package and any appropriate design requirements within which all subsequent briefing (when needed) and designing (where necessary) can take place. Accordingly, a package is defined at any point in time in the project cycle by the package information that is available:

- the brief which is progressively developed from time to time;
- the design and or specifications;
- the package programme which identifies key dates and time periods for the performance of the works and services associated with the package; and
- the package cost.

The package information in Table 11.14 is developed following the identification of a package (see Table 11.13) and is updated thereafter whenever revised information is obtained. Any change in information needs to be accepted by the client before the team moves to the next stage. At the same time, the information flow in Tables 11.13 and 11.14 can be readily audited at any point in time for compliance with the purpose for which a project is approved.

The design by employer strategy, with or without early contractor involvement, will require the services of a consulting firm to lead and manage the project. Such a firm needs not only to act as the interface between the project team and the client but also to be responsible for obtaining all the necessary approvals or acceptances relating to the end of the stage deliverables.

The introduction of contracting strategies other than design by employer requires that the design process or the finalisation of the design be managed by a contractor. This does not mean that the services of consulting firms will not be required, as very few contractors have internal design capacity. In most instances, their services will be required but for a client with a different focus and set of objectives.

The adoption of cost-based strategies involving target cost contracts requires consulting firms to:

1 develop a cost model to forecast project costs, appropriate to each stage for which services are required, and to monitor costs against a control budget;
2 develop and maintain the cost plan for all stages;
3 monitor the development of the design and assist the project team in the development of the project within budgetary constraints;
4 schedule based on the elemental cost model for inclusion in the pricing data or develop bills of quantities as necessary;
5 confirm the reasonableness of the target price where such contracts are negotiated or provide advice during the evaluation of tender offers where tenders are called for; and
6 review the contractor's accounts, payroll and programme relating to a contract, prepare forecasts of the contractor's cost and monitor expenditure in accordance with the requirements of the contract.

This is a significant departure from the traditional role and responsibilities played by the quantity-surveying profession.

The adoption of new procurement strategies also opens up business opportunities for consulting firms to provide specialist services relating to the development of procurement strategies and appropriate contract documentation. It also permits consulting firms to offer a more specialised service to manage and administer contracts associated with a package in terms of a specific contracting system.

Overcoming constraints facing consulting firms in developing countries

Capability and capacity considerations

Through the application of information technology, some aspects of the work performed by consulting firms can be delinked from the locality in which construction works are to be constructed. As a result, the consulting industry is a global industry. However, the design and procurement of

Table 11.13 Key deliverables and principal actions associated with the planning stages

Stage	Key deliverable at end of stage	Principal actions associated with the key deliverable
1: Infrastructure planning	Approved infrastructure plan which identifies long-term needs and links prioritised needs to a forecast budget for the next few years	• Identify the policy drivers, strategies and long-term objectives of national, provincial and local government which impact upon the institution's infrastructure mandate • Produce a portfolio infrastructure plan for the long-term acquisition, refurbishment, rehabilitation and maintenance of infrastructure, which provides a projected list of work items described by category, location, type, economic classification and function needs and links prioritised needs to a forecast budget for the next three to five years
2: Procurement planning	Accepted procurement strategy for implementing the infrastructure plan in the medium term	• Analyse the medium-term expenditure infrastructure plan and identify spatially located work items in the infrastructure plan grouped into categories of spend with common attributes • Perform an organisational and market analysis, formulate primary and secondary procurement objectives and make certain strategic management decisions • Package the works • Allocate risks and deciding on a suitable pricing strategy for each package and establish requirements for outsourced professional services and the manner in which such resources are to be contracted • Decide on the high-level procurement arrangements • Document the choices made in relation to the delivery management strategy, the contracting strategy and the procurement arrangements in each category and sub-category of spend

3:	Package preparation	Client accepted strategic brief setting out the package information	• Define the package objectives, business need, acceptance criteria and client priorities and aspirations • Confirm the scope of the package • Establish the project criteria, including the function, mix of uses, scale, location, quality, value, time, safety, health. environment and sustainability • Where necessary, conduct preliminary investigations or desktop studies to obtain data • Identify procedures, organisational structure, key constraints, statutory permissions and strategies to take the package forward • Establish the control budget package • Develop a strategic brief which sets out the package information including the procurement strategy to implement the package.
4:	Package definition	Client-accepted concept report setting out the integrated concept for the package	• Establish the feasibility of satisfying the strategic brief for the package with or without modification • Investigate alternative solutions • Establish the detailed brief, scope, scale, form, and cost plan for the package, including, where necessary, the obtaining of site studies and construction and specialist advice • Recommend the preferred design option • Determine the initial design criteria, design options and cost plan for the package and produce a site development plan or other suitable schematic layouts of the works • Develop a concept report which sets out the integrated concept for the package

Table 11.14 Actions and deliverables associated with implementation stages

Stage	Key deliverable at end of stage	Principal actions associated with the key deliverable
5: Design development	Client-accepted design development report setting out the integrated developed design for the package	• Designer: develop in detail the accepted concept to finalise the design and definition criteria; establish the detailed form, character, function and cost plan, defining all components in terms of overall size, typical detail, performance and outline specification, as relevant; and confirm or revise the cost plan included in the concept report • Client's representative: review the Design Development Report prepared by the contractor's project team for general conformity with design intent and conformance with the relevant requirements of the works information
6A: Design documentation (production information)	Completed and client-accepted production information	• Designer: produce the final detailing, performance definition, specification, sizing and positioning of all systems and components enabling either construction (where the contractor is able to build directly from the information prepared) or the production of manufacturing and installation information for construction • Client's representative: review the production information prepared by the contractor's project team for general conformity with design intent and conformance with the relevant requirements of the works information
6B: Design documentation (manufacture, fabrication and construction information)	Client-accepted manufacture, fabrication and construction information	• Client's representative: review the manufacture, fabrication and construction information prepared by others, based on the production information for design intent and conformance with scope of work • Contractor: produce the manufacture, fabrication and construction information based on the production information

Stage	Output	Activities
6C: Design documentation (logistics information)	Client-accepted logistic support plan for operation and maintenance	• Client: identify additional organisational structure required for operation and maintenance over life span, and office, stores, furniture, equipment, IT and staff-training requirements to run operation and maintenance facilities as well as engineering infrastructure • Professional team: establish logistic requirements in respect of facilities and/or engineering infrastructure and specify requirements, if any, for contractor to provide a servicing and maintenance plan for all facilities and engineering infrastructure
7: Works	Completed works which are capable of being occupied or used and accepted by the client	• Provide temporary and permanent works in accordance with the contract; manage risks associated with health and safety on the site • Correct notified defects which prevented the client or end user from using the works and others from doing their work
8: Handover	Works which have been taken over by the user complete with record information	• Finalise and assemble record information including drawings, specifications, manuals, guarantees and statutory certificates which accurately reflect the infrastructure that is acquired, rehabilitated, refurbished or maintained • Hand over the works and record information to the user and, if necessary, train end-user staff in the operation of the works
9A: Close out (asset data)	Archived record information and updated asset register	• Archive record information • Update the portfolio asset register
9B: Close out (package completion)	Completed contract or package order	• Correct all defects that are detected during the defects liability period • Complete the contract by finalising all outstanding contractual obligations including the finalisation and payment of amounts due after the defects correction period • Evaluate package outcomes and then compile a completion report for the package, outlining what was achieved in terms of key performance indicators and suggestions for improvements on future packages of a similar nature and enter relevant data in a database

construction works, like the challenge of sustainable development, although global in nature, requires strategies for its successful implementation which are essentially local and differ in context and content from region to region.

Local knowledge of political, social and environmental conditions as well as indigenous technologies, materials, practices and customs is immensely important in arriving at suitable and acceptable solutions and can impact significantly on project outcomes. However, the skills set required in developing countries to implement a project may not be available or may be below a critical mass within a region or a country. This is particularly true where design requirements relating to the sustainable development objectives are superimposed on the normal fitness for purpose requirements, where out of the ordinary technologies are required to satisfy requirements or services within recent specialisations or emerging fields of engineering such as fire engineering. Accordingly, what is commonly referred to as capacity building may be necessary; that is, what a recent UNESCO report describes as a process of assisting people to develop the technical skills to address their own needs for improving the living standards and prosperity of their own people and, from there, building a sustainable society (Clinton and Jones, 2010).

At the other end of the spectrum, local consulting firms, although having the capability, may lack the capacity to perform the required service. As a result, they may be overlooked in the tender process.

It is also important in developing countries that local dynamic professionals as opposed to transactional professionals be identified and afforded the opportunity to work alongside dynamic professionals from other countries. Such exposure is essential in building a leader group within the consulting fraternity that is capable of developing and putting in place systems to enable the local industry to grow and flourish. Such persons are also required to contextualise emerging and current practices within a particular society.

When procuring the services of consulting firms, clients can formulate as a secondary procurement objective (i.e. an objective additional to those associated with the immediate objective of the procurement itself) to build local capacity, engage local consulting firms or employ local professionals using a range of targeted procurement procedures and the strategic approach to procurement outlined in Figure 11.3. This will enable clients to build up or strengthen their supply chain for outsourced professional services.

Securing the participation of local consulting firms in contracts

Key performance indicators (KPIs) relating to the engagement of local enterprises, local joint venture partners, local resources and local labour in contracts are needed to set targets or to measure procurement outcomes. The flow of money from the contract to the target group (contract participation goal) is a KPI which can be used to measure the participation of targeted enterprises or targeted labour, that is, local consulting firms or professionals. The ISO 10845 family of standards for construction procurement contains performance-based specifications (see Figure 11.5) which can readily be referenced in contracts and enable such KPIs to be defined, quantified and verified in the performance of the contract. Different specifications are available for satisfying contract participation goal obligations by means of sub-contracting work to, joint venturing with or employing the target group in the performance of the contract (see Table 11.7).

The objective (level 1), the performance description (level 2) and the performance parameters (level 3) as set out in Figure 11.5 can collectively be viewed as a KPI. The contract participation goal enables targets to be set and evaluation (level 4) establishes the measurement arrangements. A contract participation goal may be used to measure the outcomes of a contract in relation to the engagement of the target groups as sub-consultants, joint venture partners or employees or to establish a target level of performance for the contractor to achieve or exceed in the performance of a contract.

A contract participation goal relating to a venture is a good vehicle for the transfer of knowledge and experience in terms of technical and management skills from one company to another as joint ventures, by their very nature, require integrated management and pooling of resources in the provision of services. Furthermore, they are good integrators of project teams between the partners and facilitate the acquisition of business skills as all partners share directly in the fortunes of the consortium. A contract participation goal relating to sub-consultancy opportunities is effective in ensuring local participation. A contract participation goal relating to the use of local labour provides an opportunity for improving the skills and deepening or broadening the experience of targeted local professionals.

A range of targeted procurement procedures are contained in ISO 10845, which enable contract participation goals to be linked to a particular procurement transaction. These include the specifying of minimum contract participation goals and the offering of incentives to attain or improve upon a stated contract participation goal through the granting of tender evaluation

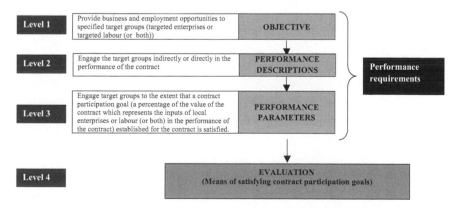

Figure 11.5 Structure of a four-level performance-based resource specification.

points during the evaluation of tender offers and the linking of performance to payments in terms of the contract.

Since contract participation goals are based on monetary flows, cost-based contracts (cost reimbursable and target cost contracts) which involve open book accounting are well suited to the monitoring and evaluation of contract participation goals as they enable monetary flow and people resources to be readily and transparently tracked.

Training and mentoring requirements for local professionals

Competencies equate to capability in a consulting firm. Competence is a function of a person's education, training, experience and contextual knowledge. Entry to a profession is frequently linked to a person's education (accredited programmes and courses) and the attainment of a number of outcomes during the first few years after graduation, ideally under the guidance of a mentor, that is an experienced and trusted advisor. Sponsor reports, interviews, the presentation of a portfolio of work and examinations, as relevant and appropriate, are used to gauge whether or not a candidate has reached a level of proficiency that warrants his or her admission to a profession.

Graduates, particularly in developing countries, often struggle to gain suitable postgraduate experience to become registered professionals either on a national register or on an international one such as that of the Engineering Council, which has its headquarters in London. International or large consulting firms can be required in terms of their contracts to provide persons identified by the client with suitable workplace opportunities in the performance of the contract as may be necessary and appropriate for them to make progress towards registration in a professional category, such as architect, engineer, engineering technologist or technician. The contracted consulting firm may be required to develop, in consultation with each identified person, an appropriate training and mentorship plan to assist such person to obtain suitable occupational or professional knowledge and experience towards registration, nominate a suitably qualified staff member to mentor or coach such persons, and submit quarterly progress reports.

When the client identifies persons who are studying towards a degree or diploma issued by an accredited institution of learning which will lead to registration, contracted consulting firms can also be required to provide them with experiential work opportunities in the performance of the contract as may be necessary and appropriate for them to satisfy requirements for admission to the next period of study. Again, contracted consulting firms can be required to mentor such persons and to submit quarterly progress reports.

Where specific skills are identified as being lacking, contracted consulting firms can be required to provide suitable experiential work opportunities and training to enable persons identified by the employer to acquire such skills and competencies in the performance of the contract. Contracted firms can be required to develop appropriate workplace plans to assist such persons to be proficient over time and to submit quarterly progress reports to the employer's representative.

The contractual obligations mentioned above produce the best result on contracts that have a sufficient duration to enable the required outcome to be achieved. Framework contracts are ideally suited to this as long-term relationships are entered into to execute a number of projects over a period of time, typically between three and five years. Framework contracts also offer flexibility in attaining secondary procurement objectives as requirements can be adjusted from one task order to another, thus allowing new key performance indicators to be introduced or improved upon over time. Thus, framework contracts are ideal vehicles for the mentoring, training and upskilling of professionals.

Financial incentives can be provided to encourage beyond minimum performance.

Balanced scorecards

A balanced scorecard approach can be adopted in terms of which a wide range of KPIs are defined in absolute terms or are qualitatively or quantitatively measured in terms of an indicator. Such indicators need to be objective, verifiable and reproducible, and, wherever possible, linked to predetermined benchmarks, reference levels or scales of value which are within levels acceptable to the client. A weighting, which reflects importance, is then assigned to each KPI and the total score measures the performance achieved. This approach enables an overall KPI to be developed. Minimum scores can also be linked to procurement transactions using a number of targeted procurement procedures.

Balanced scorecards are an effective tool in allowing a consulting firm to respond to a wide range of client objectives including provision of work opportunities to local consulting firms and capacity building at both corporate and individual levels. Balanced scorecards enable competing KPIs to be considered and matched to opportunities presented by actual projects.

Identifying suitably qualified professionals

The key considerations in engaging consulting firms in construction-related projects are:

1 The quality of the outputs or deliverables of the service satisfy client requirements and expectations.
2 The service is provided with the reasonable skill and care that is normally used by professionals providing such services.
3 The advice is independent from any affiliation, economic or otherwise, which may cause conflicts between the professional service provider's interests and those of the client.

The qualification of consulting firms can be established by assessing credentials in specific contexts (service areas) at a key personnel level (for example, professional registrations, assessed competencies and/or levels of experience) and at a corporate level (the possession of certain internal systems and demonstrated abilities to implement systems). Assessments may include interviews to discuss samples of work, police record checks and financial assessments. Performance reports from clients may also be used, where they exist.

Clients, particularly those who have little or no technical staff, have difficulty in identifying suitably qualified consulting firms to perform the required services, particularly in jurisdictions where the market is unregulated or merely regulated through codes of conduct. Reliance on persons being registered with a council in a broad discipline is not necessarily sufficient for procurement purposes. For example, registration within the broad disciplines of engineering outlined in Table 11.4 does not imply competence to perform services in the areas listed in Table 11.5. It is too coarse a screen.

Clients are seldom in a position to determine, in relation to a particular service, what are the 'generally accepted professional techniques and standards' and what is 'normally used by professionals providing services similar to the services'. However, this can easily be established by means of a well-understood benchmarking technique, namely a peer review. The phrases mentioned above frame the question that needs to be answered by people of similar standing or abilities, that is peers. Accordingly, a client can always call upon peers to establish whether or not the service is in accordance with the provisions of the contract.

The same 'test' can be applied to the assessment of a person's qualifications to do the work or supervise the work that is required in a particular service area. It should be noted in this regard that registration with a council governing a profession confirms that a person has the necessary education and training to practise independently in a broad practice area. It does not necessarily signify that such a person has the necessary experience and contextual knowledge to provide a service to a client. A peer review process can be put in place to determine that a person has the necessary experience and contextual knowledge to provide services in narrow practice areas (Watermeyer, 2009).

In developing countries, peer assessments can be made by learned societies and consulting associations. The listing of peer-assessed persons on a publicly accessible website will not only enable consumers of services to identify who is likely to be competent to perform a service but also enable gaps in capacity to be identified and addressed. Such assessments and listing can facilitate the development and growth of specialisations within a developing country.

References

Barnes, M. (1999) *Smeaton to Eaton: The Extraordinary History of Civil Engineering Management*. Smeaton Lecture, Institution of Civil Engineers.

Clinton, D. and Jones, R. (2010) *Engineering: Issues, Challenges and Opportunities for Development – section 7.2.2 Technical Capacity-Building and WFEO*. UNESCO: Paris.

cidb (Construction Industry Development Board) (2010a) *Inform Practice Note #22: cidb Infrastructure Gateway System*. www.cidb.org.za/knowledge/publications/practice_notes/default.aspx

cidb (2010b) *Delivery Management Guidelines Practice Guide 2: Construction Procurement Strategy*. Construction Industry Development Board. www.cidb.org.za/_layouts/toolkit/data/ai_docs/1555.pdf

Consulting Engineers South Africa (2010) *Constitution with Effect from 1 January 2010*. www.saace.co.za/public_downloads/constitution.pdf

Engineers Registration Board (undated) *Requirements for Registration of Engineering Consulting Firm*. www.erb.go.tz/registration.htm

Foster, V. (2008) *Overhauling the Engine of Growth: Infrastructure in Africa*. World Bank: Washington, DC.

Gereffi, G., Wadhwa, V., Rissing, B., Kalakuntla, K., Cheong, S., Weng, Q. and Lingamneni, N. (2005) *Framing the Engineering Outsourcing Debate: Placing the U.S. on a Level Playing Field with China and India*. Duke University: Durham, NC.

International Federation of Consulting Engineers (2007) *Statutes and By-Laws*. http://www1.fidic.org/about/statutes.asp

ICE (2005) *Council Report of the Study Group on Licensing, Registration, and Specialist Lists Prepared by Gordon Masterton, 24 May*.

International Organization for Standardization (2004) ISO 6707-1: 2004, Building and Civil Engineering – Vocabulary – Part 1: General Terms. International Organisation for Standardisation: Geneva.

International Organization for Standardization (2010) ISO 10845-1:2010, Construction Procurement – Part 1: Processes, Procedures and Methods. International Organisation for Standardisation: Geneva.

Jowitt, P. (2009) *Now Is the Time*. ICE Presidential address, November. http://www.ice.org.uk/downloads//pjaddress.pdf

Langehoven, H. (2010) Personal communications with Henk Langenhoven, economist, South African Federation of Civil Engineering Contractors.

Lawless, A. (2005) *Numbers and Needs: Addressing Imbalances in the Civil Engineering Profession*. South African Institution of Civil Engineering: Halfway House.

Lichtig, W. A. (2006) The integrated agreement for lean project delivery. *Construction Law*, Summer, 25.

Marjoram, T. (2010) *Engineering: Issues, Challenges and Opportunities for Development – Section 7.2 Engineering Capacity*. UNESCO: Paris.

Office of Government Commerce (2006) *Common Minimum Standards for the Procurement of Built Environments in the Public Sector*. 18 January, London. http://www.ogc.gov.uk/documents/Common_Minimum_Standards_PDF.pdf

South African Council for the Architectural Profession (2009) *Code of Professional Conduct of the South African Council for the Architectural Profession*. www.sacapsa.com/sacap/action/media/downloadFile?media_fileid=84

UK Inter-Professional Group (2002) *Professional Regulation: A Position Statement*. www.ukipg.org.uk/publications/Prof_Reg_Booklet.pdf

Watermeyer, R. B. (2000) The use of targeted procurement as an instrument of poverty alleviation and job creation in infrastructure projects. *Public Procurement Law Review*, 5, 201–266.

Watermeyer, R. B. (2004) Facilitating sustainable development through public and donor regimes: tools and techniques. *Public Procurement Law Review*, 1, 30–55.

Watermeyer, R. B. (2006) Poverty reduction responses to the Millennium Development Goals. *The Structural Engineer*, 84 (9), 27–34.

Watermeyer, R. B. (2009) Towards a minimum standard for structural engineering work. *The Structural Engineer*, 87 (19), 12, 14, 16.

Watermeyer, R. B. (2010a) *Engineering: Issues, Challenges and Opportunities for Development – Section 1.2 Engineers, Technologists and Technicians*. UNESCO: Paris.

Watermeyer, R.B. (2010b) Alternative models for infrastructure delivery. *IMIESA (Magazine of the Institute of Municipal Engineers of South Africa)*, October, 68–77.

Watermeyer, R. B. (2010c) Report back on perspectives expressed at the Wits Symposium (August 2009). *Civil Engineering*, January/February, 34–39.

Watermeyer, R. B. and Thumbiran, I. (2009) Delivering infrastructure at scale in developing countries: numbers or systems? Paper presented at the Fourth Built Environment Conference hosted by ASOCSA, Livingston, Zambia, May.

Watermeyer, R. B. and Pham, L. (2011) A framework for the assessment of the structural performance of 21st century buildings. *The Structural Engineer*, 89 (1), 19–25.

Watermeyer, R., Larkin, D., Kee, A. and Thumbiran, I. (2009) Delivering infrastructure at scale: eThekwini Water and Sanitation experience in a pilot project. *Civil Engineering*, March, 31–37.

12 Financing construction projects in developing countries

Akintola Akintoye and Suresh Renukappa

Introduction

A construction firm's financial standing is affected by retention, undervaluation, delay in payments, interest charges and a host of other economic and financial variables (Hillebrandt and Cannon, 1990). Construction businesses constantly need funds to finance their diverse activities. The long-term survival and growth of a company depend not only upon the selection of the right projects, but also on the company's ability to raise the necessary finance to fund them. Companies need capital to finance their investment projects. Short-term projects and investment in current assets are typically financed with short-term credits such as trade debt, short-term lines of credit, and cash generated from operations. However, these funding sources are not always sufficient or adequate. Long-term projects such as the expansion of the company's activities or the addition of new product lines require more permanent sources of capital such as debentures, bonds and leases.

It is generally recognised that, in the developing countries, the construction industry is associated with high rates of insolvency of companies compared with the other industries. Lowe (1997) identified the factors which cause insolvency in the construction industry as including profitability level, cash flow, credit, fluctuating demand, competitive tendering and management problems. Since the execution of construction projects demands huge investments, construction companies rarely rely on their own reserves to execute projects. Hence, the procurement of cash, termed as financing, has always been a major concern of construction companies.

In addition, liquidity, lack of which, according to Davis (1991), is an ultimate cause of insolvency, has been a major issue in construction, as the contractors or sub-contractors fail to meet their liabilities to the financial institutions and suppliers. Companies often find it difficult to obtain the right amount of funds, and at reasonable interest rates (Edge, 1988). These characteristics exist during depressed as well as buoyant economic periods. During boom periods, over-trading has been a cause of cash problems, when the capital base of the company cannot support its rapid growth (Burnett, 1991).

According to Khosrowshahi (2000), a company may enjoy business based on a healthy workforce, secure profit and loss account and an adequate number of projects, and yet fall victim to a cash-flow crisis, as it might face a shortage of liquid cash to fund the project because much of the profit is tied up in the work in progress and in interest payments. This is particularly true for low-asset companies, which are a characteristic of the construction industry in the developing countries.

The purpose of this chapter is to explore various sources of finance available to a construction enterprise and the challenges which the construction industries in developing countries face in attracting finance. The chapter draws from the literature in the areas of the construction industry and finance. It is divided into five parts. In the first part, the different meanings of construction project financing are discussed. The second part is devoted to infrastructure investment in developing countries. In the third part, key challenges associated with attracting finance for construction projects are presented. This is followed, in the fourth part, by consideration of the main sources and forms of finance available to the construction projects in developing countries, and characteristics, merits and demerits of each form, and the major issues to consider in respect of each type of financing. In the final main section of the chapter, the recent global financial crisis and implications for financing construction projects in developing countries are discussed. Last, a conclusion on key points to be considered in raising finance for construction projects in developing countries is presented.

Construction financing

'Construction financing' can take on different meanings depending on the context in which finance is required to support construction activities. Construction financing may be used in the context of industry financing, company financing, the financing of construction projects, and project financing, as shown in Figure 12.1.

Level 1: Construction industry

The construction industry needs finance to support various research, development, training and knowledge-sharing projects whereas construction

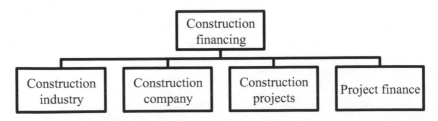

Figure 12.1 Different meanings of construction financing.

company finance concerns the financing of corporate activities and projects (Brealey *et al.*, 2007). Grants from public funds and international organisations are the main sources of finance for the construction industries in developing countries. Grants are gifts requiring no repayment. Grants from international aid agencies are attractive to developing countries, as they do not directly increase the debt burden (Robinson and Thagesen, 2004). Grants may be provided in many forms, such as support for training and education, and budget support for research and development.

Level 2: Construction company

Investors in a construction company, or the sources of corporate financing arrangements, assess the cash flow and assets of the entire company to determine its ability to service the debt and to provide security for the loan. Thus, investors carefully examine the construction company's financial track record, the viability of its expansion plans and its ability to generate sustainable cash flows. Furthermore, in the construction company finance environment, investment is financed as part of the company's existing balance sheet. Therefore, the creditworthiness of a company is primarily driven by the company's ability to service the debt.

Level 3: Construction projects

Most construction projects are paid for by the client through periodic interim payments and by the construction company. If the construction company borrows money to fund the project, it must be repaid, whether the project succeeds or fails. This type of financial arrangement is referred to as 'financing construction projects' (Venkataraman and Pinto, 2008). Hillebrandt and Cannon (1990) have noted how the finance available for construction projects can be affected by the level of retention, undervaluation, delays in payments, interest charges and a host of other economic and financial variables.

Level 4: Project finance

'Project finance' is not the same thing as 'financing projects' (Yescombe, 2002, p. 150). Project finance can be defined as 'the creation of a legally independent project company financed with equity from one or more sponsoring firms and non-recourse debt for the purpose of investing in a capital asset' (Esty, 2007, p. 213). A project finance company invests in a single-purpose capital asset, usually a long-term illiquid asset; it invests only in the particular project for which it is created. Project finance is generally used for new, stand-alone, complex projects with large risks. The project company is dissolved once the project is completed. Esty (2004a) points out that project finance involves significant costs compared with corporate finance. The term 'project finance' is generally used to refer to a non-recourse or limited recourse financing structure in which debt, equity and credit enhancement are combined for the construction and operation, or the refinancing, of a particular facility in a capital-intensive industry (Hoffman, 2007).

Project finance is designed to reduce transaction costs, in particular those arising from a lack of information on possible investments and capital allocation, insufficient monitoring and enforcement of corporate governance, risk management, and the inability to mobilise and pool savings (Nevitt and Fabozzi, 2000). The lenders pay particular attention to project performance on a regular basis because the possibility of repaying principal and interest depends on the ability of the project to generate sufficient cash flows.

The success of a construction project finance transaction will depend on various factors including (Nevitt and Fabozzi, 2000; Esty, 2004a; Esty and Sesia, 2007) (a) the size of the project; (b) the size of the company; (c) the location; (d) risks; (e) project feasibility; (f) the project team; and (g) the project's capacity to generate sufficient cash during its operating phase to match the funds which are needed to service the debt and dividends to be paid to the project sponsors. Project finance is used more extensively in developing countries as an efficient way to develop infrastructure projects (Hammami *et al.*, 2006).

As Hoffman (2007) noted, the term 'project finance' is often misused. In some circles, it refers to raising funds to pay the costs of any project. In others, it is used to describe a hopeless financial situation which can be remedied only with extreme financing options. For the purpose of this chapter, 'project finance' refers to raising funds to pay the costs of a project.

Infrastructure investment in developing countries

This section deals with project finance in the developing countries in view of its continuing importance for the funding of major capital infrastructure in those countries. The funding for project finance in developing countries comes from various sources including private-sector commercial and investment companies, commercial and investment bank syndicates, multi-national companies, equity providers and regional and international development banks.

Traditionally, and for much of the past century, in developing countries, large infrastructure[1] projects were funded and operated by governments. However, since the early 1980s, there has been increasing involvement of the private sector in the financing and management of such projects. This has taken the form of the use of 'project finance' to fund infrastructure investments (Brealey *et al.*, 1996). Private-sector finance of capital projects in the developing countries grew because the current receipts, savings and central government transfers proved to be insufficient to fund large-scale projects in most of these countries (Platz, 2009); public-sector organisations in developing countries have sought to raise private capital to deliver infrastructure services.

The total project-finance investments worldwide grew at a compound annual rate of almost 20 per cent during most of the 1990s, and reached US$165 billion in 2003 (Esty, 2004b). Most large-scale infrastructure projects, which, by nature, involve large capital investment, long completion times and high levels of risks, present formidable financial challenges to developing countries. The volume of infrastructure determined by global demographic, public-health and safety needs and economic development goals is far in excess of currently available financing resources (Fitch Ratings, 2004).

Developing countries need to spend a higher proportion of their gross domestic product (GDP) on infrastructure because of their lower infrastructure stocks and their greater growth rates. Estimates for new infrastructure investments in developing countries are in the range of 5–6 per cent of GDP. The range is higher in lower-income countries: 7–9 per cent of GDP (Delmon and Juan, 2008). For all developing countries, the infrastructure investments necessary to sustain a 4 per cent rate of economic growth are estimated to be US$500–600 billion per year. However, as Delmon and Juan (2008) noted, currently the investment is only about 40 per cent of this level.

Ofori (2006) observed that the developing countries do not form a homogeneous group. Incomes vary widely, as do determinants of economic prospects such as inflation, resource endowments, inflow of foreign direct investment (FDI) and national debt-servicing commitments. There are also differences in the level of development of their construction industries, size of industry, technologies being applied, practices and procedures, and operating environments. Therefore, the infrastructure investment required will vary across the countries. It is also pertinent to note that the costs of constructing roads, railways, pipelines and water and sanitation networks depend on terrain, weather, costs of obtaining licences and permits, and labour and material costs.

In recent years, private capital flow has fallen; investors have experienced significant distress. However, the infrastructure funding gaps in developing economies are at unprecedented levels. For example, a report by the Africa Infrastructure Country Diagnostic revealed that, in order to bring infrastructure on the continent up to a standard at which true efficiencies could be gained, some US$31 billion per year needs to be provided. The study found that bad infrastructure in 24 African countries cut national economic growth by 2 percentage points and reduced business productivity by 40 per cent (Corbett, 2010). Similarly, for Asia, it is estimated that the total infrastructure financing gap will be an average of about US$420 billion per year during 2006–2015 (ASEAN, 2008). Recognition of these funding gaps has resulted in a nearly universal acceptance that the private sector can, and should, play a larger role in the financing of infrastructure in partnership with the public sector (Fitch Ratings, 2004).

Borrowing from financial institutions has emerged as a viable option for infrastructure development in many countries. However, these loans are usually of shorter tenure of up to five or seven years, and may require sovereign guarantees (Rajivan, 2004). Hence, many developing countries are turning to domestic and international capital markets to mobilise private savings involving lengthier payback periods to finance their infrastructure projects. The level of private-sector investment in infrastructure projects in developing countries has been volatile over the last 10 years (Platz, 2009). The World Bank (2006a) estimated that, between 1970 and 2005, infrastructure-related lending oscillated between one-third and two-thirds of the bank's total lending. Investment dipped to US$50 billion in 2003 after reaching a peak of US$131 billion in 1997; it rose again to reach US$158 billion in 2007. However, with the notable exception of the telecommunications sector, investment declined in the two poorest regions (East Asia and Pacific, and sub-Saharan Africa) over the period 1990–2007 (World Bank, 2008).

Private investment in infrastructure is a relatively new trend in the developing countries (Doh and Ramamurti, 2003). Fuelled by the opening up of their political and economic systems, developing countries are increasingly turning to private investors – both local and foreign – to increase the availability of infrastructure, and hence improve access to the services they provide. The governments have moved towards market-based pricing of infrastructure.

For its operational purposes, the World Bank has classified 'developing countries' in terms of gross national income (GNI) per capita. The World Bank (2009a) defines a 'low-income' country as one with a GNI of US$995 or less in 2009, a 'lower-middle-income' country as one with a GNI of US$996–$3,945 and an 'upper-middle-income' country as one with a GNI of US$3,946–$12,195.

The World Bank Private Participation in Infrastructure (PPI) project database tracks all investment in projects with private participation. The database covers four sectors: energy, telecommunications, transport and water. The database considers projects to have private participation if a private company or investor bears a share of the project's operating risk.

Table 12.1, based on an analysis of World Bank data, shows private-sector participation in infrastructure development in some developing countries between 1990 and 2005. Table 12.1 shows the analysis of the top two countries for PPI investment in the six developing economy regions in terms of income level, total investment, type of PPI and sectors where they took place. The PPI in the 10 countries represents about 56 per cent of the total PPI investment in the 139 developing countries in the World Bank database of developing countries that had private participation in infrastructure. The analyses show that PPI investment is dominated by a small group of developing countries with relatively fast-growing markets. With the exception of the Europe and Central Asia region (33 per cent), PPI investments in each pair of top countries represent more than 50 per cent of the total PPI investment in that region. In South Asia, PPI investments in India and Pakistan represent 92 per cent of the region's total PPI investment.

Furthermore, the results of the analysis provide evidence to conclude that private-sector finance of infrastructure development is a major source of investment in developing countries. Such finance has been mainly in telecommunications in low-income developing countries rather than energy, transport, water and sewerage. Although the low-income countries constitute the bulk of developing economies, private-sector investment in infrastructure development in this category of countries is insignificant compared with middle-income developing countries. The amount of private-sector participation in infrastructure in sub-Saharan Africa is comparatively low. This could be for various reasons. The Leadership Roundtable on 'Funding African Infrastructure' identified some challenges associated with financing infrastructure projects in African countries (Corbett, 2010). These include access to long-term financing; affordability; finding reliable and credit-worthy off-takers to support the financing of projects; lack of legal and institutional frameworks to support public–private partnerships; too small a pool of talent to execute infrastructure project financing; difficulty in finding high-quality sponsors for projects; shallow domestic markets; currency risks; governments' lack of long-term perspective, capacity and experience in infrastructure development; the issue of corruption; and political instability.

Key challenges associated with attracting finance for construction projects

The construction industry in developing countries, by nature, has various special issues and challenges alongside socio-economic stress, chronic resource shortages, institutional and legal weaknesses and a general inability to deal with these key issues (Hillebrandt, 2000; Ofori, 2001; Aggarwal, 2003). There is also evidence that the problems have become greater in extent and severity after the current financial crisis, which has made the task of improving the performance of the industry even more demanding.

The key challenges in attracting finance for construction projects in developing countries include the image of the construction industry, impact of globalisation, banking sector issues, need for large up-front investment, corruption, occasional financial crises, and sustainability issues. These challenges are not peculiar to the developing countries; many developed countries also face the same challenges albeit at different levels. In the following sub-sections, each of these key challenges is discussed.

Image of the construction industry

Regardless of the level of economic development, the construction industry is a significant part of any economy because of its size and the potential role it can play in the development efforts of that economy. Ofori (1991) noted the importance of the construction industry in the national economy and attributed it to the high linkages with the rest of the economy. However, the construction industry has a poor image in both developed and developing

Table 12.1 Top two countries for Private Participation in Infrastructure (PPI) investment in developing economy regions

Region	No. of countries with PPI	Top two countries	Total investment (US$)	Sector PPI type	Energy (%) Concession (%)	Telecom (%) Divestiture (%)	Transport (%) Greenfield project (%)	Water and sewerage (%) Manage and lease (%)	% of total investment Regional	Overall total	Income level
EAP	18	China	66,955	Sector	40.9	21.6	32.8	4.6	34	8	Lower middle
				PPI type	12.2	40.0	47.7	0.1			
		Malaysia	37,845	Sector	32.6	16.3	37.7	13.5	19	4	Upper middle
				PPI type	15.0	13.8	71.2	0.0			
ECA	26	Poland	23,480	Sector	12.4	84.0	3.5	0.1	17	3	Upper middle
				PPI type	3.1	66.6	30.2	0.0			
		Russian Federation	22,485	Sector	15.1	80.0	2.3	2.6	16	3	Low
				PPI type	0.0	49.2	50.7	0.1			
LAC	28	Brazil	150,395	Sector	39.1	44.7	13.7	2.5	40	18	Upper middle
				PPI type	14.8	62.8	22.4	0.0			
		Argentina	72,575	Sector	36.2	34.3	18.1	11.4	19	9	Upper middle
				PPI type	29.2	50.8	20.0	0.0			

Region		Country			Sector						Income group
MENR	14	Morocco	15,642	Sector	58.6	41.4	0.0	0.0	37	2	Lower middle
				PPI type	41.2	30.0	28.8	0.0			
		Saudi Arabia	8,834	Sector	0.0	96.6	2.8	0.6	37	1	Upper middle
				PPI type	2.8	46.2	51.0	0.0			
SA	6	India	39,571	Sector	42.2	49.0	8.2	0.6	75	5	Low
				PPI type	2.4	11.2	86.4	0.0			
		Pakistan	8,940	Sector	66.4	28.4	5.2	0.0	17	1	Low
				PPI type	3.2	11.2	85.6	0.0			
SSA	47	South Africa	19,015	Sector	6.6	82.4	9.9	1.1	48	2	Lower middle
				PPI type	7.7	47.5	44.8	0.0			
		Nigeria	5,441	Sector	13.0	86.5	0.4	0.0	14	1	Low
				PPI type	0.4	0.0	99.6	0.0			
Total	139		471,178	Sector	35.0	44.1	16.4	4.5		56	
				PPI type	14.3	45.3	40.4	0.0			

Source: Based on an analysis of the World Bank (2005) (cited by Akintoye, 2009).

EAP, East Asia and Pacific; ECA, Europe and Central Asia; LAC, Latin America and Caribbean; MENR, Middle East and North Africa; SA, South Asia; SSA, sub-Saharan Africa.

countries. In the UK, the construction industry compares badly with other industries in terms of capital cost, product quality and client satisfaction (Egan, 1998; Latham, 1994). One of the charges often levelled against the construction industry is that it has a poor record in terms of innovation, compared with other industries. Another common feature of the industry is the high level of corruption involving construction companies and government officials.

Furthermore, in most of the developing countries, work in construction is not regarded as 'decent work' (Wells, 2001). In both developed and developing countries, the construction industry is perceived as 'dirty, dangerous, difficult' (Wells, 2001, p. 2), 'old fashioned and low performing' (Fairclough, 2002, p. 31). The lack of job security for construction workers, poor health and safety record, and lack of opportunities for training and skill formation contribute to a poor image of the construction industry in the eyes of its own workforce, clients and the public at large in developing countries. Thus, it is often difficult to attract new recruits into the industry, even while there is a shortage of skills threatening the quality of construction work.

Given the image of the construction industry and the implications for return on investments, potential investors may not be willing to finance construction projects. An enhanced reputation and brand image of the construction industry can positively affect relationships with current and potential investors, as well as attract partners and suppliers (Fombrun and van Riel, 2004). Therefore, policy makers in developing countries and the construction industry itself should institute measures to improve the industry's image.

Globalisation

Globalisation is driving major changes in the way business is undertaken in the construction industry. The global economy has been transformed in recent years by the fall of international barriers to the flow of goods, services, capital and labour, and a marked acceleration in the pace of technological and scientific progress (Jewell *et al.*, 2010). These new characteristics have increased the pace of the globalisation of the construction industry, and few construction companies find that they can be exempt from this trend (Korkmaz and Messener, 2007).

Although the globalisation of the construction industry has benefited the developing economies, challenges have also emerged (Ogunlana *et al.*, 1996). Many construction companies in developing countries face difficulties in competing against their foreign rivals in their domestic markets.

In spite of the rapid rate of globalisation, it should be noted that the construction industry is still primarily influenced by local factors. Local knowledge is critical, and local actors and institutions will continue to play a significant role. In developing countries, access to domestic and international finance to undertake construction projects that maintains activities for firms is vital (Carrillo, 1996). However, access to international finance for costly projects is beyond the control of the construction companies in developing countries. Although there have been some institutional reforms, there are still significant differences among countries that make the process of construction project development and ownership complex. Differences in business structure, demographic profiles and cultural preferences may present new opportunities and challenges for the construction industry and investors.

Banking sector issues

Hassanein and Adly (2008) noted that the lack of access to longer-term sources of finance and the lack of suitable financial resources were key barriers to sustainable growth of the construction industry in developing countries. Many construction firms in these countries still find it difficult to arrange short-term finance, and many remain critical of the banking system (Levine, 2002). Moreover, obtaining a loan from a bank is a tedious procedure which can take several months. Many construction companies do not have access to commercial funding sources at reasonable rates as compared with those available in developed countries (De Silva *et al.*, 2008). Other constraints are lack of adequate collateral, unfamiliarity with complicated procedures for raising finance from banks, high risks and transaction costs, difficulties in enforcing contracts, stringent supervisory and capital adequacy requirements, and lack of appropriate instruments to manage the risks involved (Cook and Nixson, 2000).

Most construction companies, both large and small, establish bank overdrafts to finance the cash requirements on their projects. Bankers commonly impose limits on the amount of credit provided under overdrafts. Contractors want to minimise financing costs, which are determined based on factors including the interest rate and penalties incurred on unused portions of credit. Bankers usually specify the interest rate based on the allowed credit limit. Low credit limits might cause extensions to project duration and, consequently, overruns in project costs. Thus, the credit limit allocated to a project affects the financing costs and indirect costs. Despite such problems, the banks remain the most important source of external finance for construction projects in developing countries.

Large up-front investment

Financing is a major issue in the design and development of construction projects because most construction projects require the bulk of the investments up front. Attracting large and long-tenure capital for construction projects in developing countries is a challenge (Sawant, 2010). As noted above, large construction projects have traditionally relied heavily on public expenditure for financing in developing countries. However, public funding may be less readily available in the future. Therefore, public-sector organisations in developing countries are pursuing alternative strategies for raising private capital to deliver infrastructure services.

Private sources of capital available for funding this large up-front investment include pension funds, mutual funds, hedge funds, private equity and endowments (these are discussed later in the chapter). Because of low levels of income, the financial markets in the developing countries are largely underdeveloped, informal, lacking in depth, inefficient and dominated by a few, often foreign-owned, commercial banks (Mpuga, 2010). For these reasons, it is difficult to mobilise the large up-front investment required for major construction projects in developing countries.

Corruption

Construction is frequently held up as one of the most corrupt industries worldwide (Davis, 2004). Corruption can be defined as 'dishonest or illegal behaviour, especially of people in authority, that results in distortion in the allocation of resources and performance of firms or government' (Hornby, 2000, p. 281). Kenny (2007) noted the diversity of types of corruption present in the construction sector: from bribes designed to manipulate budgeting decisions, project selection, tender specifications, procurement outcomes or contract negotiations and renegotiations, through bribes designed to cover up poor-quality construction practices and outcomes, to the theft of materials.

Transparency International drew attention to the impact of corruption in the construction industry in its 'Global Corruption Report' in 2005. In the report, the construction industry is perceived to be more susceptible to corruption because of several features, two of which stand out. First, construction projects tend to be large, and yet companies with financial and technical capabilities to implement them are few. Second, construction involves complex, non-standard production processes which foster asymmetric information stocks between clients and providers, and the latter also maintain close ties with the government. Corruption has been uncovered in World Bank-supported road projects in Vietnam and Indonesia, as well as in a large dam project in Lesotho (World Bank, 2006b).

Corruption may be a significant part of a system of decision making that leads to poor construction, limited occupational safety and low returns to investment. Therefore, many large institutional investors do not have much confidence in financing construction projects in developing countries.

Financial crisis

During the recent global financial crisis, several financial institutions in developed countries experienced a deterioration in their balance sheets due to huge losses in instruments based on sub-prime mortgages in the United States (Cali *et al.*, 2008). The most similar crisis was the Asian economic turmoil in 1997, which hit many Asian countries and triggered shock waves which had an impact on the construction industry for many years. For instance, in Hong Kong, the construction industry's gross output measured in constant prices had shrunk by almost one-third in 2007 as compared with its peak in 1998 (CSD, 2008).

The current credit crunch has caused a substantial decline in the amount of bank capital and, because of risk-based capital requirements, the banks have restricted asset growth by cutting back on lending. As a consequence, cross-border syndicated loans to developing countries and intra-bank lending have been curtailed (World Bank, 2004; Lin, 2008). The implication is that the developing countries, which have relied on borrowing from foreign banks to finance construction projects, will suffer more from the crisis (Cali *et al.*, 2008).

Sustainability issues

All industries now face the challenge posed by 'sustainability'. The impact of the construction industry on the society and the environment is significant (Renukappa, 2009). To encourage private-sector participation in construction projects in developing countries, the private-sector sponsors and, in particular, foreign investors would want to be assured that the project is technically and economically feasible, financially viable, socially and environmentally sustainable, and politically acceptable (Akintoye, 2009). Therefore, the construction industry has been called upon to become more market responsive, to reduce the number of accidents on site, to minimise waste, to minimise carbon emissions, to end its record of pollution incidents, to integrate the supply chain, to engage all of its stakeholders, and to create a more responsible and enhanced sustainability profile (Myers, 2005; Renukappa, 2009).

Large infrastructure projects, such as transport and power projects, may have adverse environmental and social impacts requiring project re-evaluation, redesign, additional investment, compensation costs and strong stakeholder engagement as well as reputation risks (World Bank, 2008). Thus, the failure of the construction industries in developing countries to address the environmental and social sustainability of construction activities could reduce investors' confidence in financing construction projects.

Sources of finance for construction projects in developing countries

The ability to borrow, or to gain access to finance, plays an important role in the construction industry (Halpin and Senior, 2009). To remain competitive, construction companies may be forced to manage liquidity by using new credit and borrowing strategies.

According to Westhead and Wright (2000), the absence of adequate funding represents a major obstacle to business development in a firm – regardless of size, location or industry or type of economic activity, although some small companies can satisfy their financial needs by taking loans from their family members, friends or acquaintances (Hussain and Matlay, 2007). However, the majority of growth-oriented companies rely on long-term funding provided by banks, financial institutions or venture capitalists.

The sources of financing available for construction projects in developing countries include (Nevitt and Fabozzi, 2000): commercial banks; local governments; institutional lenders; international agencies; government export-financing agencies and national interest lenders; money-market funds; commercial finance companies; leasing companies; investment management companies; assets funds or income funds; wealthy individual investors; companies which supply a product or raw materials; companies requiring the product or service produced by the project; contractors; trade creditors; vendor-financing sources; sponsor loans and advances; and savings and loan associations.

Figure 12.2 presents the key sources of finance for construction projects in developing countries. The first issue facing a construction company when embarking on a project, whether large or small, is whether it should be carried out using internal or external sources of finance. For a construction company, the main sources of internal finance are savings, depreciation funds and accumulated profits. Most construction companies have a strong preference for internal finance, because it involves fewer risks than external finance (Halpin and Senior, 2009). Companies might turn to external sources of finance because they do not have sufficient internal funds to finance projects.

There are two fundamental sources of external finance from which all financial engineering originates; these are equity and debt (Pretorius *et al.*, 2008). Equity and debt form the main sources of financing for construction companies, and internally generated funds form another source. The external sources may include borrowing from domestic and international markets.

This section presents possible main sources of finance available for construction projects in developing countries, their characteristics, merits and disadvantages.

The basic distinction in long-term construction project financing is that between debt and equity (Sawant, 2010). Debt refers to funds lent by financiers. It is any instrument that creates a liability, including the right to demand cash or shares at some future date. Lenders receive payments according to a predetermined rate. The repayment of debt does not depend on the profit of the company. Equity is provided by owners, also called shareholders. Owners include project sponsors, who are the driving force behind a project or a company, and 'passive' investors, who hold equity positions without being involved in the promotion of the project. Owners' return on their equity investments depends on the company's profit. These are spilt into equity and non-equity interests (Pike and Neale, 2006).

In the following sub-sections, some of the instruments available for raising finance for construction projects in developing countries are discussed.

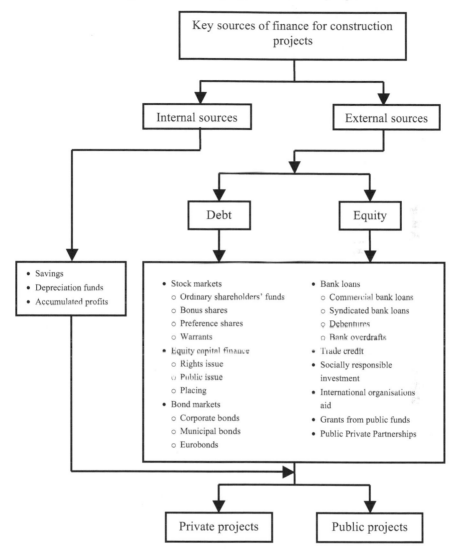

Figure 12.2 Key sources of finance for construction projects in developing countries.

Stock markets

Stock markets are where government, industry and companies can raise long-term capital and investors can buy and sell securities such as shares and bonds (Arnold, 2010). Among the various types of shares available in stock markets, ordinary shareholders' funds, bonus shares, preference shares and warrants are the most common instruments used for raising funds for construction projects in developing countries (Smith *et al.*, 2006; Halpin and Senior, 2009).

Ordinary shareholders' funds

Ownership interests of shareholders arise primarily from two sources: the amounts invested by the shareholders in the company and the amounts earned by the business on behalf of its shareholders. These two sources are reported in the balance sheet as called-up share capital and retained profits. Ordinary share capital represents the risk capital of the business. There is no fixed rate of dividend, and ordinary shareholders receive a dividend only if there are profits available for distribution after other investors (such as preference shareholders) have received their dividend or interest payments (Arnold, 2010). Ordinary shares constitute the most important source of funding for the corporate private sector. These funds are permanent, in the sense that the business makes no undertaking to repay the original investment. The owners' equity in a corporation is divided into shares, represented by stock or share certificates.

Lerner and Schoar (2004) attribute the increase in equity-financing activity in developing countries to radical reforms undertaken by governments in various areas, such as the reduction in trade and financial barriers, improvement of financial and regulatory systems, and commitment to technological innovations. A study by Deloitte (2007) found that 40 per cent of the respondents indicated that they had been attracted to developing countries' equity-financing activity partly by tax incentives, free trade zones or other incentives while 69 per cent indicated that more than half of their investments in developing countries were meeting or exceeding their revenue expectations. Therefore, in developing countries, a significant amount of equity financing can be raised for construction projects from investors by means of either floating the project company on the stock market or raising funds by private placement (Smith *et al.*, 2006). However, market inefficiencies, institutional constraints, the shallow nature of equity markets, their consequent vulnerability to shocks, and lack of relevant management expertise and professional support mean that this form of financing has been absent in most of the developing countries.

By selling (issuing) these stocks (shares) to the public, a construction company can raise relatively large amounts of owners' equity. Equity shareholders are the owners of the business who bear the greatest risk. If the company trades unsuccessfully, the ordinary shareholders are the first to suffer in terms of lack of dividends and probably a fall in the market value of their shares. If the firm is put into liquidation, it is the ordinary shareholders who will be at the bottom of the list with a claim for repayment of their investment (Watson and Head, 2007). Because of the high risks associated with this form of investment, ordinary shareholders will normally require a higher rate of return from the company (Pike and Neale, 2006). However, if the firm is successful, it is the ordinary shareholders who benefit more than any of the other groups that have an interest in the business such as employees, suppliers and creditors. Once all the obligations of these other groups are met, the balance accrues to the shareholders. Ordinary shareholders will also have control over the company. They are given voting rights and have the power both to elect the directors and to remove them from office.

Bonus shares

A bonus share or script involves the issuing of a company's shares to existing shareholders (Reuvid, 2002). Companies convert retained profits into ordinary shares, which are then distributed to existing shareholders free of charge. Retained profit is the major source of long-term finance for many construction businesses (Halpin and Senior, 2009). The reinvestment of profits rather than the issue of new shares can be a useful way of raising capital from ordinary share investors. There are no issue costs associated with retained profits, and the amount raised is certain, once the profit has been made. The retention of profits is determined by the directors of the company. They may find it easier simply to retain profits rather than ask investors to subscribe to another new share issue (Arnold, 2010).

Preference shares

It is not uncommon for a company to have more than one class of shares, each with different rights and limitations. To attract investors, companies have devised quite a variety of ownership shares. An issue of shares with certain preferences or features that distinguish it from the ordinary shares may be assigned the label 'preference'. Preference shares also constitute part of shareholders' funds – designated 'non-equity shareholders' funds', as they are hybrids falling between pure equity and pure debt (Pike and Neale, 2006). Holders receive an annual dividend, usually a fixed percentage of the par value. Preference shares offer investors a lower level of risk. They have some priority over the ordinary shares as regards the distribution of the dividend or the proceeds from the liquidation of the company.

The dividend rate on preference shares is usually set at the time of issue. This dividend must be paid first, if any dividend is paid at all, and sometimes it accumulates if it is not paid. In case of liquidation, the claims of preference shareholders are paid ahead of those of the ordinary shareholders. Because of the lower level of risk associated with this form of investment, investors will be offered a lower level of return than that offered for ordinary shares. In addition, preference shareholders are usually not given voting rights, although these may be granted where the preference dividend is in arrears (Arnold, 2010).

The three main types of preference shares are now presented.

- *Cumulative preference shares* give the holders the right to receive arrears of dividends that have arisen as a result of the company having insufficient profits in previous years. The unpaid dividends will accumulate and will be paid when the company has generated sufficient profits.

- *Non-cumulative preference shares* holders have no rights to receive dividends in arrears if the company's profit in a given year is either too low to be distributed, or negative.
- *Redeemable preference shares* allow the company to buy back the shares from the shareholders at an agreed future date. Redeemable preference shares are generally seen as a low-risk investment.

Preference shares are no longer an important source of finance for most companies. This form of fixed-return capital has declined in popularity because dividends paid to preference shareholders are not allowable against taxable profits whereas interest on loan capital is an allowable expense. This lack of a tax shield explains why preference shares are relatively unattractive to companies compared with other forms of fixed-rate security (Pike and Neale, 2006).

The dividends of preference shares tend to be fairly stable over time and there is usually an upper limit on the returns which can be received. As a result, the share price, which reflects the expected future returns from the shares, will normally be less volatile than that of ordinary shares. Factors to consider with regard to share financing include (Pike and Neale, 2006) issuing costs, servicing costs, obligation to pay dividends, tax deductibility of dividends, and effect on the control of the business.

Warrants

A warrant is another type of capital issue. It is a special security issued by a company which gives a holder the right to be allotted ordinary shares in the company on terms fixed on the issue of the warrant. That is, holders of warrants have the right, but not the obligation, to acquire ordinary shares in the company at a price and future date which are specified. Warrants are issued as a 'sweetener' to accompany the issue of loan capital or debentures (Reuvid, 2002).

Equity capital finance

A company may raise capital by operating at a profit or by borrowing. It can sell ownership in the form of shares of stock and ownership rights in order to raise capital (Spiceland *et al.*, 2005). An investor can sell his or her ownership interest at any time without affecting the company or its operations. To raise capital, a company may issue shares in a number of ways. These may involve direct appeals to investors or the use of financial intermediaries. When a company sells shares, it should make use of legal, promotional (by advertising a prospectus) and accounting services necessary to effect the sale. The cost of these services reduces the net proceeds from selling the shares. A company can increase its ordinary share capital in a number of ways: rights issue; public issue; and placing.

Rights issue

A rights issue is an offer of a company's shares to its existing shareholders. Existing shareholders have the priority to purchase a certain number of issued shares for cash. The new shares are allocated to existing shareholders in proportion to their existing shareholding. In order to make the issue attractive to shareholders, the new shares are often offered at a price significantly below the market value of the shares. In the UK, the law requires, in normal circumstances, that any new equity issue must be offered first to existing shareholders. The existence of this point in law is sometimes seen as a restriction on the ability of directors to take advantage of some other sources of equity. Similarly, to safeguard the interests of existing shareholders, in the year 2000, the Securities and Exchange Board of India formulated the Disclosure and Investor Protection Guidelines. These guidelines apply to all public issues being made by a company and govern all offers of sale and rights issues and state measures to ensure that promoters do not make extravagant promises to investors which they might subsequently be unable to fulfil.

Rights issues are now the most common form of share issue. For many companies, it is a relatively cheap and straightforward way of issuing shares. The issuing costs are quite low and the issuing procedures are simpler than for other forms of share issue. The advantage of the rights issue is that control of the company by existing shareholders will not be diluted provided they take up the rights offer. The offer does not have to be accepted personally by the existing shareholders. They may sell the rights separately from the shares, or sell the shares with the rights offer attached.

Public issue

A public issue is a form of offer for sale. The public is offered shares at an agreed price directly by the company. Such issues are relatively rare in practice but when they occur they tend to be large in size. For example, in August 2010 the world's largest ever public issue of shares was completed by the Agricultural Bank of China; the total amount raised was about US$22.1 billion.

There are two ways of making public issues (Pike and Neale, 2006). First, the issuing firm can sell the shares to an issuing house, usually a merchant bank which specialises in such work; the issuing house then sells the shares to the public. This is known as an offer for sale. Second, the issuing firm sells the shares directly to the public; often, such firms are advised by a merchant bank on such matters as the pricing of the issue. Here, the offer is known as an offer by prospectus.

When making an issue of shares, the company or the issuing house will usually set a price for the shares. However, establishing a share price may not be a straightforward task, especially when the company has specific characteristics that do not compare to those of any other business that might be used as benchmark. The problem then is, if the share price is set too high, the issue will be undersubscribed and the company or the issuing house acting on its behalf will not raise the funds expected. On the other hand, if the share price is too low, the issue will be oversubscribed and the company or the issuing house will receive less than could have been anticipated. This method is normally used only for large issues by well-known companies because the costs involved are usually high.

Placing

Placing does not involve an invitation to the public to subscribe to shares. Instead, the shares are 'placed' with selected investors such as large financial institutions (Reuvid, 2002). This can be a quick and relatively cheap form of raising funds as savings can be made in advertising and legal costs. However, this form of issuing shares can result in the ownership of the company being concentrated in a few hands. This form is mostly used by small companies which are seeking relatively small amounts of cash.

Bond markets

In recent years, the use of bonds as a vehicle for obtaining long-term debt funds has increased. Bond financing is characterised as a direct tool used in financial markets comprising a broad range of investors. In order to issue bonds, a firm's financial conditions are rated and the information gathered in the process of rating is shared with potential investors (Yescombe, 2002). Given the nature of construction projects, local bond markets could play an important role in mobilising supplementary funds for large-scale projects and in minimising double mismatch problems. The large financing gaps and advantages of bond financing for long-term construction projects constitute an impetus for the development of long-term, local currency-denominated bond markets (Hyun *et al.*, 2008). However, instituting a local bond market is a relatively new activity for many developing countries, and insights from experience are limited.

Developing a bond market can be more complicated than developing an equity market. Bond markets need supporting pricing infrastructure. They operate best when they have money-market and longer-term benchmarks. However, many developing countries lack these benchmarks. Moreover, bond markets need sophisticated market participation. Market participation results when a fairly comprehensive range of economic, technical, political and 'behavioural' factors come together (Harwood, 2000). However, many of these factors are not well developed in many developing countries and it can take a considerable period of time to set up such a market.

Bonds are issued by corporations, governments or regional entities. Government bonds generate tax-free income for investors. Thus, these bonds can be issued at lower rates than corporate bonds (Van Horne and

Wachowicz, 2005). This benefit provides governments with the opportunity to use bonds to finance projects. Among the types of bonds available in markets, corporation bonds, municipal bonds and Eurobonds are most popular for raising funds for construction projects in developing countries.

Corporate bonds

Many construction companies have started issuing bonds in order to raise funds. Corporate bonds are long-term debt securities issued by corporations. The corporate bond market now attracts more diverse investor groups than it did 10 years ago (Dietze *et al.*, 2009). Corporations promise the owner coupon payments on a semi-annual basis. The interest paid by the construction company to investors is tax-deductible to the company, which reduces the cost of financing with bonds. Since equity financing does not involve interest payments, it does not offer the same tax advantage (Madura, 2008). This is one of the main reasons why many large construction companies rely on bonds to finance their operations. Another reason for this market's success among investors may be the fact that corporate bonds have achieved higher returns on average than government bonds (Dietze *et al.*, 2009).

Corporate bonds can be placed with investors through a public offering or a private placement; the former is commonly used. A construction company which plans to issue bonds hires an investment bank to underwrite the bonds. The underwriter assesses market conditions and attempts to determine the price at which the corporation's bonds can be sold, and the appropriate size of the offerings. Some corporate bonds are privately placed rather than being sold in a public offering. Small companies borrowing relatively small amounts may consider private placements rather than public offerings, since they may be able to find an institutional investor which will purchase the entire offering. Since privately placed bonds do not have an active secondary market, they are more desirable to institutional investors which are willing to invest for long periods of time. The institutional investors which commonly purchase private placements include insurance companies, pension funds and bond mutual funds.

The last two decades have seen significant internationalisation of firms from developing countries as they have participated in international trade, being responsible for outflows of FDI, and in cross-border mergers and acquisitions. However, it is not easy for a company from a developing country to issue bonds abroad. This is largely because bonds issued by companies from developing countries require more effort in placement and sales to investors compared with a similar undertaking for a company in a developed country. Investors generally prefer to purchase securities of companies they are familiar with. The economic and political situation of the country in question is also important and often needs to be described in detail for the benefit of potential investors. Furthermore, the company's accounts and other information are typically not disclosed in a sufficiently detailed manner to meet the requirements of institutional investors in the major capital markets. In addition, accounting standards used in many developing countries differ significantly from those used in the industrialised countries (Kumiko, 1991). However, over the past few years, the International Finance Corporation (IFC) has become increasingly involved in helping companies in developing countries to raise financing through international offerings of investment funds and individual corporate bonds.

Municipal bonds

The bonds issued by state and local governments are known as municipal bonds. Kidwell and Koch (1983) noted that municipal borrowers can use long-term debt for current operations only when a constitutional amendment has been made, or a public referendum has been held, making it difficult for municipalities to effect substitutions between long- and short-term debt. Many investors like municipal bonds because their coupon interest payments are free of income tax. Municipal bonds are of two types: general obligation bonds and revenue bonds. They differ in where the money comes from to pay them. General obligation bonds are to be paid for with money raised by the issuer from a variety of different tax revenue sources. Revenue bonds are supposed to be paid off with money generated by the project which the bonds were issued to finance. For example, water fees can be used to pay for the bonds issued to finance the municipal water system.

Several features distinguish municipal bonds from the corporate bond market. The number of issues and issuers in the municipal market makes it unique. Peng and Brucato (2004) noted that many municipal bonds are sold by first-time and small issuers who are not well known in the investment community. Moreover, individual investors are a large part of the municipal market. Their capacity for security analyses is not as well developed as those of institutional investors. Also, although some progress has been made about disclosure practices in the municipal market, they continue to be inadequate; the timeliness of reporting is of major concern. Furthermore, there are no organised exchanges for municipal securities as there are for corporate securities.

Eurobonds

The Eurobond market provides an important source of funding for financial institutions, corporations, governments and supranational agencies (Kollo and Sharpe, 2006). Eurobonds are unsecured loans denominated in a currency other than the home currency of the firm which made the issue. They are foreign currency loans usually issued by governments and large companies in various countries. The Eurobond market is global and involves a range of currencies (Haste, 2008).

Eurobonds are the most widely used type of international bonds in developing countries. Since the early 1990s, developing countries have tapped Eurobonds markets by private placement and, in a few cases, by public issuance of bonds. For example, Mockler and Adams (2008) observe that the National Power Corporation of the Philippines issued a 15-year, US$100 million Eurobond in 1994. The issue was guaranteed by the World Bank and placed in the US and European markets.

Eurobonds are a vital source of finance for large companies with strong credit ratings. Eurobonds are medium- to long-term instruments (five or more years) not subject to the rules and regulations which are imposed on foreign bonds, such as the requirement to issue a detailed prospectus (Arnold, 2010). The attraction of Eurobonds, in contrast to the more conventional forms of loans capital, is that they give access to the international capital market; they are less costly to service and are not subject to national restrictions regarding loan issues (Van Horne and Wachowicz, 2005). Interest on Eurobonds is paid gross without any tax deduction – which has attractions to investors keen on delaying, avoiding or evading tax.

The advantages of borrowing on the Eurobond market are (Nevitt and Fabozzi, 2000): the potential for lower-cost funding under certain market conditions; access to a large diversified group of individual lenders not otherwise available; rapid access to the market to take advantage of current market conditions; no registration requirements; ability to make loans in any of several currencies; more choices of maturities than in syndicated bank loans; and the possibility of fixed-rate financing. The disadvantages of borrowing in the Eurobond market are that the funds available are more limited than in syndicated loans; draw-downs of funds are less flexible than in a syndicated loan; and lenders are not able to understand complex credits, thus limiting the market to well-known creditors or guarantors.

Bank loans

Bank loans have been the traditional financial arrangement for many types of construction projects (including individual housing projects) in developing countries. Distinctions between bond and bank loan financing are shown in Table 12.2.

Atkin and Glen (1992) found that, for firms in developed countries, internally generated funds were dominant, whereas for firms in developing countries, externally generated funds, such as bank loans, were more important. Despite the increasing importance of bank loans in developing countries, achieving the optimum capital structure for firms in such countries presents additional complications because of market inefficiencies and institutional constraints. For example, developing country banks cannot provide the resources required by the firms in these countries, especially where demand for credit by the government crowds out the private sector, or where the macro-economic environment is too risky for long-term loans. However, financial liberalisation in developing countries has broadened the range of financial instruments available to developing country firms (World Bank, 2001).

Table 12.2 Distinctions between bonds and bank loan financing

Factor	Bond financing	Bank loan financing
Size of financing	Substantially large; no particular limit; small issues are impractical	Smaller than bond financing unless syndicated; limited by a credit line available to a borrower, industry, country or other category to which the borrower belongs
Term	Usually one year or longer	Usually shorter than bond financing and rolled over; limited by the credit policy of a bank
Repayment	Bullet or limited pre-payment patterns; generally inflexible	Generally flexible
Interest rate	Fixed or floating rates	Floating rates for long maturities
All-in cost	Normally cheaper than bank loan financing, depending on market conditions; relatively cheap	Normally more expensive than bond financing
Credit analysis	Standardised rating by rating agencies	Proprietary credit analysis by a bank
Security	Normally unsecured	Normally secured
Use of proceeds	Normally not restricted	Normally restricted
Listing	Either listed or non-listed	Non-listed
Creditors	'Unspecific', many investors including individuals, corporations, banks, insurance companies, pension funds and mutual funds	A small number of banks and some other financial institutions
Transferability and liquidity	Readily transferable, and limited liquidity except for 'major' issuers	Often not transferable and not liquid

Source: Endo (2000).

Bank loans may take many forms, including commercial bank loans, syndicated bank loans, debentures and bank overdrafts.

Commercial bank loans

Commercial banks remain the most popular and largest source of finance for construction projects. A commercial bank's willingness to provide a loan depends on the borrower's financial health, experience and number of years in business (Thumann and Woodroof, 2005). Bank loans are not spontaneous financing as is trade credit (Shim and Siegel, 2007). Bank financing may take any of the following forms: unsecured loans, secured loans, lines of credit or instalment loans. A bank loan is transacted through a direct relationship between a lender and a borrower and can be characterised as a negotiable form of financing with flexible disbursement and a possible rescheduling of repayment (Gorton and Kahn, 2000). After a loan is made, the bank monitors the borrower's business to prevent moral hazards and negligence.

Normally, the financial institutions will take into account many factors when considering a loan application including: the period of the loan; the nature of the security offered by the company for the loan; the nature of the business; the purpose for which the loan will be used; the quality of the case presented to support the loan application; the financial position of the business; the integrity and quality of the management of the business; and the financial track record of the business (Arnold, 2010).

The merits of the bank loan are usually considered carefully before a firm borrows money. The factors bearing upon the selection of the bank loan include (Shim and Siegel, 2007) cost; effect on credit; risk; restrictions; flexibility; expected financial market conditions; the inflation rate; corporate profitability and liquidity positions; and the stability of the firm's operations.

Syndicated bank loans

Syndicated loans – loans that, prior to signing, are shared among groups of banks – have long been a part of corporate and project finance (Thomas and Wang, 2004). Large bank loans are made by syndicates of international commercial banks.

The general advantages of syndicated loans are (Nevitt and Fabozzi, 2000) that large amounts of debt can be raised; the syndicated bank loan market is the largest source of international capital; loans may be made in any one of several currencies; the number of participants can be substantial; and draw-downs can be flexible and relatively quick and cheap to arrange. However, the major disadvantage of the syndicated loan market is that the interest rate is floating and relatively high compared with other market rates. Syndicated loans are generally used by governments and government agencies. However, large corporations, and water and energy utilities projects, have used this market to raise funds.

Debentures

A debenture is one of the most common methods of borrowing in developing countries. The money raised through debentures forms a part of a company's capital structure although it does not form part of the company's share capital (Campbell, 2007). For large companies, the instrument usually used is a debenture that is a secured loan. The debenture contract specifies the amount of money to be borrowed, the rate of interest, dates of interest payments and capital repayments the company must make (Van Horne and Wachowicz, 2005). Failure to pay either the interest on the loan or the outstanding amount on the promised date entitles the lender, according to the terms of the contract, to seize the assets on which the loan is secured and sell them in order to recover the amount due. As a result of the lower level of risk associated with this form of investment, debenture holders are usually prepared to accept a lower rate of return (ASIC, 2008).

It is possible to have an unsecured debenture whereby the issuing company does not require any property as security for payment of interest and repayment of principal, but in practice such a marketable loan is usually called 'unsecured loan stock'. An important advantage is the fixed claim nature of the debenture contract; that is, the debenture holders receive only the amount stipulated in the contract. In addition to this, the cost of the debt is tax deductible, and this lessens the eventual cost to the firm.

Bank overdrafts

Overdrafts are short-term facilities provided by commercial banks to enable firms to have negative balances on their current accounts. The overdraft is a form of borrowing for which the bank will charge interest and an arrangement fee (Van Horne and Wachowicz, 2005).

Trade credit

Many construction business leaders cite insufficient access to bank credit as one of the most important constraints to the operation and growth of their companies. These financing constraints are especially binding for small- and medium-sized enterprises (SMEs) in developing countries (Okpara and Wynn, 2007). Campello and colleagues (2009) noted that during times of financial crisis, as the banks become more reluctant to lend, the financing constraints are likely to be exacerbated, leading companies to cut investments in capital and bypass attractive investment projects. An alternative source of finance to contractors as against bank credit is credit provided by suppliers of materials and other inputs; this is referred to as trade credit.

Trade credit is finance obtained from suppliers of goods and services over the period between the delivery of the goods or services and the subsequent settlement of the account by the recipient (Pike and Neale, 2006). Since trade credit is predominantly based on long-term relationships and is likely to involve sunk costs, suppliers have an interest in keeping their customers in operation (Cunat, 2007). According to Love and Zaidi (2010), trade credit has a potential to serve as an important source of finance to financially constrained construction companies because suppliers might be better able to overcome information and enforcement problems than financial institutions. The length of the trade credit period which is offered depends on the custom and practices of the industry, as well as the relationship between the contractor and the supplier, and the relative bargaining power of the contractor.

Trade credit is the least expensive form of financing inventory. The benefits of trade credit are (Nevitt and Fabozzi, 2000) that it is readily available, since suppliers want business; no collateral is required; interest is typically not demanded or, if so, the amount is minimal; it is convenient; and trade creditors are frequently lenient in the event of corporate financial problems. However, the drawbacks of trade credit are that the contractor must give up any cash discounts that would have been offered. There is also the probability that the firm's credit rating might be lowered.

Socially responsible investment

Socially responsible investment (SRI) is defined as 'investments enabling investors to combine financial objectives with social values' (Munoz-Torres *et al.*, 2004, p. 201). SRI, sometimes also called sustainable investing or ethical investing, refers to investment strategies that seek to maximise financial returns while maximising social good and the minimising environmental footprint of the investment. The general method of investment for SRI is 'engagement' – doing what they can to encourage business to do more with respect to sustainability. There are different approaches to SRI but, in general, socially responsible investors favour investments that promote environmental stewardship, protect the users of the services or products, promote human rights, prevent corruption and promote diversity. Thus, such investors avoid investments in projects that have negative social or environmental impacts. SRI is a relatively new source of financing but it is an important catalyst for mobilising other sources of debt (Head, 2000).

The Robeco and Booz company (2009) predicts that the responsible investing market will become mainstream within asset management by 2015, reaching 15–20 per cent of total global assets under management. A World Economic Forum (2003) survey revealed that over 70 per cent of CEOs believe that mainstream investors will have an increased interest in sustainability issues. Some new opportunities are arising in the SRI space to work with pension funds or other institutional investors which are managing their own money and incorporating specific social or environmental criteria in projects involving new asset classes such as water and power infrastructure projects in developing countries including Costa Rica, Mozambique and Vietnam. Sustainability intersects with the banking, insurance and financial industries in many other ways. These include issues in products (such as green mortgages), project finance, lending practices (such as responsible lending, microfinance), commodities trading (including renewable energy credit trading, carbon trading) and community development projects (such as projects aimed at developing sustainable communities).

International aid

The major sources of international assistance for construction project financing in developing countries are multi-lateral organisations such as the World Bank and the regional institutions such as the African Development Bank; and bilateral agencies such as the United States Agency for International

Development. These funds are available through government relations and are targeted towards capital projects such as development of infrastructure projects.

Some construction projects in developing countries that have recently been funded by the World Bank (2009b) are shown in Box 12.1.

In recent times, the provision of financial assistance to developing countries has been weighed down by a number of problems, prominent among which are the limited availability and excessive concentration on a few recipient countries (Ikhide, 2004). The increasing scarcity of development finance is considered as one of the most critical development problems that many low-income countries face.

Chipungu (2005) noted that decisions concerning foreign aid are normally attached to political circumstances. Failure of a recipient to agree on political ideologies or principles can result in the aid being withdrawn. It has also been known for governments of developing countries to divert international aid funds to other projects, thereby defeating the whole purpose of the funds. There are also instances when the beneficiaries misuse the funds. The lack of ownership, proliferation of aid sources and lack of co-ordination of such assistance have been major obstacles to aid effectiveness in developing countries. The myriad funding agencies in developing countries have their own priorities and the current aid system provides all of them with outlets to pursue those priorities. In many cases, this results in a duplication of functions and activities by funding institutions.

Grants from public funds

Public funds are sources of finance which a company may tap into with little or no direct cost. The funding is usually available to those companies to act in a particular way and in a particular location in the country. Examples of such action include funds provided to a construction company to invest in a new plant or equipment, the training or retraining of staff, research and development, and operations relating to the protection of the environment such as energy conservation.

Public–private partnerships (PPPs) finance source

Project finance through private participation in infrastructure investment in developing countries has been discussed previously. This sub-section considers public–private partnerships (PPPs) as a source of construction project financing in developing countries. Funding major construction projects is a problem for many developing countries, which usually rely on the government's annual capital investment budget or foreign aid for such purposes (Akintoye, 2009). Given the shortage of funding, it has become crucial and popular to tap into the vast capital and expertise of the private sector through various forms of PPP (OECD, 2007). Box 12.2 shows some examples of PPP projects that have been implemented in developing countries.

Box 12.1 World Bank-supported key construction projects in developing countries

- US$14 million Small Towns Water Supply and Sanitation Project in Ghana
- US$238 million Road Sector Development Program and Fourth Adaptable Program Loan Project to restore and expand Ethiopia's road network
- US$1.2 billion for the India Infrastructure Finance Company Ltd (IIFCL) to catalyse private financing for public–private partnerships in in infrastructure in India
- US$1 billion for the Fifth Power System Development Project to help address India's deficit of power
- US$150 million for the Andhra Pradesh Rural Water Supply and Sanitation Project, to improve water supply and sanitation services in India
- US$2.1 billion Kazakhstan South West Roads Project undertaken together with other development partners such EBRD, ADB, JICA, and Islamic Development Bank in Kazakhstan
- US$500 million for the Turkey Private Sector Renewable Energy and Energy Efficiency Project
- US$300 million for the Colombia Integrated Mass Transit Systems Projects
- US$20 million for the Solid Waste Management Program Project in Colombia
- US$167 million for the São Paulo Feeder Roads Project to improve the efficiency of the paved municipal road network in Brazil
- US$212 million for the Second Rio de Janeiro Mass Transit Project in Brazil
- US$5.5 million for the Energy Efficiency Project in Tunisia
- US$250 million Third Infrastructure Development Policy Loan to support Indonesia's effort to increase the level and effectiveness of infrastructure financing
- US$100 million Indonesia Infrastructure Finance Facility Project
- US$300 million for the NanGuang Railway Project in China
- US$129 million for the Jiangsu Water and Wastewater Project in China

Source: World Bank (2009b)

Box 12.2 Examples of PPP projects in developing countries

* Modernisation of Delhi International Airport in India
* Delhi–Mumbai–Chennai–Kolkata–Delhi Golden Quadrilateral Road in India
* Bangalore International Airport in India
* Kochi International Airport in India
* N4 Toll Road from South Africa to Mozambique
* Maputo Port in Mozambique
* Water and sanitation provision in the Dolphin Coast/Ilembe District Municipality in South Africa

---

As reported by the European Investment Bank (EIB, 2005), PPPs are risk-sharing investments in the provision of public goods and services, seen by governments as a means for launching investment programmes which would not have been possible under the public-sector budget within a reasonable period of time.

Given the changing economic, social, environmental and political circumstances, coupled with globalisation and budgetary constraints, PPPs have become a main source of finance for many infrastructure development projects in developing countries (Akintoye, 2009). However, according to Doh and Ramamurti (2003), the level of finance attracted by developing countries from private investors to fund construction projects is relatively low; and it is a new phenomenon. Akintoye (2009) suggests that there is a need for institutional development and more effective regulatory frameworks to facilitate greater involvement of private capital in construction projects in these countries.

Engaging the private sector in construction projects has been easier for some governments than others. The obstacles can be major. For example, in some countries, private participation in certain infrastructure sectors (such as water) is possible only through a change to the constitution. However, many of the obstacles are not quite as insurmountable, and it is often a case of ensuring that the legal framework is sufficiently flexible to enable private investors to be attracted, and ensure sufficient project volume to make large investors' opportunity costs worthwhile (OECD, 2007).

As shown in Table 12.1, some developing countries have financed infrastructure projects through public participation. The involvement of the private sector in the development and financing of construction projects can take many forms, ranging from the traditional procurement processes and concessions to more complex partnership arrangements in which the different segments of the process (including design, planning, finance, construction, operation and maintenance) are redistributed between the public and private parties involved (OECD, 2007).

Summary of sources of project finance

Table 12.3 presents a summary of sources of finance available for construction activities in developing countries. The table shows that grants from public funds and international aid are the main sources of finance for the

Table 12.3 Sources of finance for construction projects in developing countries

Sources of finance	Construction financing			
	Construction industry	Construction company	Construction projects	Construction project finance
Ordinary shareholders' funds		×	×	×
Bonus shares		×		
Preference shares		×	×	×
Warrants		×	×	×
Rights issue		×	×	×
Public issue		×	×	×
Placing		×	×	×
Corporate bonds		×		
Municipal bonds			×	×
Eurobonds		×	×	×
Commercial bank loans		×	×	×
Syndicated bank loans		×	×	×
Debentures		×	×	×
Bank overdrafts		×		
Trade credit		×	×	×
Socially responsible investment		×	×	×
International organisations aid	×	×	×	×
Grants from public funds	×	×	×	
Public–private partnerships source				×

construction industry whereas bank loans, bond markets and share markets are available to construction companies. To raise finance for construction projects in developing countries, companies can use any of these different sources, or a combination of them.

Implications of the global financial crisis for construction project financing in developing countries

The construction industries in the developing countries must be prepared for the volatility of the financial markets, as evinced by the current global crisis, and take appropriate measures and financial engineering strategies to mitigate against the risks and uncertainties associated with such crises.

Economic and financial crises tend to affect developing countries through declining private capital flows such as FDI, portfolio flows and international lending; official flows such as those from development finance institutions; and capital and current transfers such as official development assistance and remittances (Massa and te Velde, 2008). By the end of 2009, developing countries may have lost incomes of at least US$750 billion – more than US$50 billion of it in sub-Saharan Africa (ODI, 2009). This disruption led to a decrease in the lending capacity and risk appetite of lenders and investors with respect to construction projects in developing countries.

The current financial crisis has increased the pressure of liquidity constraints on bank and non-bank intermediaries (i.e. institutional investors such as mutual funds and hedge funds) in developed economies, with adverse consequences for developing countries (Cali *et al.*, 2008). In an environment where the economy is volatile, large banks, which are dominant sources of capital for large-scale development projects, would have little appetite for transactions that stray from generic, wholesale-type financing. Consequently, this might make it difficult for construction firms, particularly in developing countries, to secure funding, especially for difficult development projects, from or through them.

An increase in the volatility of economic activities may impair the smooth functioning of the financial system and adversely affect the performance of commercial entities including those in the construction industry (Eizaguirre *et al.*, 2004). Volatility may create uncertainty about future profits, which would impair long-term investment decisions. Bond and Devereux (1988) have identified three main routes by which a fall in a company's share evaluation during a financial crisis may be reflected in low investment. First, the fall may be associated with a decline in business confidence and expectations of future profitability. Second, there may be an indirect influence on the firm's expectations of future demand resulting from the reduced value of shareholders' wealth. Finally, the increased cost of issuing new shares in order to finance investment may have some effect.

Summary and conclusion

The facilities that are provided by the construction industry play a critical and highly visible role in the process of socio-economic development of any country. In spite of this, there are major challenges for construction firms and governments in developing countries in their efforts to obtain the finance to provide these facilities.

This chapter has explored various sources of finance available and the challenges for developing countries in attracting finance for construction projects, as well as the different meanings of construction project financing and infrastructure investment and the context in which they are used.

There are internal and external sources of finance available for construction projects in developing countries. Construction firms must have some equity finance but when they are considering increasing their long-term financing base they must assess the merits of all the options open to them.

Construction companies in developing countries have relied on traditional sources of financing (primarily through internal funds and bank loans or those obtained from family members). There is a need for them to explore alternative sources of financing for their businesses and projects. Private equity, PPPs, sustainable investment funds and sovereign wealth funds will be major areas of growth in the future and are underexploited in developing countries (Sawant, 2010). Corporate bonds are also becoming increasingly common as a means to raise finance without dependence on banks while many companies are looking to raise equity in overseas markets. There is an opportunity for the construction industries in developing countries to enhance their reputation and attract more socially responsible investment through the integration of issues of sustainability into their business operations.

To attract more funds for public-sector projects and infrastructure development, governments need to improve the investment climate (World Bank, 2010). This might involve legal reforms, risk mitigation mechanisms, innovative approaches to mobilising private investment, and broadening and deepening local financial markets in order to reduce the costs of pursuing infrastructure investment opportunities to private investors.

Well-developed financial markets can stimulate the quality of capital in several ways (Levine, 1997). They contribute to transparency because firms release information to attract capital; this ultimately improves the efficiency with which investment is allocated (World Bank, 2010). Moreover, it improves resource allocation and allows easier access to capital for enterprises, thus lowering their financing costs (Boyd and Prescott, 1986). Moreover, markets' mobilisation and pooling of savings should be considered in addition to the markets that facilitate the pooling and sharing of risks. Through well-developed financial markets, investors can diversify their portfolios (Levine, 1997). Therefore, countrywide and sectoral analysis is a necessary component of infrastructure investing, and it makes sense for investors to develop sectoral as well as geographical expertise before allocating capital. This will require that the special characteristics of construction are considered and understood in order to develop the right financing portfolio for the industry.

Financial institutions should support the development of capital markets, fulfilling the need for underwriting various loans and securitisation of the debts for getting the necessary funds for construction projects in developing countries (Gupta and Sravat, 1998). The participation of the multi-lateral financial institutions by way of equity would enhance investor confidence and the success of construction project financing. In developing countries, financial institutions can best support the construction industry by sharing the right information with the right individuals at the right time. They can also support them in the identification of value-creating opportunities by being closely involved in the execution of the project, providing robust financial information and insight, combined with financial rigour and being active business partners of the firms.

Difficulties in accessing knowledge on project financing and lack of awareness of the alternative sources of financing for construction projects are hampering the growth of the construction businesses in developing countries. An industry-wide awareness-raising programme on the concept of construction financing should be implemented. Education and training programmes should be reorientated; the syllabuses should cover aspects of financial engineering that are relevant to the construction industry in developing countries, as well as innovative financing arrangements, challenges in attracting finance and how to overcome them, and main sources of finance available for construction industries and projects and their characteristics.

Note

1 'Infrastructure' means physical structures that provide or permit transportation; energy generation and transmission; water distribution and sewage collection; and the provision of social services such as health and education; this underpins the quality of life as well as the ability of economies to function effectively (Grigg, 2010).

References

Aggarwal, S. (2003) *Challenges for Construction Industries in Developing Countries*. Proceedings of the 6th National Conference on Construction, 10–11 November, New Delhi, CDROM, Technical Session 5, Paper No. 1.

Akintoye, A. (2009) PPPs for physical infrastructure in developing countries. In A. Akintoye and M. Beck (eds) *Policy, Finance and Management for Public–Private Partnership*. Blackwell Publishing: Oxford.

Arnold, G. (2010) *Investing: The Definitive Companion to Investment and the Financial Markets*. Prentice Hall: London.

ASEAN (2008) *Estimate of the ASEAN Infrastructure Financing Mechanism*. Task Force: Kuala Lumpur.

ASIC (2008) *Debentures: Improving Disclosure for Retail Investors.* Regulatory guide 69. Australian Securities and Investments Commission: Adelaide.

Atkin, M. and Glen, J. (1992) Comparing corporate capital structures around the globe. *International Executive,* 5, 369–387.

Bond, S. and Devereux, M. (1988) Financial volatility, the stock market crash and corporate investment. *Fiscal Studies,* 9 (2), 73–80.

Boyd, J. H. and Prescott, E. C. (1986) Financial intermediary-coalitions. *Journal of Economic Theory,* 38, 211–232.

Brealey, R. A., Cooper, I. A. and Habib, M. A. (1996) Using project finance to fund infrastructure investments. *Journal of Applied Corporate Finance,* 9 (3), 25–39.

Brealey, R. A., Myers, S. C. and Allen, F. (2007) *Principles of Corporate Finance.* McGraw Hill Higher Education: London.

Burnett, R. G. (1991) *Insolvency and the Sub-contractor.* CIOB Paper no. 48. Charted Institute of Building: Ascot, UK.

Cali, M., Massa, I. and te Velde, D. W. (2008) *The Global Financial Crisis: Financial Flows to Developing Countries Set to Fall by One Quarter.* Overseas Development Institute: London.

Campbell, C. (2007) *Legal Aspects of Doing Business in Asia and the Pacific.* Yorkhill Law Publishing: Salzburg, Austria.

Campello, M., Graham, J. R. and Harvey, C. R. (2009) *The Real Effects of Financial Constraints: Evidence from a Financial Crisis.* http://ssrn.com/abstract=1357805

Carrillo, P. (1996) Technology transfer on joint venture projects in developing countries. *Construction Management and Economics,* 14, 45–54.

Chipungu, L. (2005) *Institutional Constraints to Land Delivery for Low Income Housing: A Case Study of Harare, Zimbabwe.* Our Common Estate Paper Series, 7–26. RICS Foundation: London.

Cook, P. and Nixson, F. (2000) *Finance and Small and Medium-Sized Enterprise Development.* Working Paper Series, 14 (4). Finance and Development Research Programme: IDPM, University of Manchester.

Corbett, C. (2010) Funding African infrastructure. *The Banker,* May, 1–5.

CSD (2008) *Report on the Quarterly Survey of Construction Output.* Census and Statistics Department, Government of Hong Kong Special Administrative Region: Hong Kong.

Cunat, V. (2007) Suppliers as debt collectors and insurance providers. *Review of Financial Studies,* 20, 491–527.

Davis, R. (1991) *Construction Insolvency.* Chancery Law Publishing: London.

Davis, J. (2004) Corruption in public service delivery: experience from South Asia's water and sanitation sector. *World Development,* 32 (1), 53–71.

Delmon, J. and Juan, E. (2008) *The Role of Development Banks in Infrastructure Finance: World Bank View.* Euromoney Books: London.

Deloitte (2007) *The Big Picture: Private Equity Trends.* Deloitte: London.

Dietze, L. H., Entrop, O. and Wilkens, M. (2009) The performance of investment grade corporate bond funds: evidence from the European market. *European Journal of Finance,* 15 (2), 191–209.

De Silva, N., Rajakaruna, R. and Bandara, K. (2008) Challenges faced by the construction industry in Sri Lanka: perspective of clients and contractors. *CIB Building Education and Research International Conference,* 10–15 February, Kandalama, Sri Lanka.

Doh, P. and Ramamurti, R. (2003) Reassessing risk in developing country infrastructure. *Long Range Planning*, 36 (4), 337–353.

Edge, C. T. (1988) *Solving Business Cash Problems*. Kogan Page: London.

Egan, J. (1998) *Rethinking Construction: Report of the Construction Task Force on the Scope for Improving the Quality and Efficiency of UK Construction*. Department of the Environment, Transport and the Regions: London.

EIB (2005) *Evaluation of PPP Projects Financed by the EIB*. http://www.eib.org/projects/publications/evaluation-of-ppp-projects-financed-by-the-eib.htm

Eizaguirre, C. J., Gomez, B. J. and de Gracia, H. F. P. (2004) Structural changes in volatility and stock market development: evidence for Spain. *Journal of Banking and Finance*, 28 (7), 1745–1773.

Endo, T. (2000) *Corporate Bond Market Development*. IFC working paper. International Finance Corporation: Washington, DC.

Esty, B. (2004a) *Modern Project Finance: A Casebook*. John Wiley and Sons: New York.

Esty, B. (2004b) Why study large projects? An introduction to research on project finance. *European Financial Management*, 10, 213–224.

Esty, B. C. (2007) *An Overview of Project Finance and Infrastructure Finance: 2006 Update*. Harvard Business School Publishing: Boston, MA.

Esty, B. C. and Sesia Jr., A. (2007) *An Overview of Project Finance and Infrastructure Finance: 2006 Update*. Harvard Business School Teaching Note, 9-207-107.

Fairclough, J. (2002) *Rethinking Construction Innovation and Research: A Review of Government Research and Development Policies and Practices*. DTI: London.

Fitch Ratings (2004) *Public–Private Partnerships: The Next Generation of Infrastructure Finance*. Special Report. Fitch Ratings: New York.

Fombrun, C. J. and van Riel, C. B. M. (2004) *Fame and Fortune: How Successful Companies Build Winning Reputation*. Financial Times Prentice-Hall: New York.

Gorton, G. and Kahn, J. (2000) The design of bank loan contracts. *Review of Financial Studies*, 13 (2), 331–364.

Grigg, N. S. (2010) *Infrastructure Finance: The Business of Infrastructure for a Sustainable Future*. John Wiley and Sons: New York.

Gupta, J. P. and Sravat, A. K. (1998) Development and project financing of private power projects in developing countries: a case study of India. *International Journal of Project Management*, 16 (2), 99–105.

Hammami, M., Ruhashyankiko, J. F., Yehoue, E. B. (2006) *Determinants of Public–Private Partnerships in Infrastructure*. IMF Working Paper No. 06/99, April.

Halpin, D. W. and Senior, B. A. (2009) *Financial Management and Accounting Fundamentals for Construction*. John Wiley and Sons: New York.

Harwood, A. (2000) *Building Local Bond Markets: An Asian Perspective*. International Finance Corporation: Washington, DC.

Haste, D. (2008) *Management Accounting: Financial Strategy, Financial Management*. Study notes, paper P9. Chartered Institute of Management Accountants: London.

Hassanein, A. and Adly, S. (2008) Issues facing small construction firms: the financing barrier. *Journal of Small Business and Entrepreneurship*, 21 (3), 307–320.

Head, C. (2000) *Financing of Private Hydropower Projects*. World Bank Publications: Washington, DC.

Hillebrandt, P. M. (2000) *Economic Theory and the Construction Industry*. Palgrave Macmillan: London.

Hillebrandt, P. M. and Cannon J. (1990) *The Modern Construction Firm*. Macmillan Press: London.

Hoffman, S. L. (2007) *The Law and Business of International Project Finance: A Resource for Governments, Sponsors, Lawyers, and Project Participants*. Cambridge University Press: Cambridge.

Hornby, A. S. (2000) *Oxford Advanced Learners Dictionary*. Oxford University Press: Oxford.

Hussain, J. and Matlay, H. (2007) Financing preferences of ethnic small businesses in the UK. *Journal of Small Business and Enterprise Development*, 14 (3), 487–500.

Hyun, S., Nishizawa, T. and Yoshino, N. (2008) *Exploring the Use of Revenue Bond for Infrastructure Financing in Asia*. JBICI Discussion Paper No. 15. Japan Bank for International Cooperation (JBIC): Tokyo.

Ikhide, S. (2004) Reforming the international financial system for effective aid delivery. *World Economy*, 27, 127–152.

Jewell, C., Flanagan, R. and Anac, C. (2010) Understanding UK construction professional services exports: definitions and characteristics. *Construction Management and Economics*, 28 (3), 231–239.

Kenny, C. (2007) *Construction, Corruption, and Developing Countries*. World Bank Policy Research Working Paper 4271. World Bank: Washington, DC.

Khosrowshahi, F. (2000) A radical approach to risk in project financial management. In *Proceedings, 16th Annual Conference*, Association of Researchers in Construction Management.

Kidwell, D. S. and Koch, T. W. (1983) Market segmentation and the term structure of municipal yields. *Journal of Money, Credit, and Banking*, 15 (1), 40–55.

Kollo, G. M. and Sharpe, I. G. (2006) Relationships and underwriter spreads in the Eurobond floating rate note market. *Journal of Financial Research*, 29 (2), 163–180.

Korkmaz, S. and Messener, J. I. (2007) Competitive positioning and continuity of construction firms in international markets. *Journal of Management in Engineering*, 24 (4), 207–216.

Kumiko, Y. (1991) Accessing the international capital markets. *Finance and Development*, 1, 1–3.

Latham, M. (1994) *Constructing the Team: Final Report of the Government/Industry Review of Procurement and Contractual Arrangements in the UK Construction Industry*. HMSO: London.

Lerner, J. and Schoar, A. (2004) *Transaction Structures in the Developing World: Evidence from Private Equity*. Working paper. Massachusetts Institute of Technology (MIT), Sloan School of Management: Cambridge, MA.

Levine, R. (1997) Financial development and economic growth: views and agenda. *Journal of Economic Literature*, 35, 688–726.

Levine, R. (2002) Bank based or market-based financial systems: which is better? *Journal of Financial Intermediation*, 11, 398–428.

Lin, J. (2008) *The Impact of the Financial Crisis on Developing Countries*. World Bank Draft: Washington, DC.

Love, I. and Zaidi, R. (2010) Trade credit, bank credit and financial crisis. *International Review of Finance*, 10, 125–147.

Lowe, J. (1997) Insolvency in the UK construction industry. *Journal of Financial Management of Property and Construction*, 2 (1), 83–110.

Madura, J. (2008) *International Corporate Finance*. Thomson South-Western: Mason, OH.

Massa, I. and te Velde, D. W. (2008) *The Global Financial Crisis: Will Successful African Countries Be Affected?* Background Note. Overseas Development Institute: London.

Mockler, R. J. and Adams, J. F. (2008) *Multinational Strategic Management*. Taylor & Francis: London.

Munoz-Torres, M. J., Fernandez-Izquierdo, M. A. and Balaguer-Franch, M. R. (2004) The social responsibility performance of ethical and solidarity funds: an approach to the case of Spain. *Business Ethics: A European Review*, 13, 200–218.

Mpuga, P. (2010) Constraints in access to and demand for rural credit: evidence from Uganda. *African Development Review*, 22 (1), 115–148.

Myers, D. (2005) A review of construction companies' attitudes to sustainability. *Construction Management and Economics*, 23 (8), 781–785.

Nevitt, P. K. and Fabozzi, F. J. (2000) *Project Financing*. Euromoney Books: London.

ODI (2009) *The Global Financial Crisis and Developing Countries: Taking Stock, Taking Action*. Briefing Paper. Overseas Development Institute: London.

OECD (2007) *Infrastructure to 2030: Mapping Policy for Electricity, Water and Transport, Volume 2*. Organisation for Economic Co-operation and Development: Paris.

Ofori, G. (1991) Programmes for improving the performance of contracting firms in developing countries: a review of approaches and appropriate options. *Construction Management and Economics*, 9, 19–38.

Ofori, G. (2001) *Challenges Facing the Construction Industries of Southern Africa*. Paper presented at regional conference: Developing the Construction Industries of Southern Africa, Pretoria, South Africa, 23–25 April.

Ofori, G. (2006) Revaluing construction in developing countries: a research agenda. *Journal of Construction in Developing Countries*, 11 (1), 1–16.

Ogunlana, S. O., Promkuntong, K. and Jearkjiran, V. (1996) Construction delays in a fast-growing economy: comparing Thailand with other economies. *International Journal of Project Management*, 14 (1), 37–45.

Okpara, J. O. and Wynn, P. (2007) Determinants of small business growth constraints in a sub-Saharan African economy. *SAM Advanced Management Journal*, 21 (1), 1–10.

Peng, J. and Brucato, P. F. (2004) An empirical analysis of market and institutional mechanisms for alleviating information asymmetry in the municipal bond market. *Journal of Economics and Finance*, 28 (2), 226–238.

Pike, R. and Neale, B. (2006) *Corporate Finance and Investment: Decisions and Strategies*. Pearson Education Limited: London.

Platz, D. (2009) *Infrastructure Finance in Developing Countries: The Potential of Sub-sovereign Bonds*. DESA Working Paper No. 76.United Nations Department of Economic and Social Affairs: New York.

Pretorius, F., Lejot, P., McInnis, A., Arner, D. and Fong-Chung Hsu, B. (2008) *Project Finance for Construction and Infrastructure: Principles and Case Studies*. Blackwell Publishing: Oxford.

Rajivan, K. (2004) *Linking Cities with Domestic Capital*. citiesalliance.com

Renukappa, S. H. (2009) *A Theoretical Framework for Managing Change and Knowledge Associated with Sustainability Initiatives for Improved Competitiveness*. PhD thesis, Glasgow Caledonian University, Glasgow.

Reuvid, J.,(2002) *The Corporate Finance Handbook*. Kogan Page: London.

Robeco and Booz company (2009) *Responsible Investing: A Paradigm Shift from Niche to Mainstream*. Robeco, Booz company: London.

Robinson, R. and Thagesen, B. (2004) *Road Engineering for Development*. Spon Press: London.

Sawant, R. J. (2010) *Infrastructure Investing*. John Wiley and Sons: Hoboken, NJ.

Shim, J. K. and Siegel, J. G. (2007) *Financial Management*. McGraw-Hill: New York.

Smith, N. J., Merna, T. and Jobling, P. (2006) *Managing Risk in Construction Projects*. Blackwell Publishing: Oxford.

Spiceland, J. D., Sepe, J. and Tomassini, A. (2005) *Intermediate Accounting*. McGraw-Hill: New Delhi.

Thomas, H. and Wang, Z. (2004) The integration of bank syndicated loan and junk bond markets. *Journal of Banking and Finance*, 28 (2), 299–329.

Thumann, A. and Woodroof, E. A. (2005) *Handbook of Financing Energy Projects*. Fairmont Press: Lilburn, GA.

Van Horne, J. C. and Wachowicz, J. M. (2005) *Fundamentals of Financial Management*. Prentice Hall: London.

Venkataraman, R. R. and Pinto, J. K. (2008) *Cost and Value Management*. John Wiley and Sons: Hoboken, NJ.

Watson, D. and Head, A. (2007) *Corporate Finance: Principles and Practice*. Prentice Hall–Financial Times: London.

Wells, J. (2001) *The Construction Industry in the Twenty-First Century: Its Image, Employment Prospects and Skill Requirements*. International Labour Organisation: Geneva.

Westhead, P. and Wright, M. (2000) Introduction. In P. Westhead and M. Wright (eds) *Advances in Entrepreneurship, Vol. 1*. Edward Elgar Publishing: Cheltenham.

World Bank (2001) *Developing Government Bond Markets: A Handbook*. World Bank: Washington, DC.

World Bank (2004) *Global Development Finance 2004*. World Bank: Washington, DC.

World Bank (2005) *Private Participation in Infrastructure Database*. http://ppi.world-bank.org/reports/customQueryAggregate.asp

World Bank (2006a) *Infrastructure at the Crossroads: Lessons from 20 Years of World Bank Experience*. World Bank: Washington, DC.

World Bank (2006b) *Doing Business in 2006*. World Bank: Washington, DC.

World Bank (2008) *Attracting Investors to African Public–Private Partnerships: A Project Preparation Guide*. World Bank:, Washington, DC.

World Bank (2009a) *World Development Indicators 2009 Report*. World Bank: Washington, DC.

World Bank (2009b) *Summary of Key INFRA Projects by Region*. http://siteresources.worldbank.org/INTSDNETWORK/Resources/SummaryofKeyINFRAProjectsbyRegion_FinalCR.pdf

World Bank (2010) *Global Economic Prospects: Crisis, Finance, and Growth*. World Bank: Washington, DC.

World Economic Forum (2003) *Why Global Corporate Citizenship Matters for Shareholders: A Survey of Leading CEOs*. World Economic Forum: Geneva.

Yescombe, E. (2002) *Principles of Project Finance*. Academic Press: London.

Author index

Subject index

Printed and bound by CPI Group (UK) Ltd, Croydon, CR0 4YY

23/10/2024

01778242-0017